AF692515

Numerical Analysis and Modeling of Floating Structures

Numerical Analysis and Modeling of Floating Structures

Guest Editors

Shengzhe Wang
Jinghua Wang
Chungkuk Jin

MDPI

Basel • Beijing • Wuhan • Barcelona • Belgrade • Novi Sad • Cluj • Manchester

Guest Editors

Shengzhe Wang
Department of Civil Engineering
University of Colorado Denver
Denver
USA

Jinghua Wang
Department of Civil and Environmental Engineering
The Hong Kong Polytechnic University
Hong Kong
China

Chungkuk Jin
Department of Ocean Engineering and Marine Sciences
Florida Institute of Technology
Melbourne
USA

Editorial Office
MDPI AG
Grosspeteranlage 5
4052 Basel, Switzerland

This is a reprint of the Special Issue, published open access by the journal *Journal of Marine Science and Engineering* (ISSN 2077-1312), freely accessible at: https://www.mdpi.com/journal/jmse/special_issues/07MT0634GJ.

For citation purposes, cite each article independently as indicated on the article page online and as indicated below:

Lastname, A.A.; Lastname, B.B. Article Title. *Journal Name* **Year**, *Volume Number*, Page Range.

ISBN 978-3-7258-7715-7 (Hbk)
ISBN 978-3-7258-7716-4 (PDF)
https://doi.org/10.3390/books978-3-7258-7716-4

Cover image courtesy of Shengzhe Wang

© 2026 by the authors. Articles in this reprint are Open Access and distributed under the Creative Commons Attribution (CC BY) license. The reprint as a whole is distributed by MDPI under the terms and conditions of the Creative Commons Attribution-NonCommercial-NoDerivs (CC BY-NC-ND) license (https://creativecommons.org/licenses/by-nc-nd/4.0/).

Contents

About the Editors

Shengzhe Wang

Shengzhe Wang is an Assistant Professor in the Department of Civil Engineering and Construction at the University of Colorado Denver. He holds a BEng in Structural Engineering from the University of Auckland, an MPhil in Civil Engineering from the University of Sydney, and a Ph.D. in Civil and Environmental Engineering from Princeton University. Dr. Wang's work encompasses the disciplines of ocean, structural, and mechanical engineering to foster innovative solutions for climate change adaptation. Through this interdisciplinary perspective, his research explores the interaction between extreme hydrodynamic events with onshore, offshore, and floating structures; architecture- and nature-inspired solutions to coastal resilience; and the geometric mechanics of thin-shell structures across different scales and fields of application. He received the 2021 Moisseiff Award conferred by the American Society of Civil Engineers and is the author of a monograph "Applied Ocean Mechanics for Structural Engineering and Climate Adaptation" published by Springer Nature.

Jinghua Wang

Jinghua Wang is an Assistant Professor in Ocean and Coastal Engineering at the Hong Kong Polytechnic University (PolyU). Prior to joining PolyU, he worked as a Research Fellow at the National University of Singapore and City University of London. He has been engaged in research on nonlinear water wave modeling and wave–structure interactions for many years.

Chungkuk Jin

Chungkuk Jin is an Assistant Professor at the Florida Institute of Technology, specializing in fluid–structure interaction for marine civil infrastructures. With a Ph.D. from Texas A&M University, he has authored over 140 scholarly works. His excellence in research is recognized by the prestigious Early-Career Research Fellowship from the U.S. National Academies, through which he contributes to a diverse array of funded projects. His work develops numerical tools for hydro-elasticity, advancing the resilience of next-generation coastal and offshore systems.

Journal of
*Marine Science
and Engineering*

MDPI

Editorial

Numerical Analysis and Modeling of Floating Structures

Shengzhe Wang [1,*], Jinghua Wang [2] and Chungkuk Jin [3]

1 Department of Civil Engineering and Construction, University of Colorado Denver, Denver, CO 80204, USA
2 Department of Civil and Environmental Engineering, The Hong Kong Polytechnic University, Hong Kong, China; jinghua.wang@polyu.edu.hk
3 Department of Ocean Engineering and Marine Sciences, College of Engineering and Science, Florida Institute of Technology, Melbourne, FL 32901, USA; cjin@fit.edu
* Correspondence: shengzhe.2.wang@ucdenver.edu

Over recent decades, offshore activities have increasingly migrated into deeper waters, where conventional fixed-bottom infrastructures are becoming increasingly impractical or uneconomical. In response, floating structures have emerged as a versatile solution capable of supporting a broad range of maritime activities. Examples include energy production [1], coastal protection [2], transportation [3], aquaculture [4], and ocean colonization [5]. Relative to fixed infrastructure, the engineering complexity associated with floating structures is usually considerably higher given their inherently dynamic nature. This complexity often results from the high degree of coupling between numerous physical processes spanning hydrodynamics, aerodynamics, structural dynamics, mooring dynamics, and at times even impact dynamics from collisions with floating sea ice or other types of debris. These diverse interaction mechanisms frequently lead to highly stochastic or nonlinear responses that are difficult to capture using simplified analytical approaches in isolation.

In response, numerical analysis and computational modeling have become indispensable tools for understanding and designing floating marine systems. Recent advances in computational fluid dynamics (CFD) and multiphysics simulation frameworks have significantly expanded the ability of researchers and engineers to predict the behavior of floating structures exposed to a broad range of environmental conditions. Examples include fatigue evaluation under realistic sea states [6], large-amplitude motions induced by breaking wave impact [7], and complex mooring line dynamics [8]. This Special Issue, "Numerical Analysis and Modeling of Floating Structures", was conceptualized to highlight recent developments in this rapidly evolving discipline within the broader context of ocean engineering. The ten articles showcased in this Special Issue can be broadly classified into four distinct themes: (1) energy infrastructure, (2) transportation systems, (3) coastal resilience, and (4) aquaculture facilities. When taken as the whole, these contributions reflect both the breadth of modern offshore research and the growing importance of numerical modeling in the design and analysis of floating structures across diverse applications.

The most common theme involves the numerical modeling of floating energy infrastructure, spanning wind, tidal, and nuclear sources, as discussed in four separate contributions. Lyu et al. (Contribution 1) employed OpenFAST with SIMDYN to model the large-amplitude motions of the semi-submersible 5 MW OC4 floating offshore wind turbine (FOWT) exposed to coupled wind and wave loading. This work revealed the importance of nonlinearity, particularly in relation to the heave and pitch motions of the FOWT under extreme environmental conditions. Srinivasamurthy et al. (Contribution 2) explored the collapse mechanisms of spar-type floating wind turbines during blade pitch malfunction within a coupled aero-hydro-elastoplastic modeling framework. It was discovered that damage in the plastic region (at the tower root) associated with floating turbines were eight times greater than onshore turbines under the same wind conditions. Lee et al. (Contribution 3) investigated the dynamic response of a dual vertical-axis tidal turbine under combined wind, wave, and current loads. In particular, the authors employ time-domain simulations via OrcaFlex to understand both the motions and mooring line tensions associated with both taut and catenary mooring systems. It was revealed that sway and heave motions are significantly influenced by wave excitation, with the catenary system exhibiting larger responses. On the other hand, taut lines were observed to increase mooring tensions but proved more effective in restricting dynamic motions. The implications of mooring line failure were additionally revealed to significantly exacerbate yaw movements of the floating structure. Yuan et al. (Contribution 4) developed novel time-domain design stress (TDDS) method for the fatigue analysis of reactor pressure vessels (RPVs) within floating nuclear power plants. The technique systematically incorporates the effects of wave loads, internal pressure, and thermal effects on the fatigue life of RPVs, yielding an average error of only 0.033% relative to traditional time-domain methods, while being 30% faster. Finite element modeling (FEM) further revealed the locations of stress concentrations that would be more susceptible to fatigue.

Beyond energy infrastructure, three contributions discuss the numerical modeling of different transportation systems. Gong et al. (Contribution 5) employed an arbitrary Lagrangian Eulerian (ALE) method via LS-DYNA to study the collision dynamics of floating sea ice impacting bridge piers. It was discovered that the maximum impact force increases linearly with sea ice mass but decreases when the spacing between blocks of sea ice becomes larger. The influence of different sea ice geometries across various stages of collision is also evaluated. Lee et al. (Contribution 6) leveraged nonlinear static analysis to study the stability and seakeeping performance of a submerged floating tunnel module (SFTM) during wet tow operations. In particular, numerical static offset tests were performed at varying tow speeds to determine the equivalent system stiffness in surge, sway, roll, and yaw. Statistical analyses further revealed the sea states most likely to result in the most extreme motions of the SFTM, thereby decreasing towing stability. Qu et al. (Contribution 7) used CFD to model the hydrodynamic response of a return capsule for manned space missions to ensure safe water landings. Taking the form of a conical structure, the study evaluated how environmental factors, including waves, currents, and wind, affected the overall stability of the floating capsule. It was revealed that severe sea states can exacerbate nonlinear phenomena, such as subharmonic resonance, thereby amplifying rotational motions that can seriously compromise safety margins.

Two contributions focus on the numerical modeling of infrastructure for wave attenuation to enhance the resilience of coastal communities. Smith et al. (Contribution 8) leveraged FEM in conjunction with Lagrangian smoothed particle hydrodynamics (SPH) to study wave attenuation offered by free-surface breakwaters (FSBWs) adopting hyperbolic paraboloid (hypar) surfaces. It was discovered that hypar-faced FSBWs enhance wave attenuation when compared to traditional flat-faced designs, particularly for shorter wave periods and smaller drafts. The principle stresses within hypar FSBWs were also observed to be three times lower than that of their flat counterparts, which could potentially yield more economical designs. Jia et al. (Contribution 9) studied the wave attenuation performance of a box-type floating breakwater incorporating flexible wave-dissipating structures. The immersed boundary method was used in conjunction with a finite element structural solver to facilitate the accurate simulation of fluid–structure interaction (FSI) under regular waves. The results demonstrated that structural flexibility played a crucial role in modifying wave reflection, transmission, and dissipation mechanisms. A lower stiffness was observed to enhance energy dissipation through large deformation and vortex generation, while higher stiffness promoted reflection with reduced dissipation.

https://doi.org/10.3390/jmse14070640

The sole discussion on floating aquaculture facilities is provided by Kim et al. (Contribution 10). Here, a numerical model was developed using the Cummins equation and a lumped-mass line model to describe the hydrodynamic performance of a floating fish cage equipped with a spread mooring system comprising eight mooring lines separated into four station-keeping groups. In particular, studies were performed to assess the effects of varying mooring line lengths, waves, current speeds, and biofouling on cage behavior. It was revealed that heavier chains reduced excursions but increased tension, while high current speeds increased the line tension and mooring line stiffness by up to 66%. The presence of biofouling was further noted to increase both the maximum excursion and line tensions. Ultimately, it was concluded that the spread mooring system can maintain reliable station-keeping, even under severe environmental conditions with a 50-year return period, thus validating its suitability for the deployment of floating fish cages to promote offshore aquaculture activities.

The articles presented in this Special Issue collectively demonstrate the growing sophistication of numerical modeling techniques applied to floating structures. By incorporating a diverse range of systems, including energy platforms, breakwaters, aquaculture facilities, and transportation infrastructure, these contributions highlight both the versatility of floating structures and the wide range of environmental challenges that must be addressed to ensure the robustness of their design. Table 1 provides a summary of the different structures and forcing mechanisms pertaining to each contribution.

Table 1. Summary of all contributions within the Special Issue.

Contribution	Theme	Structure Studied	Forcing Mechanisms
1	Energy	Semi-submersible floating offshore wind turbine	Wind and waves
2	Energy	Spar-type floating offshore wind turbine	Wind and waves
3	Energy	Dual vertical-axis tidal turbine	Wind, waves, and currents
4	Energy	Floating nuclear reactor pressure vessel	Waves, internal pressure, and thermal effects
5	Transportation	Bridge piers	Floating sea ice
6	Transportation	Submerged floating tunnel module	Wet tow operations
7	Transportation	Return capsule for manned space missions	Wind, waves, and currents
8	Coastal resilience	Free-surface breakwater with hypar faces	Waves
9	Coastal resilience	Floating breakwater with flexible wave-dissipating structures	Waves
10	Aquaculture	Floating fish cage with spread mooring system	Waves, currents, and biofouling

The outcomes of this Special Issue also offer crucial insight into a range of future research directions pertaining to the numerical modeling of floating structures. Namely, the continued development of robust numerical models that integrate multiple excitation sources (e.g., coupled aero-hydrodynamics) will be essential in ensuring the survivability of floating systems exposed to realistic marine environments. The expansion of research efforts into the modeling of non-rigid floating bodies will further shed light into the relationship between structural dynamics (e.g., superstructure flexibility) and hydrodynamics (e.g., characteristics of wind/wave loading). A particular area of research not explicitly addressed by the contributions summarized in Table 1 relates to the modeling of floating structures excited by highly impulsive breaking waves. Wave breaking is known to impart

https://doi.org/10.3390/jmse14070640

short-duration loads up to 50 times higher than their nonbreaking counterparts [9]. Understanding the response of floating structures under impulsive conditions will be of critical importance in ensuring their operability within increasingly severe ocean environments exacerbated by a changing climate.

The Guest Editors thank all authors for their contributions, the reviewers for their careful and constructive feedback, and the *JMSE* editorial staff for their support throughout the review process.

Funding: This research received no external funding.

Acknowledgments: The Editors gratefully acknowledge all authors, reviewers, and technical editors who contributed to the success of this Special Issue.

Conflicts of Interest: The authors declare no conflicts of interest.

List of Contributions:

1. Lyu, W.; Falzarano, J. Numerical Analysis and Modeling of a Semi-Submersible Floating Wind Turbine Platform with Large Amplitude Motions Subjected to Extreme Wind and Wave Loads. *J. Mar. Sci. Eng.* **2025**, *13*, 243. https://doi.org/10.3390/jmse13020243.
2. Srinivasamurthy, S.; Pandit, A.; Yoshida, S. Collapse Behavior of Onshore and Spar-Floating Wind Turbine Towers During Blade Pitch Malfunction. *J. Mar. Sci. Eng.* **2026**, *14*, 378. https://doi.org/10.3390/jmse14040378.
3. Lee, Y.; Park, J.; Chung, W.C. Dynamic Response of a Floating Dual Vertical-Axis Tidal Turbine System with Taut and Catenary Mooring Under Extreme Environmental Conditions in Non-Operating Mode. *J. Mar. Sci. Eng.* **2025**, *13*, 1315. https://doi.org/10.3390/jmse13071315.
4. Yuan, J.; Ma, F.; Zhang, M.; Shen, K.; Tang, J. The Time-Domain Design Stress Method for Fatigue Analysis of the Reactor Pressure Vessel in Floating Nuclear Power Plants. *J. Mar. Sci. Eng.* **2025**, *13*, 235. https://doi.org/10.3390/jmse13020235.
5. Gong, L.; Cui, Y.; Du, Y.; Qin, L.; Zhao, X. Morphology and Effect of Load on Bridge Piers Impacted by Continuous Sea Ice. *J. Mar. Sci. Eng.* **2024**, *12*, 1871. https://doi.org/10.3390/jmse12101871.
6. Lee, I.; Jin, C.; Kim, S.-J.; Kim, M. On the Static Stability and Seakeeping Performance of a Submerged Floating Tunnel Module in Wet Tow. *J. Mar. Sci. Eng.* **2025**, *13*, 77. https://doi.org/10.3390/jmse13010077.
7. Qu, F.; Chen, S.; Zhang, K. The Numerical Analysis of Hydrodynamic Response and Structural Stability of an Eccentric Conical Floating Structure. *J. Mar. Sci. Eng.* **2025**, *13*, 1032. https://doi.org/10.3390/jmse13061032.
8. Smith, S.; Wu, G.; Pawitan, K.A.; Garlock, M. Hyperbolic Paraboloid Free-Surface Breakwaters: Hydrodynamic Study and Structural Evaluation. *J. Mar. Sci. Eng.* **2025**, *13*, 245. https://doi.org/10.3390/jmse13020245.
9. Jia, X.; Guo, S.; Wang, K.; Fu, S.; Yu, X.; Peng, W. Wave Attenuation Performance of a Floating Breakwater Integrated with Flexible Wave-Dissipating Structures. *J. Mar. Sci. Eng.* **2026**, *14*, 97. https://doi.org/10.3390/jmse14010097.
10. Kim, S.-J.; Jeong, S.-J.; Park, S.-J. Hydrodynamic Performance and Mooring Safety Assessment of an Offshore Floating Movable Fish Cage. *J. Mar. Sci. Eng.* **2024**, *12*, 2351. https://doi.org/10.3390/jmse12122351.

References

1. Zeng, X.; Shao, Y.; Feng, X.; Xu, K.; Jin, R.; Li, H. Nonlinear Hydrodynamics of Floating Offshore Wind Turbines: A Review. *Renew. Sustain. Energy Rev.* **2024**, *191*, 114092. [CrossRef]
2. Dai, J.; Wang, C.M.; Utsumomiya, T.; Duan, W. Review of Recent Research and Developments on Floating Breakwaters. *Ocean Eng.* **2018**, *158*, 132–151. [CrossRef]
3. Di Pilato, M.; Perotti, F.; Fogazzi, P. 3D Dynamic Response of Submerged Floating Tunnels under Seismic and Hydrodynamic Excitation. *Eng. Struct.* **2008**, *30*, 268–281. [CrossRef]

4. Dal Bo Zanon, B.; Roeffen, B.; Czapiewska, K.M.; de Graaf-Van Dinther, R.E.; Mooij, P.R. Potential of Floating Production for Delta and Coastal Cities. *J. Clean. Prod.* **2017**, *151*, 10–20. [CrossRef]
5. Wang, G.; Rosenfeld, Y.; Drimer, N.; Goldfeld, Y. Occupant Comfort Analysis for Rigid Floating Structures—Methodology and Design Assessment for Offshore Dwelling Module. *Ships Offshore Struct.* **2021**, *16*, 184–199. [CrossRef]
6. Pimenta, F.; Ribeiro, D.; Roman, A.; Magalhaes, F. Predictive Model for Fatigue Evaluation of Floating Wind Turbines Validated with Experimental Data. *Renew. Energy* **2024**, *223*, 119981. [CrossRef]
7. Wang, S.; Chuang, W.-L. Dynamic Analysis of Breaking Wave Impact on a Floating Offshore Wind Turbine via Smoothed Particle Hydrodynamics. *Mar. Struct.* **2025**, *100*, 103731. [CrossRef]
8. Chen, H.; Hall, M. CFD Simulation of Floating Body Motion with Mooring Dynamics: Coupling MoorDyn with OpenFOAM. *Appl. Ocean Res.* **2022**, *124*, 103210. [CrossRef]
9. Cuomo, G.; Allsop, W.; Bruce, T.; Pearson, J. Breaking Wave Loads at Vertical Seawalls and Breakwaters. *Coast. Eng.* **2010**, *57*, 424–439. [CrossRef]

Disclaimer/Publisher's Note: The statements, opinions and data contained in all publications are solely those of the individual author(s) and contributor(s) and not of MDPI and/or the editor(s). MDPI and/or the editor(s) disclaim responsibility for any injury to people or property resulting from any ideas, methods, instructions or products referred to in the content.

https://doi.org/10.3390/jmse14070640

Journal of
Marine Science and Engineering

MDPI

Article

Numerical Analysis and Modeling of a Semi-Submersible Floating Wind Turbine Platform with Large Amplitude Motions Subjected to Extreme Wind and Wave Loads

Weishan Lyu * and Jeffrey Falzarano

Department of Ocean Engineering, Texas A&M University, College Station, TX 77840, USA; jfalzarano@tamu.edu
* Correspondence: lyuweishan@tamu.edu

Abstract: The objective of this study is to predict the large amplitude motions of floating wind turbine platforms and to emphasize the significance of nonlinear forces when these platforms are subjected to combined wind and wave loads. The analysis utilizes the 5 MW OC4 semi-submersible model. First, we couple the OpenFAST v3.1.0 with SIMDYN, validate the effectiveness of the coupled program, and highlight the considerable impact of nonlinearity on the results, particularly in relation to the heave and pitch motions of offshore wind platforms under extreme environmental conditions. We then discuss the primary reasons for this phenomenon. Ultimately, this study proposes an optimized model aimed at mitigating the nonlinear effects associated with such conditions.

Keywords: large amplitude motion; semi-submersible floating wind turbine; blended time-domain method; nonlinearity; optimization

1. Introduction

Wind power is an endlessly renewable energy source that can be harnessed both onshore and offshore. Fixed wind and floating wind are two distinct types of offshore wind energy systems that harness the power of wind at sea to generate electricity. Fixed wind refers to traditional offshore wind turbines that are directly anchored to the seabed. These turbines are generally installed in shallower waters, typically less than 50–60 m deep. Fixed wind technology is the more established and widely deployed method of offshore wind generation, with numerous operational offshore wind farms, particularly in Europe, such as those located in the North Sea. However, its application is constrained by the depth of the water. Floating wind technology refers to offshore wind turbines that are not anchored directly to the seabed but are instead mounted on floating platforms tethered to the ocean floor. This innovative approach allows for deployment in much deeper waters than traditional fixed turbines, thus unlocking vast new areas for offshore wind development. Typically, they are utilized in waters deeper than 60 m and can be positioned further offshore. Common types of floating wind platforms include spar and semi-submersible designs, and one of the popular models is OC4 semi-submersible [1]. Floating platforms can be arranged in larger and more expansive wind farms, enhancing overall energy production capacity. However, it is important to note that floating wind technology is still in the early stages of development compared to its fixed counterparts. Currently, offshore wind platforms are receiving increasing attention due to the stronger and more predictable wind conditions. The number of floating offshore wind turbines (FOWTs) has grown rapidly in recent years, highlighting the increasingly critical role of offshore wind energy in global energy systems. Despite the commercialization of FOWT

https://doi.org/10.3390/jmse13020243

technology, there are still numerous technical challenges to be addressed. As offshore wind farms are installed in deeper water depths, wind velocity increases, necessitating longer wind blades and leading to stronger wind loads, potentially causing significant platform motion. This motion, in turn, may exert higher forces on the blades and tower, creating complex and fluctuating aerodynamic conditions [2]. Therefore, ensuring the stability of the platform is of the utmost importance.

Researchers have employed various methods to investigate the response of floating platforms. Vorpahl [3] demonstrated the environmental loads, the relevant tools, and the standards used on offshore wind turbines. Ferrandis investigated the influence of the viscosity and the nonlinearities for FOWTs in regular waves through the CFD method [4]. Wang and Chen used computational fluid dynamics (CFD) to study the hydrodynamic response of FOWT semi-submersibles and the influence of viscosity and nonlinearities in regular waves [5], respectively. They also validated the responses of a semi-submersible under random waves via the CFD method at the same time [6]. Wan and Xia used the viscous flow solver naoe-FOAM-SJTU based on OpenFOAM to analyze semi-submersible platform motions [7]. Teng proposed a method to simulate the nonlinear response of moored floating platforms by dividing Quasi-Static Transfer Functions (QTFs) into different components [8]. Additionally, Katarzyna studied the dynamic response of the spar model by analyzing the instantaneous center of rotation of FOWTs [9]. Meanwhile, Jameel conducted a dynamic analysis of coupled spar platforms in deep water [10]. It is evident that researchers employ diverse approaches to study the predictions of floating platform motions. Wang and Roberston enhanced OpenFAST to predict the nonlinear, low-frequency responses of the semi-submersible model [11]. Wind loads often do not receive adequate attention in research, which typically emphasizes wave loads instead. FOWT systems are complex, requiring a holistic approach to analysis. To effectively evaluate FOWT systems, it is essential to consider aerodynamic and hydrodynamic calculations and mooring design, as each component interacts with the others. However, many studies overlook the critical importance of examining the system as a whole.

Considering the above issues, the most popular software is FAST v8.16, updated as OpenFAST and with more functions, which the National Renewable Energy Laboratory developed [12] (Golden, CO, USA). This multi-physics solver can simulate the coupled dynamic responses of FOWTs. Many researchers coupled OpenFAST or FAST with their own programs. Bae and Kim coupled Charm3D-FAST [13,14] to study the performance of the floating wind platform. Yang coupled FAST with AQWA to examine the interaction between the floaters of the multi-body platform [15]. However, the above works are based on normal conditions and do not account for extreme conditions. Due to the stable motions observed, nonlinearity was not considered in these studies.

In recent years, extreme conditions have become more frequent due to climate change. Rising temperatures are causing more intense weather events such as storms, heatwaves, and large waves. Additionally, melting glaciers and the expansion of seawater are raising sea levels, increasing the risk of coastal flooding and storm surges, which is posing a serious threat to the safety of coasts and marine structures. In this paper, we mainly focus on the high waves and storms in the ocean. These extreme conditions are typically driven by severe weather events, including hurricanes, typhoons, and cyclones, and are characterized by powerful winds and large waves. In Journee and Massie's book, they define the extreme sea state [16]. Understanding the response of FOWTs in extreme environments has posed a significant challenge in recent years. When external loads are substantial, it is inadequate to consider only linear forces. Jang and Kim emphasized the importance of nonlinear Froude–Krylov forces and nonlinear hydrostatic restoring forces in the motions of arctic spars [17]. Similarly, Bandyk, Rodríguez, and Rajendran conducted comparable analyses on ship performance by incorporating nonlinear forces [18–20]. However, there remains a limited amount of research focused on predicting platform motions under the combined impacts of wind and wave loads while considering these nonlinear forces. Further investigation is necessary to thoroughly explore the effects of nonlinear forces on FOWTs.

Floating structures interact with waves, currents, and winds in complex ways. When it comes to extreme conditions, nonlinear forces offer a more precise representation of these interactions than linear models. The dynamics of these nonlinear forces can impact the stability of the structure. In cases where structures respond nonlinearly to hydrodynamic forces, they may exhibit behaviors not accounted for in linear models, such as increased response amplitudes. Nonlinear interactions can result in resonant phenomena, where the structure's natural frequencies align with wave frequencies, leading to amplified responses. This is a crucial consideration when designing systems to mitigate such effects. Accurately predicting nonlinear forces is vital for assessing structural integrity and ensuring safety, as well as for optimizing design and reducing the risk of catastrophic failures. The novelty of this study is to analyze the nonlinear response of motions under extreme wind and wave loads in floating offshore wind turbine systems, which is seldom considered currently.

Based on previous research conducted at the Texas A&M University Marine Dynamic Lab, we utilized the SIMDYN program, which was developed by the lab (College Station, TX, USA). This program operates in the time domain and is used in solving both linear and nonlinear governing equations. Its nonlinear capabilities account for the nonlinearity of Froude–Krylov and hydrostatic forces. A notable distinction of SIMDYN is the derivation of nonlinear forces, specifically the Froude–Krylov force and restoring force, which are obtained from the platform's instantaneous position and orientation. In contrast, other numerical methods derive the Froude–Krylov force and the restoring force from the averaged positions of the platform.

Therefore, this paper begins by demonstrating the trend and background of FOWTs and illustrating the current numerical methods used in this field and then pointing out the inadequate parts of the current study and briefly introducing the importance and breakthrough of our work. Then, it explains how to integrate SIMDYN with OpenFAST and details the evaluation of linear and nonlinear hydrodynamic loads. Following this, the program's accuracy will be established, and several cases under varying conditions will be presented and discussed. Ultimately, the paper will propose an optimized platform shape model based on the preceding calculations. The conclusions and implications of this work are discussed last.

2. SIMDYN-OpenFAST Program

There are various tools available for evaluating the performance of FOWTs, with OpenFAST being the most widely used. OpenFAST is a multi-physics solver capable of simulating the coupled dynamic responses of FOWTs. It effectively models the interactions between wind turbines, wind, their supporting structures, and control systems. The OpenFAST framework encompasses several modules, each tailored to simulate different aspects of a wind turbine system. As a comprehensive toolset, OpenFAST enables users to investigate wind turbine performance across a range of scenarios. Its modular design allows for a focused analysis of specific subsystems, which can then be integrated for full-system simulations.

OpenFAST is comparable to other simulation tools such as FAST (its predecessor), ANSYS, Bladed, and OpenWind. However, it distinguishes itself by being open source, which fosters greater transparency and customization. Additionally, OpenFAST supports both land-based and offshore wind turbine systems, providing flexibility in the models it can integrate. Therefore, OpenFAST is an advanced tool designed for simulating the performance of wind turbines across a diverse array of conditions. With its flexibility, comprehensive capabilities, and high accuracy, OpenFAST has become an indispensable resource in the wind turbine industry.

Our research group extensively studied different model types using SIMDYN and verified its accuracy across various fields. Somayajula and Falzarano demonstrated the effectiveness of SIMDYN through an example of parametric roll simulation of a container ship [21], showcasing its ability to accurately analyze nonlinear time-varying phenomena. In a separate study, Jose coupled FAST with SIMDYN to study the classic instability of negative damping that occurs in FOWTs [22]. Additionally, Wang and Falzarano integrated SIMDYN with MAP to predict the motions of a wave energy converter with mooring lines [23].

In our current work, we have coupled SIMDYN with OpenFAST to examine the large amplitude platform motions of a semi-submersible floating wind turbine under extreme wind and wave loads. This involved replacing HydroDyn, a time-domain hydrodynamics module in OpenFAST, which offers hydrodynamic simulations specifically for offshore wind turbines, modeling the interaction between turbine platforms (such as floating structures) and water and incorporates the effects of wave motion, currents, and the platform's movement in its analysis. The modules are illustrated in Figure 1.

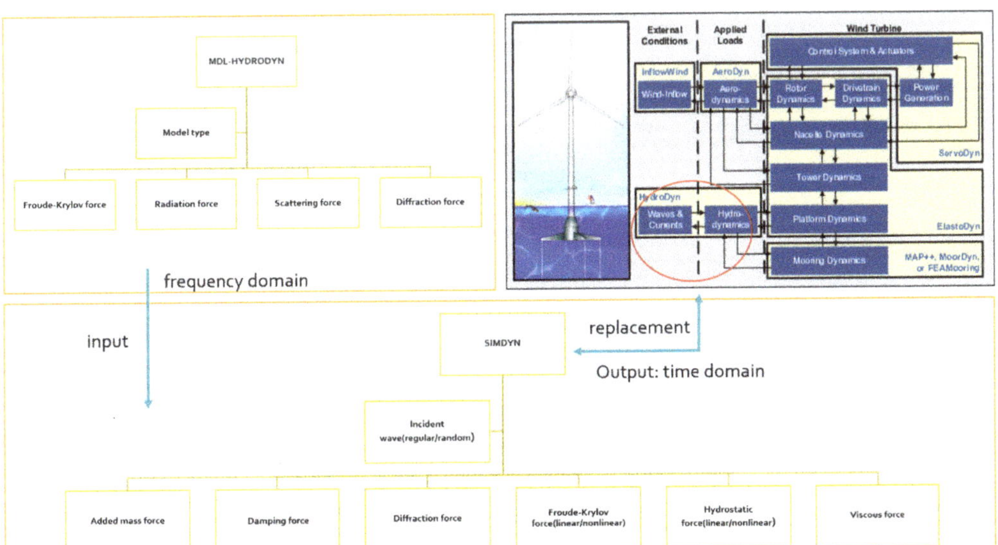

Figure 1. The modules of OpenFAST coupled with SIMDYN.

HydroDyn has the capability to calculate the linear hydrostatic force, added mass, and damping terms arising from linear wave radiation. The wave excitation term is derived from first- and second-order diffraction. In the context of SIMDYN, the real-time inertial force, damping force, hydrostatic force, and incident wave force can also be assessed. These forces are input into the SIMDYN-OpenFAST program, which then provides the acceleration at a subsequent time for the next calculation step. It is important to note that HydroDyn is based on the Wamit output, while SIMDYN relies on MDLHydroDyn [24], a research product developed by the Marine Dynamics Laboratory at Texas A&M University (College Station, TX, USA), and has been verified. SIMDYN-OpenFAST offers two calculation methods. The first is based on linear analysis, utilizing an algorithm similar to the HydroDyn module. This method will be used to verify the accuracy of our program against results from OpenFAST. The second method is based on nonlinear analysis, which will be the primary focus of this paper.

3. Theoretical Background

The general parameters of the semi-submersible platform are present in Table 1, and the model is shown in Figure 2.

Table 1. The semi-submersible platform general parameters.

Water Plane area (m^2)	368
Displacement (m^3)	13,464
Center of Buoyancy (z) (m)	−13
Center of Gravity (z) (m)	−5
Metacentric Height GM (m)	2
Water Depth (m)	320
Wave direction (deg)	180
Wave type	Regular/Random Wave
Depth of platform base below SWL (total draft) (m)	20
Elevation of main column (tower base) above SWL (m)	10
Diameter of main column (m)	6.5
Diameter of offset (upper) columns (m)	12
Diameter of base columns (m)	24

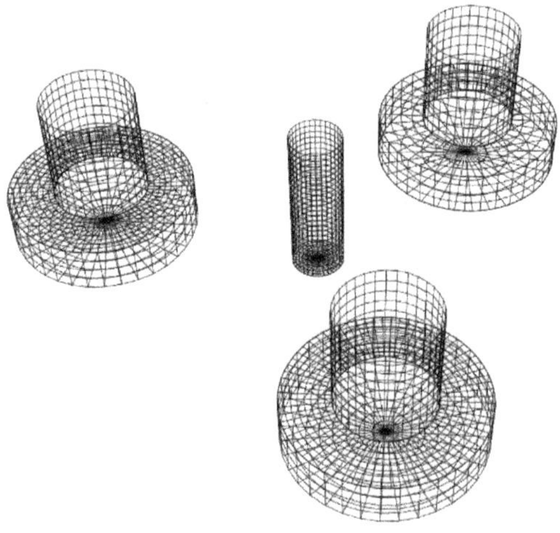

Figure 2. The semi-submersible platform model.

The frequency domain analysis is the first step in conducting a time domain analysis. The wave force and the radiation force are frequency dependent. We assume the fluid should be inviscid, incompressible, irrotational, and then we obtain the velocity potential as follows:

$$\varnothing(x,t) = \left(\varnothing_I(x,\beta,\omega_I) + \varnothing_D(x,\beta,\omega_I) + \sum_{j=1}^{6} \eta_j \varnothing_j(x,U,\omega_e) \right) e^{i\omega_e t} \tag{1}$$

where

$\varnothing_I$ is the incident wave potential;

$\varnothing_D$ is the diffracted wave potential;

$\varnothing_j$ is the radiation potential;

ω_e is the encounter frequency, which should be the same as wave frequency ω for FOWTs.

When we obtain the velocity potential, the hydrodynamic coefficients, added mass coefficients, and damping coefficients, can be given by the following:

$$A_{jk}^{0} = -\frac{\rho}{\omega_e}\int Im(\varnothing_k)n_j\, ds \tag{2}$$

$$B_{jk}^{0} = -\rho\int Re(\varnothing_k)n_j\, ds \tag{3}$$

Then, we obtain the nondimensional frequency domain properties of the semi-submersible shown in Figure 3 from MDLHydroDyn.

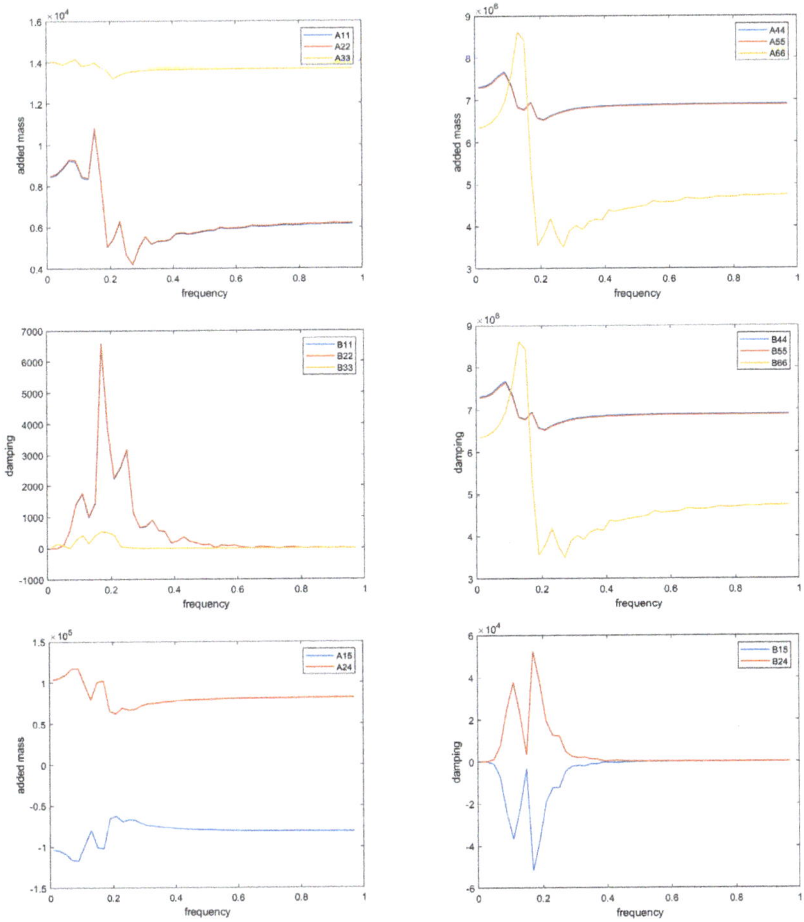

Figure 3. The nondimensional frequency domain of added mass and damping.

The governing equation for the whole system is the following:

$$(M + A)\ddot{x}(t) + \int_0^t K(t - \tau)\dot{x}(\tau)d\tau = F(t) \tag{4}$$

The motion vector x(t) represents the displacements of six degrees of freedom (DOF). M is the mass or inertia term, and A is the added mass or added inertia term. K is the damping term. $K(t - \tau)$ is the memory kernel, which is related to the radiation damping and restoring forces of the system. F(t) represents the external forces and moments, including hydrodynamic loads from waves and aerodynamic loads from wind, acting on the FOWT. The main loads applied to the system are hydrodynamic loads and wind loads. The hydrodynamic load consists of diffraction force and Froude–Krylov force induced by incoming waves. The total force in the hydrodynamic system can be expressed as follows:

$$F_{total} = F_{\text{FK}} + F_{diffraction} + F_{radiation} + F_{viscous} + F_{restoring} + F_{scattering} \tag{5}$$

The radiation forces are created by the motion of the floating body, which makes waves. The viscous forces are generated by the fluid's viscosity (seawater). The restoring forces are caused by the difference in the weight and buoyancy and the effect of the mooring system. The linear radiation and scattering forces are obtained from MDLHydroDyn. Based on the instantaneous position and orientation, the nonlinear Froude–Krylov forces, the nonlinear hydrostatic forces, and the nonlinear inertial forces can be computed.

For this paper, we calculated the nonlinear hydrodynamic forces and Froude–Krylov forces by SIMDYN, incorporating Wheeler stretching up to the instantaneous position and orientation. The underwater pressure is given by the following:

$$p(t, x, y, z) = -\rho \frac{\partial \varnothing_I}{\partial t}(t, x, y, \hat{z}) - \frac{\rho}{2}|\nabla \varnothing_I(t, x, y, \hat{z})|^2 \tag{6}$$

where

$$\hat{z} = z - \eta(t, x, y) \leq 0$$

The first term is the linear term, and the second term is the nonlinear term. The forces and moments are obtained by integration of the pressure over the instantaneous wetted surface area S_B:

$$F_I(t) = \int P(t, x, y, z) \cdot \boldsymbol{n} dS \tag{7}$$

$$M_I(t) = \int P(t, x, y, z) \cdot (\boldsymbol{x} \times \boldsymbol{n}) dS \tag{8}$$

The frequency domain program calculates the scattering wave force based on the wave force RAO. The Fourier transform of the incident wave is used to obtain the force in the time domain. In the case of a single-wave excitation frequency, the radiation force has two components, one due to the added mass and the other due to radiation damping.

$$F_{radiation} = -A(\omega)\ddot{x} - B(\omega)\dot{x} \tag{9}$$

Restoring forces are obtained using the instantaneous buoyancy and weight of the platform.

$$F_{restoring} = \int -\rho g z \cdot \boldsymbol{n} dS + \boldsymbol{W} \tag{10}$$

$$M_{restoring} = \int -\rho g z \cdot (\boldsymbol{x} \times \boldsymbol{n}) dS + (\boldsymbol{x_G} \times \boldsymbol{W}) \tag{11}$$

In order to calculate the viscous force, the Morison equation is used, which is the most important part of damping forces.

4. Simulation Results and Discussion

4.1. The Program Verification

In this paper, we will start by applying the random wave without wind load; the significant wave height is 5 m, and the wave period is 7 s. The results are presented in Figures 4–7.

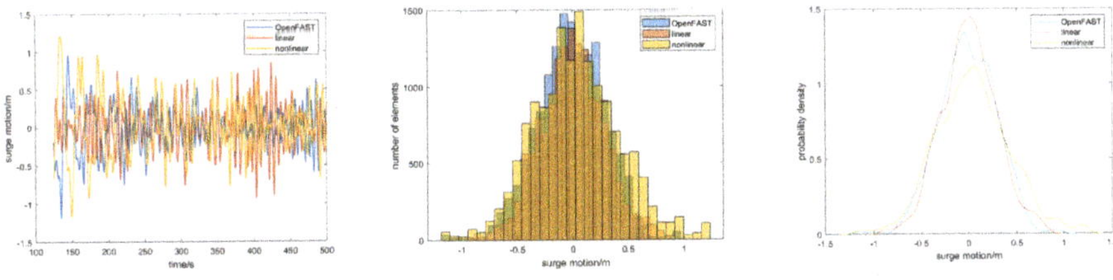

Figure 4. The response of the surge motion.

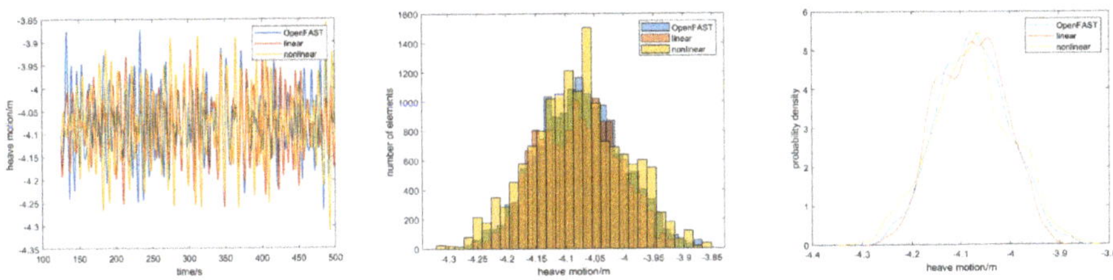

Figure 5. The response of the heave motion.

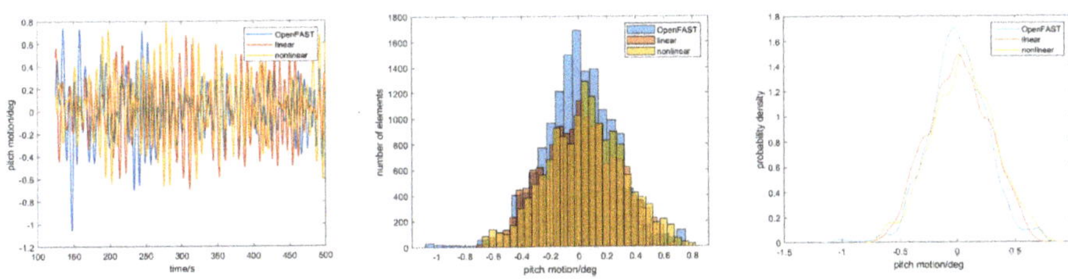

Figure 6. The response of the pitch motion.

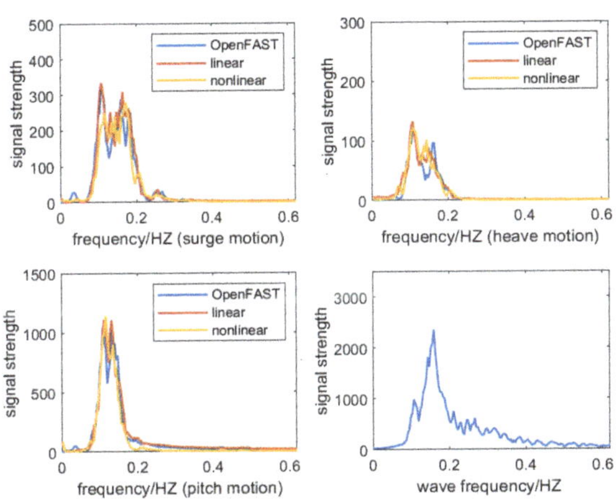

Figure 7. The response in the frequency domain.

Then, we apply regular waves to the platform, as these waves can accurately estimate much of the platform's response and serve as a clear benchmark for assessing the accuracy of the simulation. In our test model, we examine a typical scenario where the wave height is 10 m, and the wave period is 7 s. The results are presented in Figure 8.

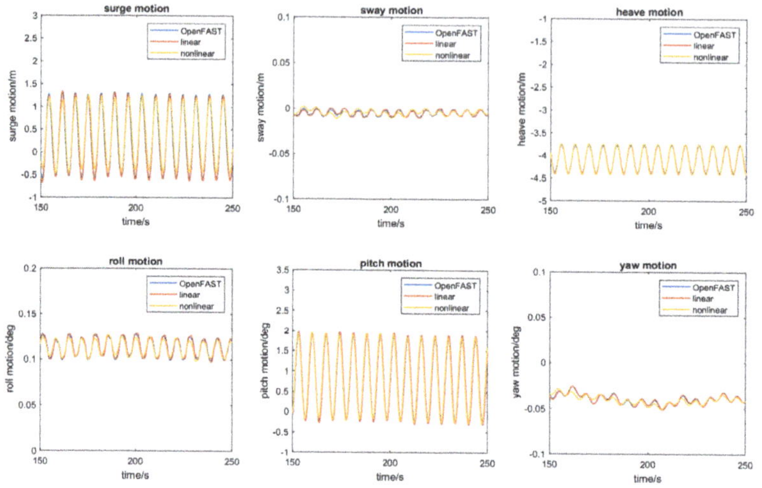

Figure 8. A comparison of the results of OpenFAST and OpenFAST coupled with SIMDYN.

From Figure 6, it is obvious that, for random cases, the motions' range generated by OpenFAST and the linear and nonlinear modules of SIMYDYN-OpenFAST are the same, and the distributions are very similar. Random cases cannot be entirely consistent, not only because of the randomness of the external loads but also because of the nonlinearity of the system, including the interactions from other modules. The program is capable of accurately predicting response ranges in all directions. Additionally, frequency domain analysis reveals that the majority of responses from the three methods fall within the same frequency range, which can be obtained from Figure 7. However, due to the limitations of the Fast Fourier Transform (FFT), converting non-stationary motions from the time domain to the frequency domain in a nonlinear system lacks precision, which means that the results from the three methods cannot be identical. To enhance clarity in our study, we utilized a filtering technique to minimize noise, leading to a smoother dataset. The close clustering of most response peaks is quite evident. Furthermore, we have included the wave amplitude in the frequency domain as a reference to support the responses of those motions. To fully explore the nonlinear characteristics of the response—such as the bifurcation diagram, chaos, or the nonlinear magnification curve—our current analysis cannot delve deeply into these aspects due to the numerous nonlinear relationships present in the system, which also introduce various noises. While we can employ filters to smooth the data, it is important to note that this process may obscure certain "special cases" arising from these nonlinear relationships. Therefore, at this stage, we can only ascertain the range of frequencies involved. For a more thorough investigation into the nonlinear characteristics, it would be advantageous to analyze specific motions or pairs of coupled motions independently [25]; however, this falls outside the current scope of our work, which should be categorized under nonlinear dynamic control theory; more details can be found in Ref [26].

To illustrate our findings clearly, we also perform a regular case study; the results are shown in Figure 8. The responses in six directions are very closed, especially the results from OpenFAST and the linear module of SIMDYN-OpenFAST, and there are some 'coupled errors', which are very small and can be ignored. The nonlinear module differs slightly from other results due to the different calculation theories, which we explained in Section 2.

From the above analyzed cases, we observe that, when platform motion is minimal, there is no significant difference between the results generated by OpenFAST and the linear and nonlinear modules of SIMYDYN-OpenFAST. This observation aligns with our previous research on response analysis in ships [21], barges [22], and wave energy converters [23]. These figures highlight the accuracy of our program and validate the consistency between OpenFAST and the linear module of SIMYDYN-OpenFAST. Furthermore, it indicates that, when wave loads remain within normal ranges, nonlinear forces do not significantly impact the calculations due to the relatively small variations in instantaneous positions. Since the sway, roll, and yaw responses are so small, they will not be discussed in this paper.

4.2. The Impacts of Nonlinear Forces

In order to examine the impacts of the nonlinear forces, extreme conditions are applied. Table 2 outlines several extreme wave conditions that will serve as the reference points for this study.

Table 2. The wave heights and periods corresponding to different hurricanes.

Hurricanes Name	Fran	Lili	Georges	Floyd
Wave height(m)	11.64	11.20	10.88	14.20
Periods (s)	14.29	13.25	13.2	15.4

In the random case analyzed, the significant wave height is measured at 15 m, with a wave period of 14 s. The random wind generated by TurbSIM v2 [27] has an average speed of 18 m/s. The results are presented in Figure 9. We chose to conduct the random case first, as it holds greater significance in real-world scenarios. It is evident that the range of responses is larger in nonlinear situations. The surge direction response remains relatively unchanged compared to the other directions due to the mooring system. As external random loads increase, the heave response becomes larger, and the pitch angle tilts more in the load direction, which is expected. However, such phenomena are not identifiable in linear modes. We also performed a frequency domain analysis, as completed previously, to determine if there were any notable frequencies that might impact our platforms. In our case, we did not observe any. However, other notable frequencies could arise in other scenarios, contingent on the relevant design factors of the system.

We then examine a regular wave with a height of 13 m, a period of 14 s, and a consistent wind speed of 18 m/s. The results are depicted in Figure 10. The trends observed are quite similar to those in the random case. In the time domain, it is evident that, in the surge motion, the response did not change much, and the responses in the heave and pitch motions became larger.

Therefore, the results from the random and regular load cases indicate that the primary responses of concern are heave and pitch. These two motions are closely coupled during instantaneous calculations, which highlights the importance of nonlinear forces, especially when the motion amplitude is large under extreme conditions.

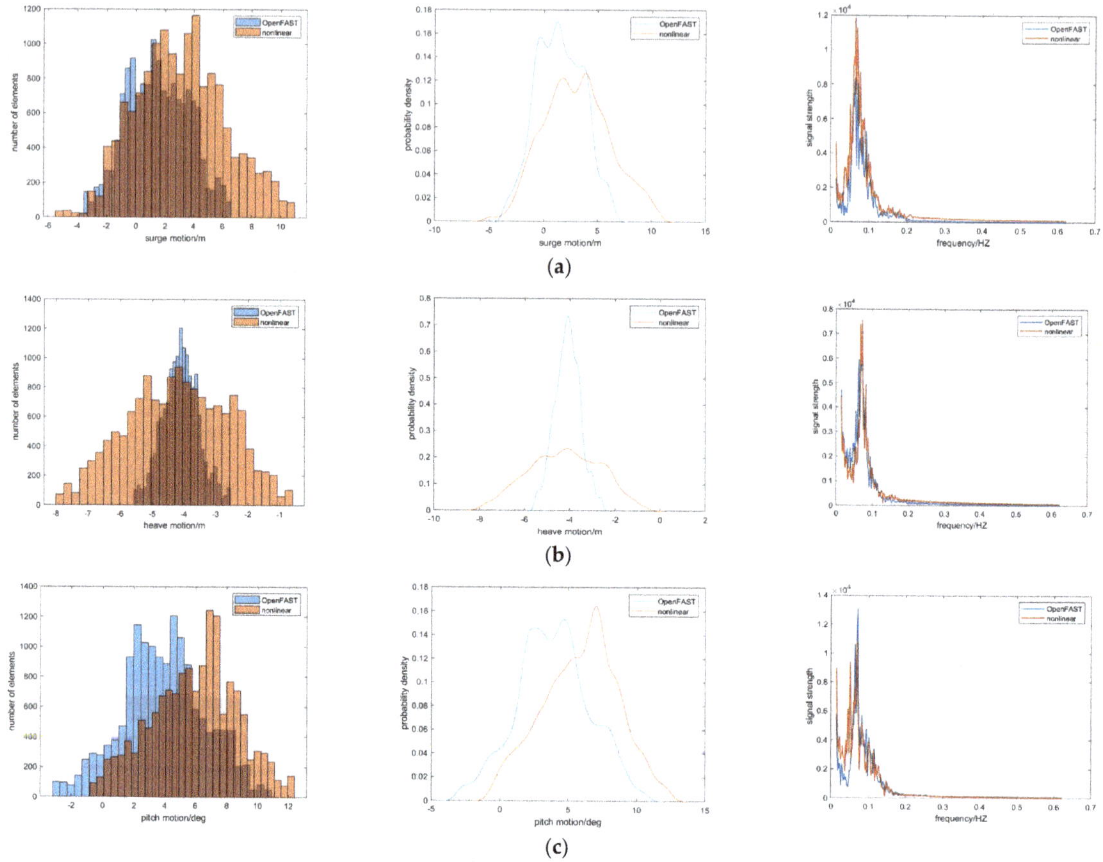

Figure 9. A comparison of the results of OpenFAST and OpenFAST coupled with SIMDYN. (**a**) Responses in surge direction. (**b**) Responses in heave direction. (**c**) Responses in pitch direction.

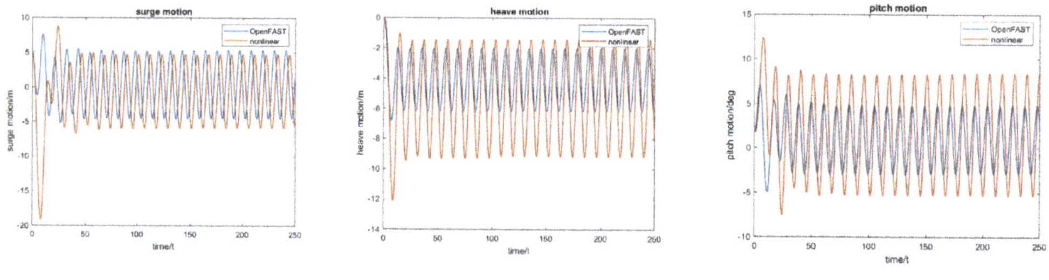

Figure 10. A comparison of the results of OpenFAST and OpenFAST coupled with SIMDYN.

4.3. The Reason Discussion

The nonlinearity arises from both the Froude–Krylov force and the restoring force. To assess which of these forces plays a more significant role in the overall nonlinear behavior, we conducted a probability density estimate; using a normal kernel function helps to estimate the probability distribution of a set of data points. This method is part of the Kernel Density Estimation (KDE), which is a way to estimate the probability density function (PDF) of a random variable without making specific assumptions about the data. In this method, a normal distribution, also known as a Gaussian kernel, is placed at each data point. These normal distributions are then added together to create a smooth estimate of the overall distribution. The results of this analysis are presented in Figures 11 and 12.

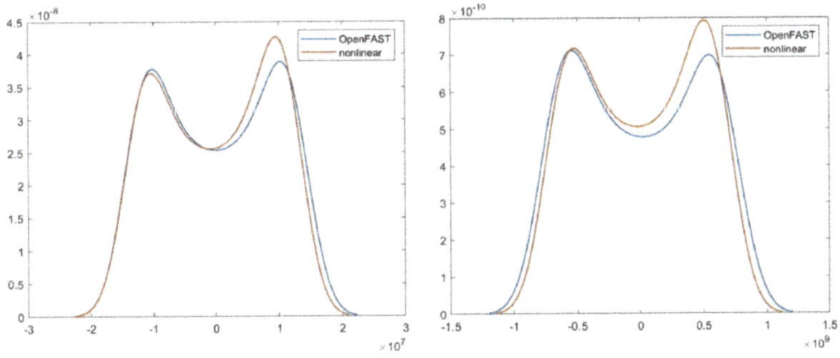

Figure 11. A comparison of the density estimate for the Froude–Krylov force in heave and pitch directions. Note: x axis presents Froude–Krylov force; y axis presents probability density.

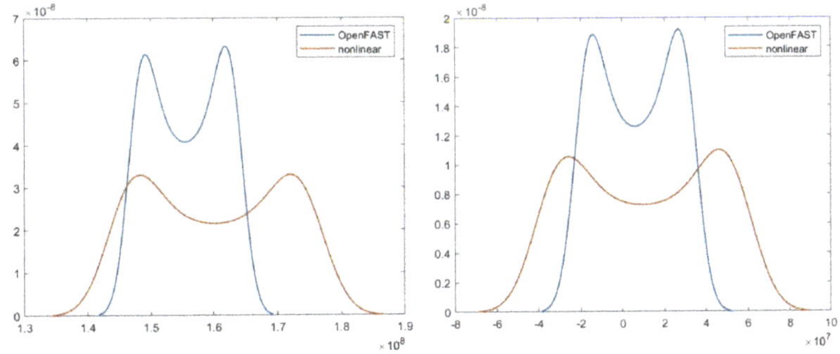

Figure 12. A comparison of the density estimate for the hydrostatic force in heave and pitch directions. Note: x axis presents hydrostatic force; y axis presents probability density.

Figures 11 and 12 show the distributions of the Froude–Krylov force and hydrostatic force from different methods. It is evident that the force distributions for the Froude–Krylov force from both methods are quite comparable, especially in contrast to the findings related to the hydrostatic force. Notably, the range of the nonlinear hydrostatic force is considerably broader than that of the linear method's results. Moreover, as shown in Figure 10, the range of the heave motion is substantial, suggesting that the amplitude of the platform's instantaneous buoyancy and weight varies considerably during nonlinear analysis. This likely explains why the primary contributor to nonlinearity in these cases is the force associated with nonlinear hydrostatics. In contrast, while the nonlinear Froude–Krylov force increases, the symmetry of the platform results in minimal changes to the instantaneous wetted surface. However, for asymmetric platforms, greater attention should be given to the nonlinear Froude–Krylov force.

To stabilize the platform's motion during large amplitude movements, it is essential to enhance the restoring force. One effective approach is to increase the wetted surface area while maintaining the positions of the centers of gravity and buoyancy to stabilize the heave motion. When subjected to large amplitude motion, the instantaneous buoyancy does not change significantly, which can help reduce the impacts of the nonlinear restoring force. Our team has previously developed an optimization framework utilizing genetic algorithms for the automated parametric optimization of the Octabuoy semi-submersible design [28]. In light of this recommendation, we present one of the optimization methods from an engineering perspective, deliberately omitting any economic analysis.

4.4. Optimization Method for Large Amplitude Platform Motion

The model is shown in Figure 13, and the general parameters are listed in Table 3.

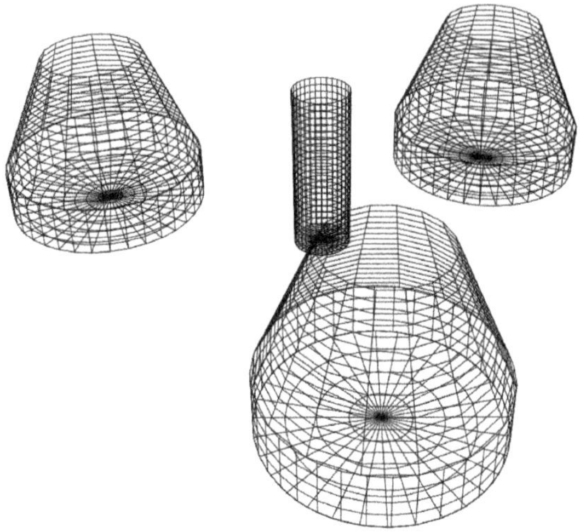

Figure 13. The optimized model.

Table 3. General parameters of the optimized platform.

Water Plane area(m^2)	368
Displacement(m^3)	19,986
Center of Buoyancy (z) (m)	−12
Center of Gravity (z) (m)	−5
Metacentric Height GM (m)	0.82
Water Depth (m)	320
Wave direction (deg)	180
Wave type	Regular wave
Depth of platform base below SWL (total draft) (m)	20
Elevation of main column (tower base) above SWL (m)	10
Diameter of main column (m)	6.5
Diameter of offset (upper) columns (m)	12
Diameter of base columns (m)	24

Then, we obtain the nondimensional frequency domain properties of the new model shown in Figure 14 from MDLHydroDyn.

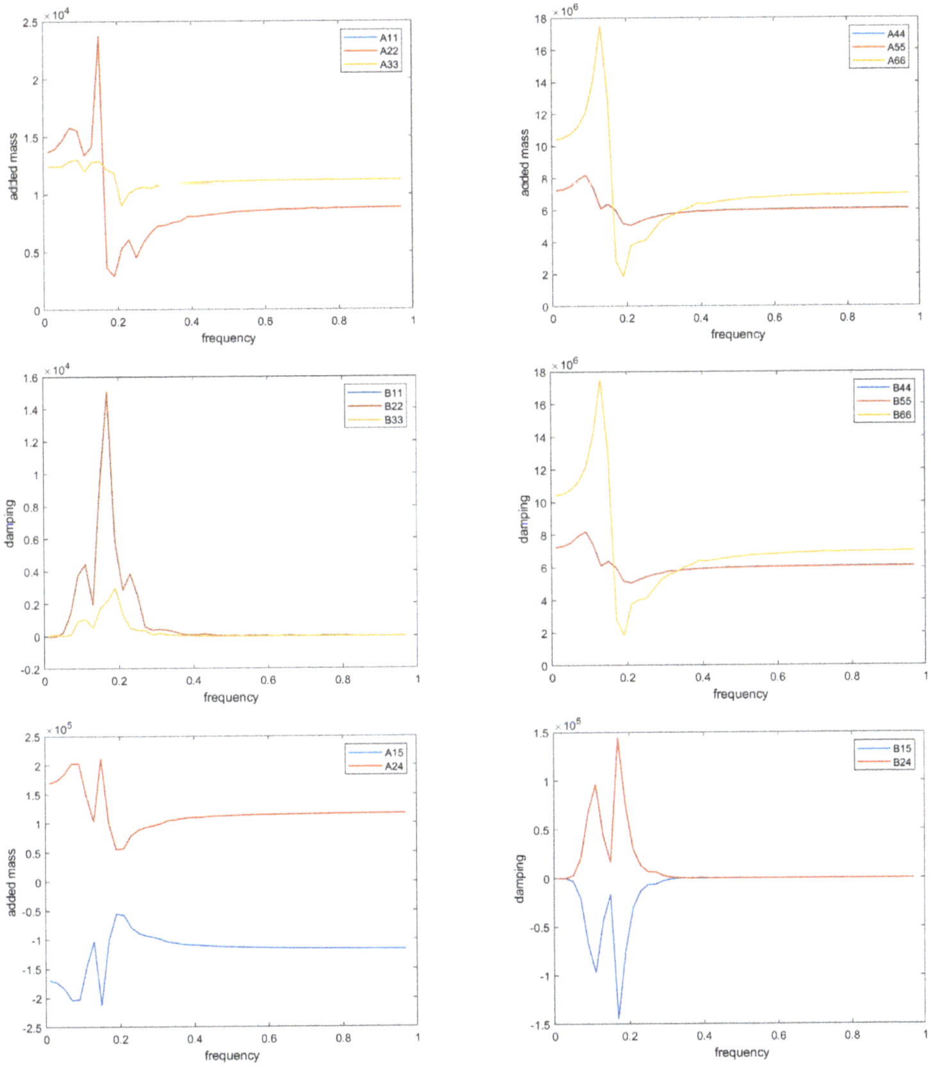

Figure 14. Nondimensional frequency domain of added mass and damping of the new model.

We utilized SIMDYN-OpenFAST and applied identical environmental conditions to the platform, with a wave height of 15 m and a wave period of 14 s. The wind speed is constant at 18 m/w. The resulting data are presented in Figure 15. It shows that the range of the responses in the heave and pitch directions becomes smaller. Our findings indicate that increasing the wetted surface area or enhancing the displacement volume, while maintaining other parameters constant and stabilizing the instantaneous buoyancy location, helps to mitigate the nonlinear effects of hydrostatic forces, particularly in the heave and pitch directions. However, there are limitations, particularly with the increased wet surface area, which introduces new requirements for materials, budget, the fabrication and installation teams. Additionally, optimizing the platform is not easy because we should consider other modules of the FOWTs, like the impacts on the mooring system. So, it is important to note that this revised model offers a design concept for platform optimization; we have not taken into account any additional factors. Additionally, there is a simpler approach to utilizing the control system to stabilize the instantaneous center of buoyancy, but this aspect belongs to control theory, which we will not address in our paper.

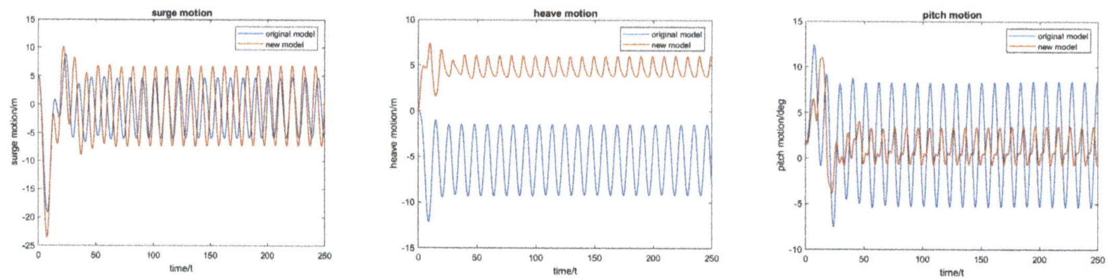

Figure 15. A comparison of the results of the original model and optimized model.

5. Conclusions

This paper examines the responses of a semi-submersible under extreme conditions through the coupled program SIMDYN-OpenFAST, which is designed to estimate the motion of floating offshore wind platforms in both linear and nonlinear contexts, particularly when subjected to varying wind and wave loads. The important results are as follows:

Under normal conditions, when the motion of a semi-submersible is characterized by small amplitudes, the impact of nonlinear forces—specifically the Froude–Krylov force and restoring force—remains minimal in the calculations.

When the amplitudes increase, these nonlinear forces significantly influence the platform's motion, especially in the heave and pitch directions.

Nonlinear hydrostatic forces are primarily responsible for the disparities observed when comparing linear and nonlinear scenarios.

To mitigate the nonlinear effects of hydrostatic forces, one approach is to stabilize the instantaneous buoyancy center to increase the wetted surface area or the displacement volume.

An optimized platform model has been provided that outperforms the original semi-submersible design, leading to a significant reduction in oscillation amplitude. But, it is important to note that this model does not take into account installation or economic considerations, serving primarily as a preliminary design that can be refined for future applications, particularly as environmental conditions or wind blade lengths change.

In this study, no resonance phenomena resulting from nonlinear forces were observed. However, it is possible for such phenomena to occur in extreme circumstances when the frequency of the platform aligns with that of the nonlinear forces. Furthermore, the presence of several modes may contribute to a larger platform's motion.

Therefore, understanding the nonlinear forces acting on floating offshore wind platforms is crucial for ensuring their safety and reliability. Extreme external loads, such as significant waves, wind gusts, and unpredictable environmental factors, can impose substantial forces on these structures. Investigating the nonlinear responses is essential for developing designs capable of withstanding and adapting to these severe conditions, thereby minimizing the risk of structural failure. This study can serve as a valuable reference for platform design, particularly in extreme conditions or when dealing with significant platform motions. During the design phase, especially when large-scale and costly experiments are impractical, it is advisable to analyze extreme conditions to identify performance limitations. It is important to consider the effects of nonlinear forces in such analyses. Furthermore, this study can inform the development of control systems, such as negative feedback control, to stabilize platform motion under challenging conditions. Ultimately, this research offers insights into mitigating the impact of nonlinear forces on the platform, proving beneficial not only for semi-submersible designs but also for spar platforms and other floating platforms.

Author Contributions: Writing—original draft, W.L.; Writing—review & editing, W.L. and J.F.; Project administration, J.F. All authors have read and agreed to the published version of the manuscript.

Funding: The authors would like to thank the Department of Ocean Engineering at Texas A&M University, who provided the teaching assistant funding for this work. The authors also thank Falzarano, who provided the research assistant funding through the TAMU OSSC Consortium project during the summer of 2023.

Institutional Review Board Statement: Not applicable.

Informed Consent Statement: Not applicable.

Data Availability Statement: Data are contained within the article.

Conflicts of Interest: The authors declare no conflicts of interest.

References

1. Robertson, A.; Jonkman, J.; Masciola, M.; Song, H.; Goupee, A.; Coulling, A.; Luan, C. *Definition of the Semisubmersible Floating System for Phase II of OC4*; National Renewable Energy Laboratory: Golden, CO, USA, 2014. [CrossRef]
2. Tran, T.-T.; Kim, D.-H. The platform pitching motion of floating offshore wind turbine: A preliminary unsteady aerodynamic analysis. *J. Wind Eng. Ind. Aerodyn.* **2015**, *142*, 65–81. [CrossRef]
3. Vorpahl, F.; Schwarze, H.; Fischer, T.; Seidel, M.; Jonkman, J. Offshore wind turbine environment, loads, simulation, and design. *Wiley Interdiscip. Rev. Energy Environ.* **2013**, *2*, 548–570. [CrossRef]
4. Ferrandis, J.d.Á.; Bonfiglio, L.; Rodríguez, R.Z.; Chryssostomidis, C.; Faltinsen, O.M.; Triantafyllou, M. Influence of viscosity and non-linearities in predicting motions of a wind energy offshore platform in regular waves. *J. Offshore Mech. Arct. Eng.* **2020**, *142*, 062003. [CrossRef]
5. Wang, Y.; Chen, H.-C.; Koop, A.; Vaz, G. Hydrodynamic response of a FOWT semi-submersible under regular waves using CFD: Verification and validation. *Ocean Eng.* **2022**, *258*, 111742. [CrossRef]
6. Wang, Y.; Chen, H.-C. Verification and validation of computational fluid dynamic simulations of a FOWT Semi-Submersible under bichromatic and random waves. *J. Offshore Mech. Arct. Eng.* **2022**, *145*, 062001. [CrossRef]
7. Wan, D.; Xia, K. Numerical validation and analysis of the semi-submersible platform of the DeepCwind floating wind turbine based on CFD. *J. Renew. Sustain. Energy* **2016**, *8*, 053104. [CrossRef]
8. Teng, B.; Cong, P.; Gou, Y. Nonlinear Time-Domain theory for the simulation of moored floating body motion. *J. Mar. Sci. Appl.* **2018**, *17*, 341–352. [CrossRef]
9. Patryniak, K.; Collu, M.; Coraddu, A. Rigid body dynamic response of a floating offshore wind turbine to waves: Identification of the instantaneous centre of rotation through analytical and numerical analyses. *Renew. Energy* **2023**, *218*, 119378. [CrossRef]
10. Jameel, M.; Ahmad, S.; Islam, A.B.M.S.; Zummat, M.Z. Non-linear dynamic analysis of coupled spar platform. *J. Civ. Eng. Manag.* **2013**, *19*, 476–491. [CrossRef]
11. Wang, L.; Robertson, A.; Jonkman, J.; Yu, Y.-H. OC6 phase I: Improvements to the OpenFAST predictions of nonlinear, low-frequency responses of a floating offshore wind turbine platform. *Renew. Energy* **2022**, *187*, 282–301. [CrossRef]
12. *OpenFAST Documentation, Version v3.4.1*; National Renewable Energy Laboratory: Golden, CO, USA, 2017.
13. Bae, Y.H.; Kim, M.H. Rotor-floater-mooring coupled dynamic analysis of mono-column-TLP type FOWT (Floating Offshore Wind Turbine). *Ocean. Syst. Eng.* **2011**, *1*, 93–109. [CrossRef]
14. Bae, Y.H.; Kim, M.H. Coupled dynamic analysis of multiple wind turbines on a large single floater. *Ocean. Eng.* **2014**, *92*, 175–187. [CrossRef]
15. Yang, Y.; Bashir, M.; Michailides, C.; Li, C.; Wang, J. Development and application of an aero-hydro-servo-elastic coupling framework for analysis of floating offshore wind turbines. *Renew. Energy* **2020**, *161*, 606–625. [CrossRef]
16. Journée, J.M.J.; Massie, W.W. *Offshore Hydromechanics*; Delft University of Technology: Delft, The Netherlands, 2001. Available online: https://ocw.tudelft.nl/wp-content/uploads/Introduction_Offshore_Hydromechanics.pdf (accessed on 31 August 2001).
17. Jang, H.; Kim, M. Effects of nonlinear FK (Froude-Krylov) and hydrostatic restoring forces on arctic-spar motions in waves. *Int. J. Nav. Archit. Ocean Eng.* **2020**, *12*, 297–313. [CrossRef]
18. Bandyk, P.J. Numerical methods for nonlinear hydrostatic and Froude-Krylov forces. In Proceedings of the SNAME Maritime Convention, Tacoma, WA, USA, 30 October–1 November 2019.

19. Rodríguez, C.A.; Neves, M.A.S.; Polo, J.C.F. A time-efficient approach for nonlinear hydrostatic and Froude-Krylov forces for parametric roll assessment in irregular seas. *Ocean Eng.* **2016**, *120*, 246–255. [CrossRef]
20. Rajendran, S.; Fonseca, N.; Soares, C.G. Body nonlinear time domain calculation of vertical ship responses in extreme seas accounting for 2nd order Froude-Krylov pressure. *Appl. Ocean Res.* **2015**, *54*, 39–52. [CrossRef]
21. Somayajula, A.; Falzarano, J. Large-amplitude time-domain simulation tool for marine and offshore motion prediction. *Mar. Syst. Ocean Technol.* **2015**, *10*, 1–17. [CrossRef]
22. Jose, A. A Study of Negative Damping Phenomenon in Floating Offshore Wind Turbines and the Role of Non-Linear Wave Forces. Ph.D. Thesis, Texas A&M University, College Station, TX, USA, 2018. Available online: https://hdl.handle.net/1969.1/174034 (accessed on 31 August 2018).
23. Wang, H.; Somayajula, A.; Falzarano, J.; Xie, Z. Development of a blended Time-Domain program for predicting the motions of a wave energy structure. *J. Mar. Sci. Eng.* **2019**, *8*, 1. [CrossRef]
24. Guha, A. Development of a Computer Program for Three Dimensional Frequency Domain Analysis of Zero Speed First Order Wave Body Interaction. 2012. Available online: http://oaktrust.library.tamu.edu/bitstream/1969.1/148193/1/GUHA-THESIS-2012.pdf (accessed on 31 May 2016).
25. Falzarano, J.M.; IEsparza Mulk, M.T.U.I. "A Combined Steady-State and Transient Approach to Study Large Amplitude Ship Rolling Motion and Capsizing. *J. Ship Res.* **1995**, *39*, 213–224.
26. Palazzolo, A.; Shin, D.; Falzarano, J. *Introduction to Engineering Nonlinear and Parametric Vibrations with MATLAB and MAPLE*; Wiley: Hoboken, NJ, USA, 2025.
27. Jonkman, B.J. *TurbSim User's Guide (Technical Report NREL/TP-500-39797)*; National Renewable Energy Laboratory: Golden, CO, USA, 2006.
28. Xie, Z.; Falzarano, J. An optimization framework of a parametric Octabuoy semi-submersible design. *Int. J. Nav. Archit. Ocean Eng.* **2020**, *12*, 711–722. [CrossRef]

Disclaimer/Publisher's Note: The statements, opinions and data contained in all publications are solely those of the individual author(s) and contributor(s) and not of MDPI and/or the editor(s). MDPI and/or the editor(s) disclaim responsibility for any injury to people or property resulting from any ideas, methods, instructions or products referred to in the content.

Journal of
Marine Science and Engineering

MDPI

Article

Collapse Behavior of Onshore and Spar-Floating Wind Turbine Towers During Blade Pitch Malfunction

Sharath Srinivasamurthy [1,*], Amiya Pandit [2] and Shigeo Yoshida [1,3]

1 Institute of Ocean Energy, Saga University, Saga 840-8502, Japan
2 Sustainability Institute, University College Cork, T12 K8AF Cork, Ireland
3 Research Institute for Applied Mechanics, Kyushu University, Kasuga 816-8580, Japan
* Correspondence: sharath@ioes.saga-u.ac.jp

Abstract

Blade pitch control is one of the most important control systems for a wind turbine: blade pitch controller malfunction can lead to increased vertical bending moment at the tower base, which may result in structural failure. This study investigated the collapse behavior mechanism at the tower root due to an extreme event of blade pitch malfunction for onshore and spar-floating wind turbines. An aero-hydro-elastoplastic coupled analysis tool previously developed and validated by one of the authors was utilized to capture the structural response at the tower root in elastic and plastic regions. Three strength models—(i) SM-01, (ii) SM-02, and (iii) SM-03—were selected to demonstrate the collapse behavior mechanism of onshore and spar-floating 5 MW wind turbines in a time-series simulation. The damage in the plastic region, termed the collapse extent, was evaluated at the collapsing section. Moment–rotational angle relationships are discussed under the same wind conditions. The tower vibrations were found to dominate the structural response of the onshore wind turbine, whereas the tower vibrations and floater response dominate the spar-floating wind turbine response during the failure event. The collapse extent of the spar-floating wind turbine was found to be 8 times larger than the onshore wind turbine under the same wind conditions. Furthermore, simulations were carried out for the spar-floating wind turbine to understand the effect of incoming waves on the collapse behavior: the collapse extent increases as the wave amplitude and period increase under the same wind conditions.

Keywords: onshore and spar-floating wind turbines; blade pitch malfunction; collapse behavior; coupled analysis

1. Introduction

To limit carbon emissions, as agreed upon in the Paris Agreement (2015) [1], wind turbines have gained global attention as a way to replace fossil fuel [2,3]. Wind turbines are complicated structures being installed in onshore and offshore regions to harness wind energy. These wind turbines are high-tech systems with advanced mechanical–electrical components and large, flexible blades easily surpassing 50 m [4,5]. The commercialization of onshore (land-based) wind turbines is consistent with technological developments. However, to reduce land-based conflicts and constraints, particularly visual impacts and noise, there has been a recent transition towards installing floating offshore wind turbines in the deep ocean [6,7].

Floating offshore wind turbines are promising and economical solutions to harvest wind energy in isolated deep-sea locations. However, the design optimization of floating platforms and mooring systems is highly complex when these tall, slender structures are installed in the harsh ocean environment. The interaction between the flexible components of the wind turbines (blade, tower) and stochastic dynamic input load can be used to quantify the true risk of failure [8,9]. Furthermore, the loading amplitude and the complexity between the structural components of the wind turbines bring uncertain failure modes [10–12]. Hence, identifying the root cause of collapse or non-linear deformation of wind turbines is crucial.

Most wind turbine collapses result from tower buckling and foundation overturn [13] during typhoons and earthquakes. Furthermore, wind turbine tower collapses during extreme events, such typhoons [14] and super typhoons [15], are analyzed when the turbine is in a parked and idling condition. Fan et al., 2018 [16], used finite-element-based ABAQUS software to determine the collapse pattern of a wind turbine tower under the coupled effect of earthquake and wind loads. Chou et al., 2018 [17], undertook a comparative analysis to generalize the collapse behavior of a wind turbine tower subjected to strong wind.

Hence, improved solutions are proposed for wind turbine safety during smooth operation while enhancing the power generation efficiency [18,19]. Sophisticated blade pitch and rotor yaw control systems are integrated into the wind turbines to improve their aerodynamic response. Therefore, designing robust controllers for wind turbines is a popular field of research. Investigations into PID [20–22] and Proportional Integral [23] conventional controllers were undertaken by Karad and Thakur, 2020, who found that PI controllers outperform PID controllers regarding blade pitch control when the blade's pitch angle is controlled by either electrical or hydraulic means. Subsequently, refs. [24–26] implemented a blade pitch control scheme to manage asymmetric loads above the rated wind speed in 5 MW wind turbines; the numerical results were compared in terms of time-domain and fatigue damage analyses. However, implementing these sophisticated control systems poses new challenges for offshore wind turbines, and mechanical faults in controller systems increases the risk of failure due to tower buckling and blade tower strike [27].

Although most wind turbines are designed to withstand 60–70 m/s wind speeds (IEC-61400 [28]), pitch control failure is a common cause for collapse compared with other triggers, such as construction problems or wind turbine blade collision. In addition to this, any fault in the blade pitch angle controller [29] affects the aerodynamic load on the wind turbines, which has impact on the platform pitch motion of floating offshore wind turbines [30,31]. Therefore, it is imperative to deeply analyze the failure risk associated with the blade pitch controller. This is even more important for floating offshore wind turbines, as they are prone to the more stochastic nature of dynamic loading coupled with the floating platform response. Notably, from a control perspective, a negative damping effect is introduced to floating offshore wind turbines by the controllers above the rated wind speed, as reported in [32]. Furthermore, blade pitch controller malfunction may inadvertently excite platform oscillations, causing structural damage [33]. To solve this issue, rather than designing rigid platforms for offshore wind turbines, flexible floating platforms may be considered as a viable solution.

Floating platforms are usually modelled as rigid in most studies. The hydroelastic response of a flexible spar platform was studied by Xiaoming at al [34], though this was limited to understanding the response in the elastic region. Since there have been limited studies that efficiently estimated the collapse nature in the plastic region for floating offshore wind turbines, this study attempted to provide a basic framework to understand the elastoplastic phenomenon. Therefore, the objectives of this study focused on three main aspects: first, to compare and understand the failure mechanism of onshore and spar-floating wind turbines under blade pitch malfunction; second, treatment of the floating

https://doi.org/10.3390/jmse14040378

platform as a flexible structure, thereby obtaining the most realistic response under wind–wave loading; and third, to understand the aero-hydro-elastoplastic phenomenon in a coupled time-series simulation. With this framework, it is possible to conduct a holistic risk assessment of floating offshore wind turbines under blade pitch malfunction.

The paper is organized as follows: Section 2 describes the analysis models of onshore and spar-floating wind turbines. Section 3 elucidates the theoretical modeling of the capacity curve and methodology of simulating blade pitch failure. Section 4 describes the results of the collapse behavior of onshore and spar-floating wind turbines towers in elastic and plastic regions. It also discusses the various collapse responses under different wave conditions. The findings of the study are summarized in Section 5.

2. Analysis Models

Onshore and spar-floating wind turbines were considered for the comparative analysis. Figure 1 illustrates the analysis models. Both the onshore and spar-floating models contain the NREL 5 MW Reference Wind Turbine [35], the details of which are summarized in Table 1. Since the main objective of the study was to understand the collapse at the tower root due to blade pitch malfunction, Figure 1 highlights the location of the tower root.

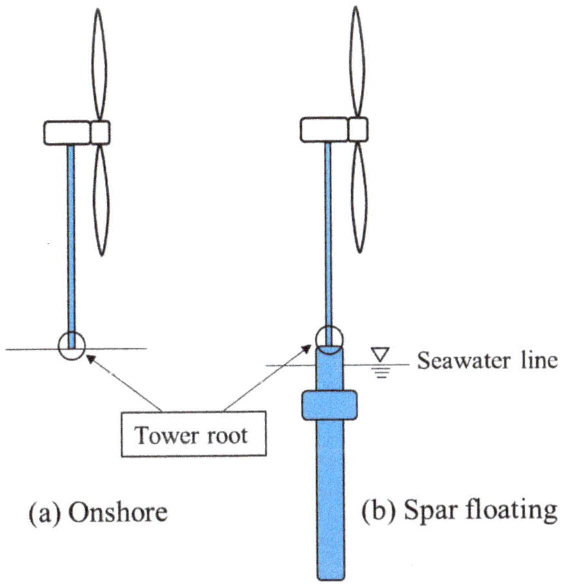

Figure 1. Illustration of onshore and spar-floating wind turbines.

Table 2 summarizes the principal particulars of the spar-floating body with a total draft of 109.3 m. It also consisted of a knob column at a distance of 10 m below the Seawater Line (SWL). The purpose of the knob column is to reduce the draft and to use the spar-floating platform in relatively shallow waters.

https://doi.org/10.3390/jmse14040378

Table 1. Principal particulars of wind turbine.

Rating	5 MW
Rotor Orientation and Configuration	Upwind and 3 Blades
Control	Variable Speed and Collective Pitch
Drivetrain	High Speed and Multiple-Stage Gearbox
Rotor and Hub Diameters	126 m, 3 m
Hub Height	90 m
Cut-In, Rated, and Cut-Out Wind Speeds	3, 11.4, and 25 m/s
Cut-In and Rated Rotor Speeds	6.9 and 12.1 rpm
Rated Tip Speed	80 m/s
Rotor Mass	110,000 kg
Nacelle Mass	240,000 kg
Tower Mass	347,460 kg

Table 2. Principal particulars of spar-floating body.

Water Draft	109.3 m
Elevation to Spar Platform Top Above SWL	10 m
Diameter of Water Plane	7 m
Diameter of Knob Column	25 m
Length of Knob Column	11.5 m
Position of the Knob Column Below SWL	10 m
Platform Mass, Including Ballast	7,940,000 kg
CM Location Below SWL	54.7 m
KG (With Wind Turbine)	54.6 m
KB	76.1 m
GM (With Wind Turbine)	28.5 m

3. Theoretical Modeling

The elastoplastic analysis was carried out to understand the collapse behavior of the onshore and spar-floating wind turbines. The tower root was selected as a critical failure location for both analysis models since the maximum bending moment is generated at the tower root section, making it the weakest section, and thus, most susceptible to collapse. The elastoplastic analysis theoretical modeling was carried out by adopting a non-linear spring model [36] (Figure 2).

Since the bending moment (BM) at the tower root is a function of the relative rotational angle between the tower side and foundation/platform side, the relations in Equations (1) and (2) hold:

$$\mathrm{BM} = M(\theta) \tag{1}$$

$$M(\theta) = K_S\theta \tag{2}$$

where K_S is the secant stiffness of the non-linear spring, and θ is the relative rotational angle assumed at the collapsing section, i.e., the tower root. Note that $\theta = \theta_T - \theta_F$ for the onshore case and $\theta = \theta_T - \theta_P$ for the spar-floating case. θ_T, θ_F, and θ_P are the rotational angle at the tower side, foundation side, and platform side, respectively. With this methodology,

https://doi.org/10.3390/jmse14040378

the sectional moment can be obtained at the collapsing section. This sectional moment at the collapsing section can be balanced with $M(\theta)$ at the tower root. This helps evaluate the bending moment at the tower root in the plastic region and gives added advantage of considering more failure sections in the future. Once the bending moment at the collapsing section exceeds the ultimate capacity of the non-linear rotational spring, it is possible to understand the collapse behavior in the plastic region.

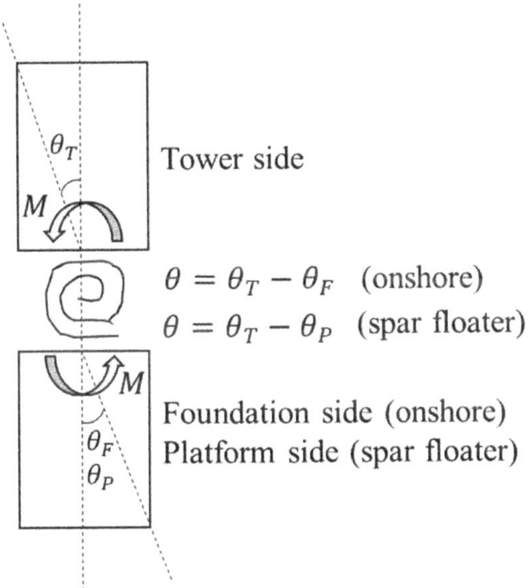

Figure 2. Theoretical model for elastoplastic analysis at tower root section.

In this study, the non-linear rotational spring was modeled as a piece-wise linear spring model that represents the relation between the relative rotational angle and the tower root moment. Figure 3 shows the idealized capacity curve assumed in this study, which includes the recovery of bending rigidity when unloaded. The path AB in the capacity curve represents the elastic region, wherein there is no failure. As we reach point B, the tower root bending moment reaches its ultimate capacity. The plastic region of the section and corresponding collapse extent is shown by the path BC. The path CD shows the bending rigidity recovery in the plastic region after partial collapse, representing the unloading path. Furthermore, it is noted here that the collapse extent is the residual relative rotational angle once the section exceeds its ultimate moment capacity M_u. In practical engineering, the collapse extent value signifies the local shell buckling at the section, while the tower regains its bending rigidity.

Three strength models were selected in this study to understand the collapse behavior of onshore and spar-floating wind turbines (Table 1). The cylindrical cross-section diameters were assumed to be 6.3, 5.8 and 5.75 m for SM-01, SM-02 and SM-03, respectively, with the same thickness of 0.03 m. The rationale for this selection was to avoid the full collapse of the section and to demonstrate the collapse extent in the plastic region such that the two analysis models could be compared for a strength model. By assuming a Young's modulus of 200 GPa, the collapsing moment for each strength model could be calculated (Table 3).

A validated coupled analysis simulation code [37] was utilized to obtain the response of the onshore and spar-floating wind turbines to the blade pitch malfunction. Validation was carried out by comparing the simulation results with the model scale experimental findings. The blade pitch malfunction was artificially induced in the spar-floating model

https://doi.org/10.3390/jmse14040378

scale experiment and the generated bending moment was measured. This was compared with the numerically simulated results from the coupled analysis code.

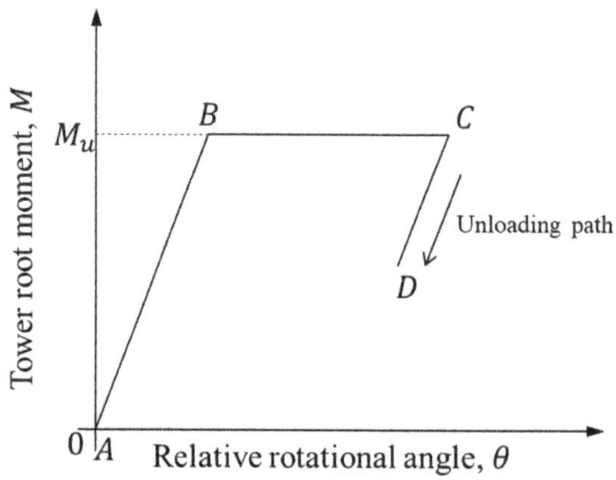

Figure 3. Theoretical moment–rotational angle relationship—capacity curve.

Table 3. Strength models for collapse behavior.

Strength Models	Cross-Section Diameter (m)	Collapsing Moment M_u ($\times 10^5$ kNm)
SM-01	6.3	1.53
SM-02	5.8	1.30
SM-03	5.75	1.28

Figure 4 shows the simulation technique to capture the response when the blade pitch control was up (simulation with controller) and down (simulation without controller). The blade pitch malfunction occurred at a certain time T_S, and a ramp function was used to simulate the malfunction. Therefore, at the end of the ramp function, i.e., at a certain time T_E, the coupled response obtained was without the controller. It was therefore possible to simulate the response in the tower root section under the same strength model before and after the malfunction. The simulations were conducted under a steady 18 mps wind speed for the three strength models under varying wave conditions; then, the resulting collapse behavior was studied.

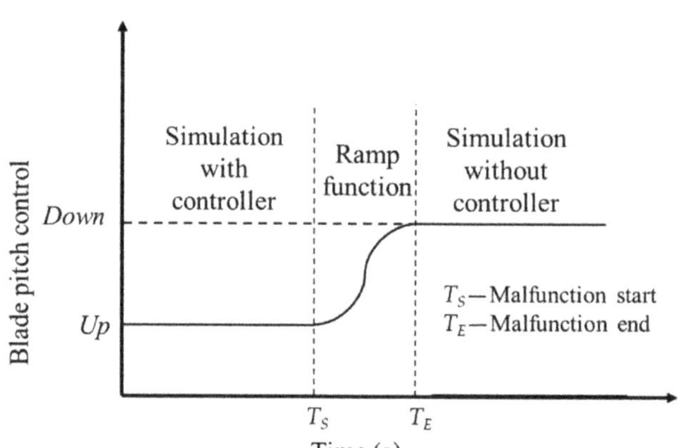

Figure 4. Simulation of blade pitch malfunction.

4. Results and Discussion

In this section, the study results are divided into four sections. First, in Section 4.1, the tower base moment time-series due to blade pitch malfunction and the corresponding moment–rotational angle relationships are summarized for the three strength models. Section 4.2 compares the partial collapse extent between the onshore and spar-floating platform cases. Sections 4.3 and 4.4 discuss the effects of the wave period and amplitude on the collapse extent for the spar-floating platform.

4.1. Tower Root Moment and Moment–Rotational Angle Relationship

First, the collapse behavior due to different strength models is discussed. A steady 18 mps wind was selected and the wave load was suppressed for all simulations discussed in this section. Three different strength models are discussed to understand the collapse behavior when the collapsing moment was varied while subjected to a blade pitch control malfunction. The time series of the tower base moment was plotted to understand the difference between onshore and spar-floating platform. The corresponding moment–rotational angle relationship is also shown to understand the elastoplastic response in the collapsing section.

Figure 5a shows the tower base moment for SM-01, which has a diameter of 6.3 m and thickness of 0.03 m. The SM-01 simulation was carried out as a baseline to show the no-collapse condition. The blade pitch malfunction occurred at 50 s, and the comparison between the onshore and spar-platform shows different behavior for the same incoming 18 mps wind. The response of the onshore platform is dominated by the structural vibrations of the tower, whereas the response of the spar-platform is dominated by both structural vibrations and rigid body response of the floating platform.

The natural period of the spar in the pitch direction is about 15 s, as clearly confirmed from the three peaks between 0 and 50 s in the simulation results. As the blade pitch malfunction occurred at 50 s, the results show that the tower root moment increases rapidly after the extreme event. It is noteworthy that the tower root moment peaks at different times for onshore and spar-floating wind turbines. The moment for the onshore case peaks at about 51 s, soon after the malfunction occurs, whereas the spar-floating case peaks at about 60 s, following the rigid-body response of the spar platform.

It is hypothesized that soon after the blade pitch malfunction, the additional moment generated is transferred to the soil foundation in the onshore wind turbine system. The soil foundation is a rigid system and responds instantly, so the moment peak at 51 s. However, in the spar-floating wind turbine case, the additional load is transferred to the floating platform and resets the spar platform. The response is delayed, and therefore, the tower root moment increases to the maximum value at about 60 s.

The corresponding moment–rotational angle relationship is plotted in Figure 5b; it can be inferred that there is no section collapse. The tower root bending moment is always less than the collapsing moment and the value lies in the elastic region. The maximum bending moment that developed is about 1.37×10^5 kNm, which is less than the model capacity of 1.53×10^5 kNm.

To demonstrate the collapse behavior in the elastoplastic region and understand the effects of different strength models, SM-02 and SM-03 with reduced strengths were adopted. Correspondingly, the ultimate moment capacity of the section, i.e., the collapsing moments, for SM-02 and SM-03 were 1.30×10^5 and 1.28×10^5 kNm, respectively. The blade pitch malfunction was simulated at 50 s, as in SM-01, and simulations were carried out under the same wind conditions.

https://doi.org/10.3390/jmse14040378

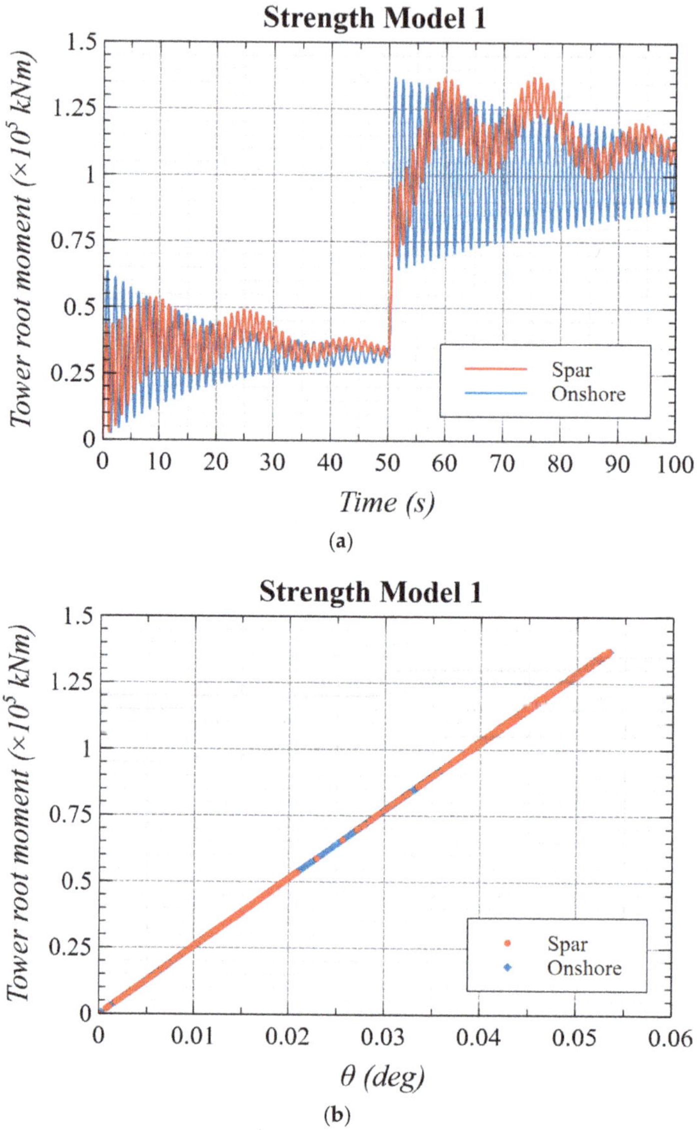

Figure 5. Strength model 1: (**a**) tower base moment; (**b**) moment–rotational angle relationship.

Figure 6a,b show the tower root moment and corresponding moment–rotational angle relationship for SM-02, respectively. The behavior until 50 s is the same as that in SM-01 and there is no collapse of the section. As the malfunction occurs and the tower root moment increases rapidly, a partial collapse at the tower root section can be observed for the onshore and spar-floating cases. The partial collapse occurs at the tower root between 52 and 60 s for the onshore case and between 74 and 78 s for the spar platform. However, the tower regains bending rigidity after a partial collapse of about 0.07 degrees in the plastic region for both cases (Figure 6b).

https://doi.org/10.3390/jmse14040378

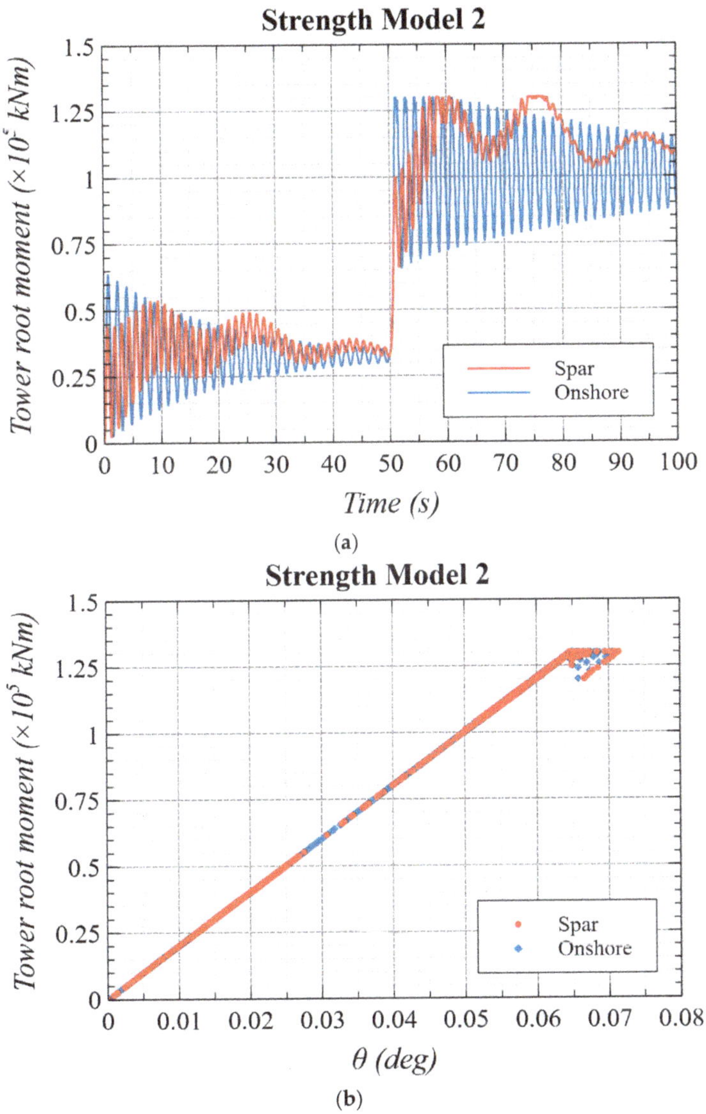

Figure 6. Strength model 2: (**a**) tower base moment; (**b**) moment–rotational angle relationship.

To further understand the sensitivity of the capacity curve, the tower root moment and corresponding moment–curvature relationship is plotted for SM-03 (Figure 7a,b). The partial collapse occurs between 52 and 60 s for the onshore case and regains the bending rigidity with partial collapse of about 0.08 degrees. However, for the spar platform case, there is a partial collapse between 58 and 62 s, followed by another partial collapse between 74 and 82 s. Even though the tower does not fail completely and regains bending rigidity, a total collapse extent of about 0.55 degrees is observed in the plastic region. The advantage of the such a simulation technique is that it can capture the collapse behavior as a time series in the elastic and plastic regions. Therefore, the sensitivities to collapse even under the same strength model for onshore and spar-floating wind turbines are different.

https://doi.org/10.3390/jmse14040378

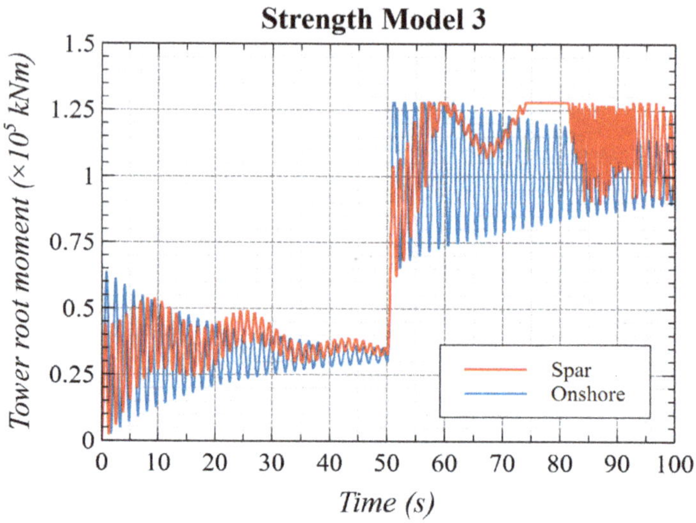

(a)

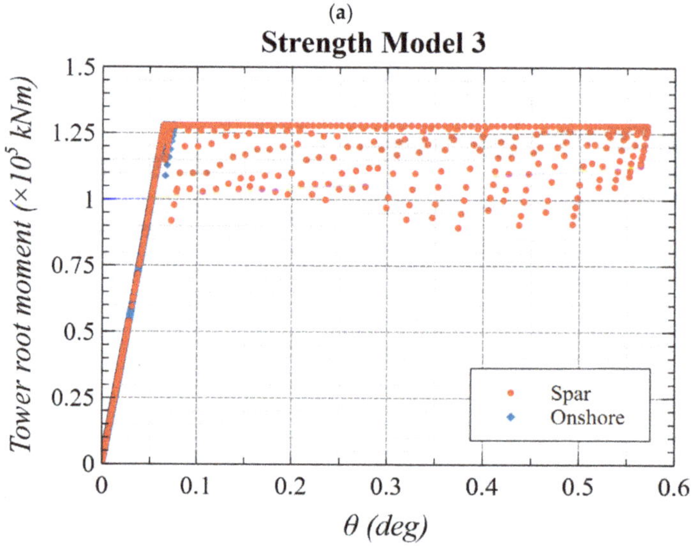

(b)

Figure 7. Strength model 3: (**a**) tower base moment; (**b**) moment–rotational angle relationship.

4.2. Collapse Extent of Onshore and Spar-Floating Wind Turbines

After confirming the effects of different strength models, this section shows plots and comparisons of the collapse extent as a time series for the onshore and spar-floating wind turbines.

Figure 8a,b summarize the collapse extent for different strength models for the onshore and spar-floating cases, respectively. The simulations carried out using SM-03 (cross-section diameter: 5.75 m) for the spar-floating case lead to a much higher collapse extent of 0.55 degrees than the onshore case, which is about 0.07 degrees. The partial collapse increases for the spar-floating case due to the load superposition effect, i.e., the platform rigid-body motion couples with tower structural vibrations at the moment of blade pitch failure. It can also be confirmed that the maximum collapse occurs at 52 s for the onshore

https://doi.org/10.3390/jmse14040378

case and 82 s for the spar-floating case. The partial collapse extent of the spar-floating wind turbine is about 8 times that of the onshore wind turbine under the same wind conditions. Therefore, spar-floating wind turbines are much more sensitive and show a larger collapse extent, even for the same strength model. Furthermore, the time at which the maximum collapse occurs is different for onshore and spar-floating wind turbines. Table 4 summarizes the collapse extent comparison for the onshore and spar-floating wind turbines.

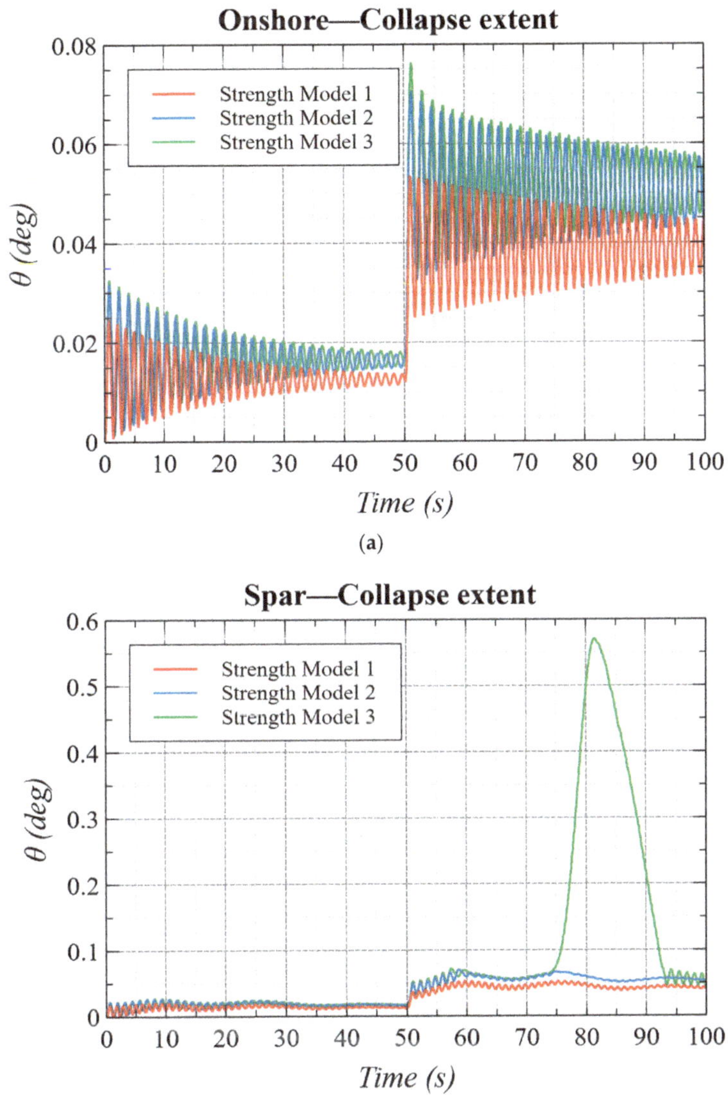

Figure 8. Collapse extent: (**a**) onshore; (**b**) spar floating.

https://doi.org/10.3390/jmse14040378

Table 4. Comparison of onshore vs. spar-floating—wind-only condition.

Strength Model	Cross-Section Diameter (m)	Collapsing Moment ($\times 10^5$ kNm)	Collapse Extent (Degrees)	
			Onshore	Spar Floating
SM-01	6.3	1.53	No collapse	No collapse
SM-02	5.8	1.30	0.07	0.07
SM-03	5.75	1.28	0.07	0.55

Note that all the results shown in Sections 4.1 and 4.2 are based on a steady 18 mps wind to provide an effective comparison of the onshore and spar-floating wind turbines. In reality, it is essential to include incoming waves for a spar platform. To understand the collapse behavior under the presence of waves, further simulations were carried out for the spar-floating wind turbine, with the results presented in the next sections.

4.3. Effect of Wave Period—Strength Model 1

In this section, the wave period's effect on the collapse extent is discussed. SM-01 was considered for all the simulations. First, the pitch tilt angle of the spar platform is plotted in Figure 9.

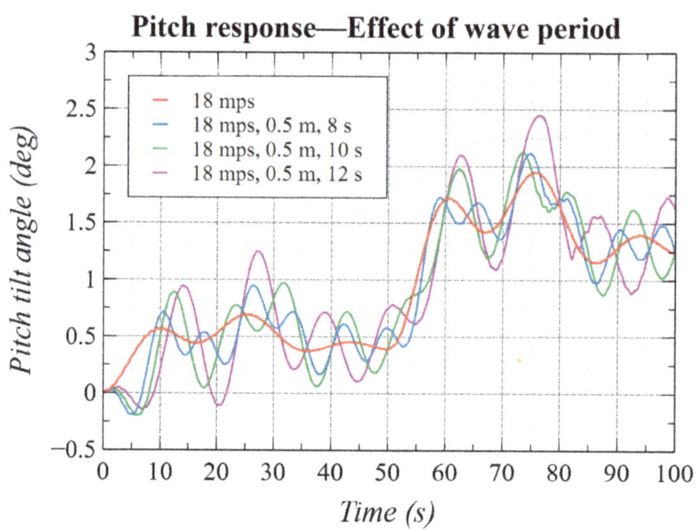

Figure 9. Pitch tilt angle—varying wave period.

The wave amplitude was set to 0.5 m and the wave period was varied for the same 18 mps wind speed. The pitch response of the spar platform under the combined wind-wave load was simulated for wave periods of 8, 10, and 12 s. From 0 to 50 s, the pitch motion response is dynamic and reflects the incoming wave period. As malfunction occurred at 50 s, the pitch response increases drastically, even more so for longer wave periods. The pitch response is the coupled response due to wind and regular waves and therefore shows a more realistic dynamic response than the wind-only response.

The tower root moment and corresponding moment–rotational angle relationship is plotted to show the wave period's effect on the collapse behavior. Figure 10a shows the tower root moment as a time series under combined wind–wave loads. There is a partial collapse, even for SM-01, under the combined loading. The partial collapse increases as the wave period increases for the same wind loading condition. The collapse extents for

https://doi.org/10.3390/jmse14040378

wave periods of 8, 10, and 12 s are about 0.1, 0.12, and 0.5 degrees, respectively. However, as observed before, the tower regains its bending rigidity. The sensitivity to the incoming wave period can be understood here based on the rapid increase in the collapse extent from 0.12 to 0.5 degrees when the wave period increases from 10 to 12 s.

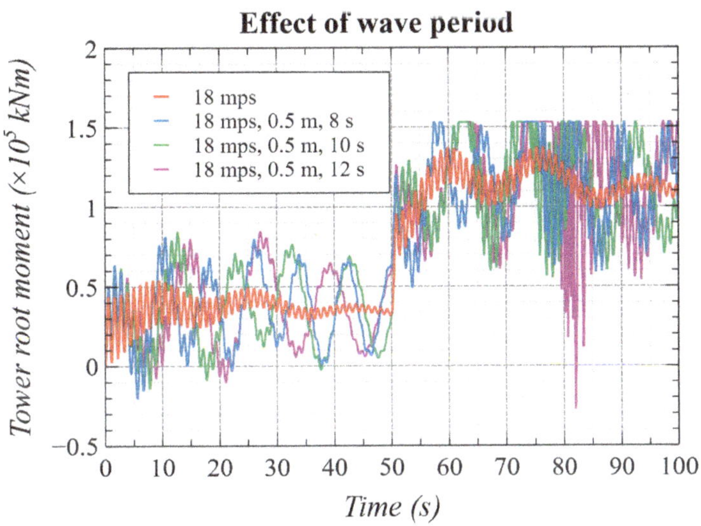

(a)

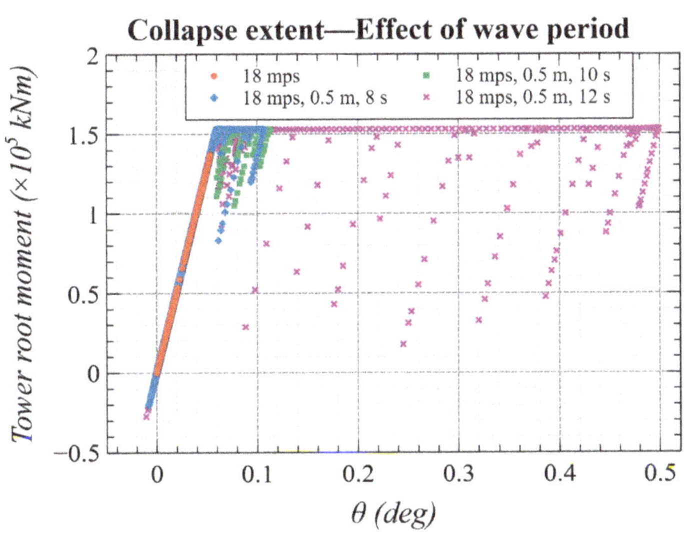

(b)

Figure 10. Effect of wave period: (**a**) tower root moment; (**b**) collapse extent.

Notably, even with SM-01, there is partial collapse for spar-floating wind turbines under the combined wind–wave loading. Therefore, it can be concluded that the design of floating offshore wind turbines needs special attention relative to onshore wind turbines. Table 5 summarizes the wave period's effect on the collapse extent.

https://doi.org/10.3390/jmse14040378

Table 5. Effect of wave period for SM-01—wind–wave condition.

Wave Amplitude (m)	Wave Period (s)	Collapse Extent (Degrees)
0.5	8	0.1
	10	0.12
	12	0.5

4.4. Effect of Wave Amplitude—Strength Model 1

In this section, the wave amplitude's effect on the collapse extent is discussed. As in Section 4.3, SM-01 was considered for all the simulations. A wave period of 8 s was selected and the wave amplitude was varied under the same steady 18 mps wind speed. Figure 11 shows the pitch response under the combined wind–wave load for various wave amplitudes. The dynamic response's amplitude slightly increases as the wave amplitude increases between 0 and 50 s. After the blade pitch malfunction, the pitch response increases rapidly and attains a maximum value of 2.8 degrees for the 0.7 m wave amplitude at about 78 s.

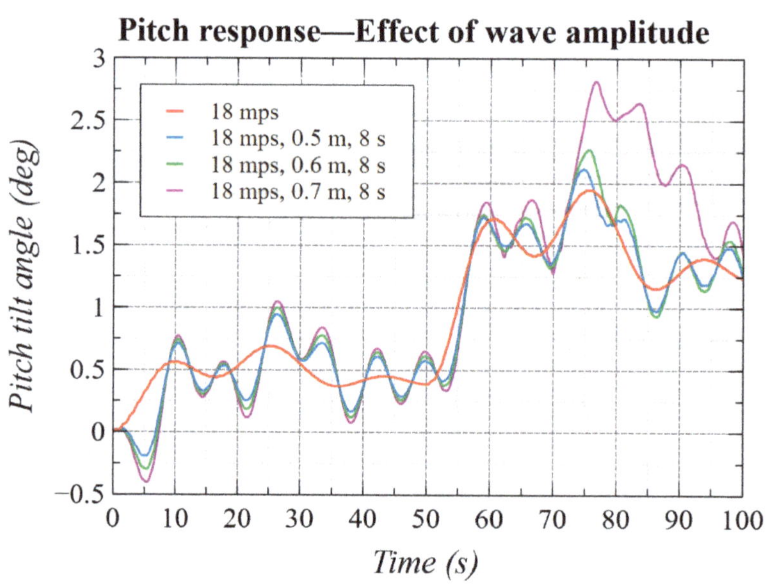

Figure 11. Pitch tilt angle—varying wave amplitude.

Furthermore, Figure 12a,b show the tower root moment and moment–rotational angle relationship, respectively, to demonstrate the wave amplitude's effect on the collapse extent. From the simulation between 50 and 100 s, the wave amplitude has a major effect on the collapse extent. The collapse extents observed for the 0.5, 0.6, and 0.7 m wave amplitudes are 0.1, 0.4, and 1.4 degrees, respectively. The collapse extent increases as the wave amplitude increases. Thus, it is important to consider the wave amplitude variation while designing the accidental limit state of an FOWT system subjected to a blade pitch control malfunction.

https://doi.org/10.3390/jmse14040378

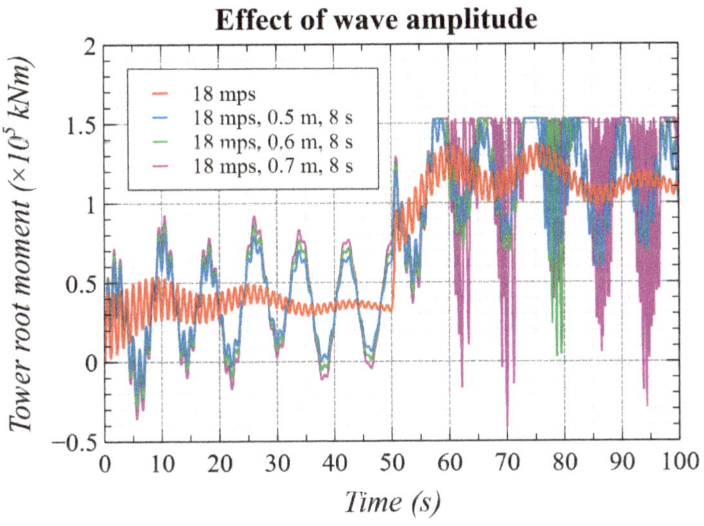

(**a**)

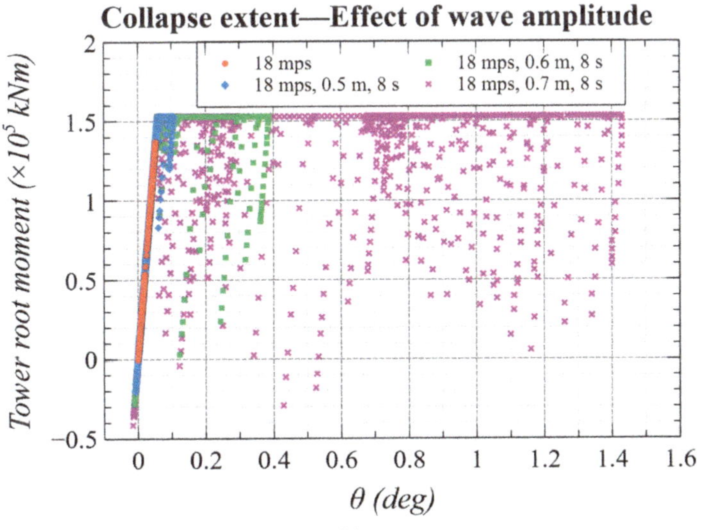

(**b**)

Figure 12. Effect of wave amplitude: (**a**) tower root moment; (**b**) collapse extent.

Table 6 summarizes the collapse extent of various simulation results discussed in this study: the collapse extent increases as the wave period and wave amplitude increases. Furthermore, the collapse extent of spar-floating wind turbines is larger than the onshore case, even under the same strength model, and needs much more attention. In all these demonstration cases, the tower regains rigidity in the plastic region, but as the wave amplitude or wave period increases further, the tower section might not regain rigidity and the tower will collapse entirely, as discussed in [35]. This study gives a framework for simulating such complex collapse scenarios, even during partial collapses under the action of combined wind–wave loading.

https://doi.org/10.3390/jmse14040378

Table 6. Effect of wave amplitude for SM-01—wind–wave condition.

Wave Period (s)	Wave Amplitude (m)	Collapse Extent (Degrees)
8	0.5	0.1
	0.6	0.4
	0.7	1.4

5. Conclusions

In the present study, the collapse behavior mechanisms of onshore and spar-floating wind turbines under blade pitch malfunction were investigated. The tower root was selected as a critical section where the partial collapse occurred in this investigation. Aero-hydro-elastoplastic coupled simulations were carried out as time series, and the tower root bending moment was evaluated for three different strength models. The collapse extent, which is a quantitative measure of the damage in the plastic region at the collapsing section, was evaluated and the moment–rotational angle relationships are discussed. The specific conclusions deduced from the study are as follows:

- The onshore wind turbine response is dominated by tower vibrations, whereas the spar-floating wind turbine response is dominated by tower vibrations and the natural period of the floating platform. Therefore, the tower root moment peaks at different times for onshore and spar-floating wind turbines under the same wind conditions.
- The collapse extent in the plastic region for the spar-floating wind turbine is 8 times larger than the onshore wind turbine, even for the same strength model. This shows that a spar-floating wind turbine is more critical than the onshore case for failures.
- Regarding the collapse behavior of the spar-floating wind turbine, the collapse extent increases with increasing wave periods and amplitudes under the same wind conditions. Therefore, special attention is needed to evaluate the safety of spar-floating wind turbines.

In the present study, the tower root was selected as the critical section for partial collapse in the plastic region. Therefore, in the future it is desired to conduct simulations with multi-section distributed plasticity models to study the partial collapses throughout the tower rather than at one section. Furthermore, a simplistic capacity curve model was utilized in this study for the plastic region, and therefore, a detailed investigation is essential to adopt an advanced capacity curve in the future.

Author Contributions: Conceptualization, S.S.; methodology, S.S. and A.P.; software, S.S.; validation, S.S.; formal analysis, S.S.; investigation, S.S.; resources, S.S. and S.Y.; data curation, S.S.; writing—original draft preparation, S.S.; writing—review and editing, S.S. and A.P.; visualization, S.S.; supervision, S.Y.; project administration, S.S.; funding acquisition, S.S. All authors have read and agreed to the published version of the manuscript.

Funding: This research was partly funded by JSPS Kakenhi, grant number 25K08153.

Data Availability Statement: Dataset available on request from the authors.

Conflicts of Interest: The authors declare no conflicts of interest.

References

1. Agreement, P. Paris agreement. In Proceedings of the Report of the Conference of the Parties to the United Nations Framework Convention on Climate Change (21st Session), Paris, France, 30 November–13 December 2015; p. 2.
2. IEA. *World Energy Outlook 2022*; International Energy Agency (IEA): Paris, France, 2022.

https://doi.org/10.3390/jmse14040378

3. Díaz, H.; Soares, C.G. Review of the current status, technology and future trends of offshore wind farms. *Ocean Eng.* **2020**, *209*, 107381. [CrossRef]
4. Naqash, T.M.; Alam, M.M. A State-of-the-Art Review of Wind Turbine Blades: Principles, Flow-Induced Vibrations, Failure, Maintenance, and Vibration Suppression Techniques. *Energies* **2025**, *18*, 3319. [CrossRef]
5. Menezes, E.J.; Araújo, A.M. Wind turbine structural control using H-infinity methods. *Eng. Struct.* **2023**, *286*, 116095. [CrossRef]
6. Edwards, E.C.; Holcombe, A.; Brown, S.; Ransley, E.; Hann, M.; Greaves, D. Evolution of floating offshore wind platforms: A review of at-sea devices. *Renew. Sustain. Energy Rev.* **2023**, *183*, 113416. [CrossRef]
7. Srinivasamurthy, S.; Iwamatsu, S.; Hashimoto, K.; Suzuki, H.; Chujo, T.; Haneda, K.; Nihei, Y. Study of slow-drift damping on wind tracking performance of a new-type FOWT 'Optiflow' with single-point mooring. *Ocean Eng.* **2021**, *242*, 110131. [CrossRef]
8. Lara, M.; Mulders, S.P.; van Wingerden, J.-W.; Vázquez, F.; Garrido, J. Analysis of adaptive individual pitch control schemes for blade fatigue load reduction on a 15 MW wind turbine. *Appl. Sci.* **2023**, *14*, 183. [CrossRef]
9. Fu, B.; Zhao, J.; Li, B.; Yao, J.; Teifouet, A.R.M.; Sun, L.; Wang, Z. Fatigue reliability analysis of wind turbine tower under random wind load. *Struct. Saf.* **2020**, *87*, 101982. [CrossRef]
10. Sun, Y.; Li, H.; Sun, L.; Soares, C.G. Failure analysis of floating offshore wind turbines with correlated failures. *Reliab. Eng. Syst. Saf.* **2023**, *238*, 109485. [CrossRef]
11. Li, H.; Díaz, H.; Soares, C.G. A failure analysis of floating offshore wind turbines using AHP-FMEA methodology. *Ocean Eng.* **2021**, *234*, 109261. [CrossRef]
12. Chou, J.-S.; Ou, Y.-C.; Lin, K.-Y. Collapse mechanism and risk management of wind turbine tower in strong wind. *J. Wind Eng. Ind. Aerodyn.* **2019**, *193*, 103962. [CrossRef]
13. Chou, J.-S.; Tu, W.-T. Failure analysis and risk management of a collapsed large wind turbine tower. *Eng. Fail. Anal.* **2011**, *18*, 295–313. [CrossRef]
14. Ishihara, T.; Yamaguchi, A.; Takahara, K.; Mekaru, T.; Matsuura, S. An analysis of damaged wind turbines by typhoon Maemi in 2003. In Proceedings of the Sixth Asia-Pacific Conference on Wind Engineering, Seoul, Republic of Korea, 12–15 September 2005.
15. Chen, X.; Xu, J.Z. Structural failure analysis of wind turbines impacted by super typhoon Usagi. *Eng. Fail. Anal.* **2016**, *60*, 391–404. [CrossRef]
16. Fan, J.; Li, Q.; Zhang, Y. Collapse analysis of wind turbine tower under the coupled effects of wind and near-field earthquake. *Wind Energy* **2019**, *22*, 407–419. [CrossRef]
17. Chou, J.-S.; Ou, Y.-C.; Lin, K.-Y.; Wang, Z.-J. Structural failure simulation of onshore wind turbines impacted by strong winds. *Eng. Struct.* **2018**, *162*, 257–269. [CrossRef]
18. Namik, H.; Stol, K. Individual blade pitch control of a spar-buoy floating wind turbine. *IEEE Trans. Control Syst. Technol.* **2013**, *22*, 214–223. [CrossRef]
19. Jonkman, J. Influence of control on the pitch damping of a floating wind turbine. In Proceedings of the 46th AIAA Aerospace Sciences Meeting and Exhibit, Reno, NV, USA, 7–10 January 2008; p. 1306.
20. Soni, U.K. Pitch control of CFVWS turbine using PID controlled internal pitch loop. *Int. J. Electr. Electron. Comput. Syst. (IJEECS)* **2015**, *3*, 3–6.
21. Gambier, A.; Nazaruddin, Y.Y. Collective pitch control with active tower damping of a wind turbine by using a nonlinear PID approach. *IFAC-PapersOnLine* **2018**, *51*, 238–243. [CrossRef]
22. Erol, H. Stability analysis of pitch angle control of large wind turbines with fractional order PID controller. *Sustain. Energy Grids Netw.* **2021**, *26*, 100430. [CrossRef]
23. Mulders, S.P.; Pamososuryo, A.K.; Disario, G.E.; Wingerden, J.W.v. Analysis and optimal individual pitch control decoupling by inclusion of an azimuth offset in the multiblade coordinate transformation. *Wind Energy* **2019**, *22*, 341–359. [CrossRef]
24. Tang, S.; Tian, D.; Wu, X.; Huang, M.; Deng, Y. Wind turbine load reduction based on 2DoF robust individual pitch control. *Renew. Energy* **2022**, *183*, 28–40. [CrossRef]
25. Yuan, Y.; Chen, X.; Tang, J. Multivariable robust blade pitch control design to reject periodic loads on wind turbines. *Renew. Energy* **2020**, *146*, 329–341. [CrossRef]
26. Zhang, S.; Wang, Y.; Liu, Y.; Wang, X.; Bai, W.; Cao, T. Improved control strategy and designed control parameters of pitch system for wind turbine considering blade load reduction. *Renew. Energy* **2024**, *232*, 121050. [CrossRef]
27. Cheng, Y.; Wang, J.; Dai, K.; Damatty, A.E. Influence of mechanical faults in pitch, brake, controller systems on fragility of offshore wind turbine under aero-hydro loadings. *Ocean Eng.* **2023**, *285*, 115386. [CrossRef]
28. *IEC CDV 61400-1*; Wind Energy Generation Systems—Part 1: Design Requirements. International Electrotechnical Commission: Geneva, Switzerland, 2016.
29. Pujana-Arrese, A.; Elorza, I.; Trojaola, I.; Arrizabalaga, I.; Irigoyen, E. On the effects of pitch system faults on a wind turbine. *Results Eng.* **2024**, *22*, 102230. [CrossRef]
30. Kakuya, H.; Shiraishi, T.; Yoshida, S.; Utsunomiya, T.; Sato, I. Experimental results of floating platform vibration control with mode change function using full-scale spar-type floating offshore wind turbine. *Wind Eng.* **2018**, *42*, 230–242. [CrossRef]

https://doi.org/10.3390/jmse14040378

31. Kakuya, H.; Yoshida, S.; Sato, I.; Utsunomiya, T. A study on the platform-pitching vibration of floating offshore wind turbines based on classical control theory. *Wind Eng.* **2020**, *44*, 610–630. [CrossRef]
32. Keflemariam, Y.A.; Lee, S. Control and dynamic analysis of a 10 MW floating wind turbine on a TetraSpar multi-body platform. *Renew. Energy* **2023**, *217*, 119194. [CrossRef]
33. Jeong, M.; Loth, E.; Qin, C.; Phadnis, M.; Pusch, M.; Pao, L. Operational Pitch Actuation Dynamics for Offshore Wind Turbines Ranging from 5 to 50 MW. *Wind Energy* **2025**, *28*, e2975. [CrossRef]
34. Ran, X.; Leroy, V.; Bachynski-Polić, E.E. Hydroelastic response of a flexible spar floating wind turbine: Numerical modelling and validation. *Ocean Eng.* **2023**, *286*, 115635. [CrossRef]
35. Jonkman, J.; Butterfield, S.; Musial, W.; Scott, G. *Definition of a 5-MW Reference Wind Turbine for Offshore System Development*; National Renewable Energy Lab. (NREL): Golden, CO, USA, 2009.
36. Srinivasamurthy, S.; Iijima, K.; Fujikubo, M.; Nihei, Y. Elastic-plastic behavior analysis for a flexible FOWT subjected to blade pitch control malfunction. *J. Jpn. Soc. Nav. Arch. Ocean Eng.* **2016**, *23*, 179–187. [CrossRef]
37. Srinivasamurthy, S.; Iijima, K.; Nihei, Y.; Hara, N. Coupled simulation between FAST and Hydro-structural code for a flexible FOWT considering blade pitch control malfunction. In *International Conference on Offshore Mechanics and Arctic Engineering, Busan, Republic of Korea, 19–24 June 2016*; American Society of Mechanical Engineers: New York, NY, USA, 2016.

Disclaimer/Publisher's Note: The statements, opinions and data contained in all publications are solely those of the individual author(s) and contributor(s) and not of MDPI and/or the editor(s). MDPI and/or the editor(s) disclaim responsibility for any injury to people or property resulting from any ideas, methods, instructions or products referred to in the content.

https://doi.org/10.3390/jmse14040378

Journal of
Marine Science and Engineering

MDPI

Article

Dynamic Response of a Floating Dual Vertical-Axis Tidal Turbine System with Taut and Catenary Mooring Under Extreme Environmental Conditions in Non-Operating Mode

Yunjun Lee [1], Jinsoon Park [2] and Woo Chul Chung [1,*]

[1] Division of Mechanical Engineering, National Korea Maritime & Ocean University, Busan 49112, Republic of Korea; yunjun9771@g.kmou.ac.kr
[2] Korea Institute of Ocean Science and Technology, Busan 49111, Republic of Korea; jpark@kiost.ac.kr
* Correspondence: wcchung@kmou.ac.kr; Tel.: +82-051-410-4361

Abstract: This study analyzes the dynamic response of a floating dual vertical-axis tidal turbine platform under extreme environmental loads, focusing on two different mooring systems as follows: taut and catenary. The analysis assumes a non-operational turbine state where power generation is stopped, and the vertical turbines are lifted for structural protection. Using time-domain simulations via OrcaFlex 11.4, the floating platform's motion and mooring line effective tensions are evaluated under high waves, strong wind, and current loads. The results reveal that sway and heave motions are significantly influenced by wave excitation, with the catenary system exhibiting larger responses due to mooring system features, while the taut system experiences higher mooring effective tension but shows more restrained motion. Notably, in the roll direction, both systems exhibit peak frequencies unrelated to the wave spectrum, attributed instead to resonance with the system's natural frequencies—0.12438 Hz for taut and 0.07332 Hz for catenary. Additionally, the failure scenario of ML02 (Mooring Line 02) and the application of dynamic power cables to the floating platform are analyzed. The results demonstrate that under non-operational and extreme load conditions, mooring system type plays a main role in determining platform stability and structural safety. This comparative analysis offers valuable insights for selecting and designing mooring configurations optimized for reliability in extreme environmental conditions.

Keywords: dual vertical-axis tidal turbine system; mooring system; mooring line failure; dynamic power cable

1. Introduction

The EU recognized the risk of greenhouse gas problems due to rapid industrialization and announced a climate policy. This climate policy aims to reduce greenhouse gas emissions by 90% by 2040 compared to 1990 levels [1]. Accordingly, many experts started research on greenhouse gas reduction in their respective fields. Among them, tidal current energy is one of the best energy sources considering high efficiency, production stability, and predictability. The difference in tidal range, caused by the gravitational pull of the moon and the sun, can be converted into energy. The kinetic energy of the fluid particles in the tide as the seawater flows in turns the turbines to produce electricity. The turbines transmit the generated electrical energy to the land through power cables. Usually, seawater is about 832 times denser than air and a relatively small turbine can produce a lot of electricity. Also, tidal current energy can be produced both during the day and at night.

Recently, floating tidal energy systems have attracted attention because they have overcome the limitations of onshore tidal energy platforms. The floating tidal platform can be installed with all equipment inside and is easy to maintain [2–9]. Therefore, many previous studies have been conducted. Jo, C. et al. analyzed the mooring system of a floating tidal power plant (TCP) using a commercial analysis program (OrcaFlex) and designed an optimal design to obtain a safe dynamic response [10]. Arini, N. R. et al. performed a dynamic response analysis on vertical and horizontal modular tidal energy platforms, and the effective tension of the mooring line under different environmental loads was analyzed [11].

In addition to tidal energy, research on floating platforms to obtain various energy sources has been conducted. Song, J. et al. analyzed the dynamic response of floating solar panels under various environmental loads and when the mooring line was broken [12]. Yang, R. Y. et al. analyzed the dynamic response of floating photovoltaics (FPV) under normal and extreme environmental conditions through experiments and numerical analysis. The commercial analysis program ANSYS AQWA 2020 was used for numerical analysis, and ANSYS FLUENT was used for wave pressure analysis [13]. Li, L. et al. designed a novel offshore floating renewable energy system combining an offshore floating wind turbine (OFWT), a wave energy converter (WEC), and a tidal turbine [14]. Muliawan, M. J. et al. performed numerical analysis in the time domain using SIMO/TDHMILL3D and calculated the wind power generation of Spar under operating conditions [15].

Furthermore, many studies related to the structure monitoring approach of floating system have also been carried out thus far. Chung, W. C. et al. created an algorithm to analyze the status of mooring lines and risers. The results of numerical analysis using Orcaflex were compared with the results derived through the algorithm, and it can be confirmed that the results are in good agreement [16–18].

Also, effective tension applied to mooring chains were analyzed using multiple approaches [19]. Guo, X. et al. investigated the effects of thrust and torque on the dynamic response of a floating tidal turbine [20]. Du, J. et al. compared the dynamic responses of six mooring systems, including those combining different buoys with mooring lines [21]. Tang, Y. et al. presented a methodology to overcome the limits of Orcaflex and obtain more accurate results by using additional software [22]. Thomsen, J. B. et al. compared the results obtained from Open FAST and Orcaflex, and the reason for this was analyzed [23]. Sobhaniasl, M. et al. evaluated axial forces and fatigue in dynamic power cables connected to a 5 MW floating offshore wind turbine using ANSYS AQWA and FAST NREL (National Renewable Energy Laboratory) codes [24]. Taninoki, R. et al. conducted behavior simulation tests on a dynamic power cable system, measuring strain on wire armor under tensile and bending loads. Forced oscillation tests confirmed that tensile-induced strain remained within safe limits, validating the cable's ability to endure stress in floating platforms [25]. Okpokparoro, S. et al. presented effective tension as a main parameter for failure risk assessment in floating platforms, linking effective tension to long-term reliability [26]. Abrahamsen, A. B. et al. presented a reference dynamic power cable design, assessing maximum cable effective tension as a key failure mode [27]. Yang, R. Y. et al. simulated two mooring system failure scenarios. The effective tension distribution range of the floating platform and anchor chain after the mooring system failure was analyzed [28].

However, previous studies have primarily focused on the structural analysis of simple floating tidal power systems under normal sea conditions. While numerous papers have addressed the feasibility and static stability of single tidal power generators, studies analyzing the dynamic response of floating platforms and the effective tension in mooring lines under extreme environmental conditions remain limited. In particular, although mooring line failure scenarios have been extensively studied for floating offshore wind platforms, such analyses are scarce for tidal power floating platforms.

This study goes beyond a basic global analysis by conducting a comprehensive simulation that encompasses reliability assessments under extreme environmental conditions, considerations of mooring line failure scenarios, and integrated system analysis including dynamic power cables with dual vertical-axis turbines. By applying traditional analysis methodologies, this study offers a valuable reference for validating the applicability of floating dual vertical-axis tidal turbine systems and provides a foundation for an initial conceptual design.

2. Theoretical Background

In this chapter, this study's theoretical background is explained briefly.

2.1. Governing Equations [29,30]

In the fluid area, the fluid flow is assumed to be inviscid, incompressible, and irrotational. If the flow is assumed to be inviscid, then it is simplified to the Euler equation; if potential flow is assumed, then it is derived to the Laplace Equation (1). In addition, since the kinetic energy and pressure change transmitted during wave oscillation are not greatly attenuated by viscosity, ignoring viscosity does not have a significant effect on the accuracy of wave analysis.

$$\nabla^2 \Phi = 0 \tag{1}$$

where velocity potential is Φ, respectively. By inputting the (1) into the Navier–Stokes equation and integrating it, the Bernoulli equation for the fluid pressure over the wetted surface, as in (2), is obtained.

$$p = -\rho \frac{\partial \Phi}{\partial t} + \frac{1}{2}(\nabla \Phi)^2 + gZ \tag{2}$$

Bottom BC (Boundary Condition) at z = $-\infty$, the vertical velocity of fluid particles at the sea bottom is zero.

$$\frac{\partial \Phi}{\partial t} = 0 \tag{3}$$

Free surface at z = η(x, y, t), the pressure at sea is equal to atmospheric pressures.

$$\rho \frac{\partial \Phi}{\partial t} + \frac{1}{2}\left(\Phi_x^2 + \Phi_y^2 + \Phi_z^2\right) + \rho g z = 0 \tag{4}$$

Kinematic BC on the surface of a rigid body moving at z = η(x, y, t), fluid particles at the sea surface are at the sea surface.

$$\frac{\partial \eta}{\partial t} + u\frac{\partial \eta}{\partial x} + V\frac{\partial \eta}{\partial y} - \frac{\partial \Phi}{\partial t} = 0 \tag{5}$$

This condition applies at the instantaneous free surface at z = η(x, y, t), which is defined by (2).

$$\eta(X, Y, t) = -\frac{1}{g}\frac{\partial \Phi}{\partial t} + \frac{1}{2}(\nabla \Phi)^2 \tag{6}$$

However, the exact solution of the governing equation is hard to obtain precisely due to the non-linearity of BC. So, the approximate solution is obtained based on the perturbation approach and used to calculate the wave load acting on the floating body. The approximate solution obtained using the perturbation approach is as follows [31]:

$$\begin{aligned} \sum \Phi &= \Phi_1 + \Phi_2 + \ldots \\ \sum p &= p_1 + p_2 + \ldots \end{aligned} \tag{7}$$

2.2. Hydrodynamic Coefficients [32–34]

Hydrodynamic coefficients are numerical values that describe how a body behaves when moving through a fluid, such as seawater. The coefficients represent the relationship between the fluid forces and the motion of the body (such as velocity or acceleration). They are essential in predicting and analyzing the performance of floating structures.

Continuously, floating platforms take on additional fluid loads as they move due to pressure changes in the surrounding fluid called added mass/moment of inertia. Also, when a floating body moves, energy is dissipated as waves are generated on the surface. The following is Equation (8) related to the added mass/moment and radiation damping:

$$(M_a)_{ij} - \frac{i}{w} C_{ij} = \rho \int_s n_i \Phi_j ds \tag{8}$$

where M_a is added mass, C_{ij} is radiation damping, n_i, Φ_j is normal velocity and fluid potential.

2.3. Wave Excitation Force

Wave excitation forces are split into two components such as Froude–Kriloff and diffraction forces as below. Froude–Kriloff forces, derived from pressure on the body due to undisturbed incident waves:

$$F_{fk} = \iint_{S_B} p_d n dS \tag{9}$$

Diffraction forces result from wave field disturbance by the structure. For example, the vertical component is:

$$F_{d,3} = M_{a,33} \cdot a_3 \tag{10}$$

where $M_{a,33}$, a_3 are vertical added mass and wave acceleration, respectively.

The complete wave excitation force, accounting for both effects in complex form, is:

$$F_{wave}^{1st\ order} = -i\omega\rho \iint_{S_B} n_i (\varnothing_0 + \varnothing_S) dS \tag{11}$$

Here, ϕ_0 and ϕ_s represent the incident and scattered wave potentials. Due to the geometrical complexity of the platform, this must be evaluated numerically.

2.4. Time Domain Analysis in Orcaflex

The floating platform has generally six degrees of freedom, including three translational and three rotational motions. The equations describing this motion are expressed based on Newton's law or Lagrange's equations. The equations can be expressed as follows:

$$(M + M_a)\ddot{u} + C\dot{u} + Ku = F_{exteral\ force} \tag{12}$$

$$F_{exteral\ force} = F_{wave} + F_{current} + F_{wind} \tag{13}$$

where $M, M_a, \mathbf{C}, \mathbf{K}$ is structure mass, added mass, damping, restoring load, respectively. $F_{exteral\ force}$ is external load including wave, current, wind load. This equation includes mass, damping and restoring force, evaluated in time steps, and considers instantaneous and time-varying geometry. F_{wave} consists of the first and second-order wave excitation force. Continuously, in random waves, wave load can be expressed using the Volterra series expansion.

$$F_{wave}^{1st\ order} = Re\left[\sum_{i=1}^{N} A_i L(\omega_i) e^{iw_i t}\right] \tag{14}$$

$$F_{wave}^{2nd\ order} = Re\left[\sum_{j=1}^{N}\sum_{k=1}^{N} A_j A_k * D(\omega_{j,} - \omega_k)\cdot e^{i(\omega_j - \omega_k)t} + \sum_{j=1}^{N}\sum_{k=1}^{N} A_j A_k S(\omega_j, \omega_k)\cdot e^{i(\omega_j + \omega_k)t}\right] \tag{15}$$

(*) represent the complex conjugate. Also, L, D, S mean linear transfer function (LTF), the difference $D(\omega_{j,} - \omega_k)$ and sum $S(\omega_j, \omega_k)$, frequency quadratic transfer function (QTF), accordingly.

2.5. Lumped Mass Method for Mooring Line [34,35]

The lumped mass method, which is implemented in OrcaFlex, is used to model the mooring line as a series of beam elements. In this approach, each mooring line—assumed to behave like a beam—is divided into nodes and segments, as illustrated in Figure 1. Forces such as weight, buoyancy, and drag are concentrated at each node (mass point), while massless springs are used to simulate the dynamic behavior of the mooring line. The Morison equation is applied to calculate the hydrodynamic forces acting on the line. Furthermore, $\rho V\dot{u}$ is Froude–Krylov force, $\rho VC_a(\dot{u} - \dot{v})$ is hydrodynamic mass force, and $\frac{1}{2}\rho C_d A(u - v)|u - v|$ is drag force, respectively.

$$F = \rho V\dot{u} + \rho VC_a(\dot{u} - \dot{v}) + \frac{1}{2}\rho C_d A(u - v)|u - v| \tag{16}$$

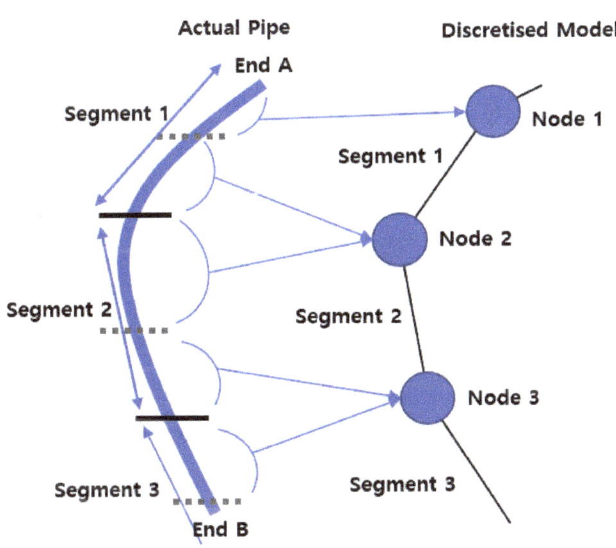

Figure 1. Lumped mass method in Orcaflex.

3. Target Model

3.1. Floating Platform

Figure 2 shows the front view and top view of the floating platform model. In the figure, two pontoons are colored red and are located 4 m deep in the water on both sides. It is assumed that the pontoons are equipped with a large-capacity battery inside. Also, the shape of the pontoons is streamlined to reduce resistance at high fluid speeds. Additionally, the yellow block connecting these two pontoons on both sides holds the system controllers, which act as generators for the turbines. In addition, Figure 3 plotted a mesh configuration for the floating platform which is used in Orcawave. Red and gray meshes represent wet and dry surfaces, respectively. The mesh size is set to 0.4 m by default, and the number of total nodes is 3944. In this numerical analysis, it is assumed the power generation is stopped and the vertical turbines are lifted to prevent failure under the extreme environments. For this reason, the gyroscopic moments generated when the turbine motor rotates, which could affect the floating platform, are not also considered. Additionally, it is assumed that the moment generated by lifting the vertical turbines is quite small compared to the entire system. As can be seen in Table 1, the mass of vertical turbines is extremely low among the total mass. In fact, the mass of the vertical turbines is only 0.39% of the total mass. This can be the basis for the insignificant effect of the turbine moment on the entire floating platform. So, it is excluded from numerical analysis. Table 1 lists the specification of the base model of the floating platform and Table 2 tabulates hydrostatic restoring stiffness. To analyze the dynamic response of the floating platform in time domain, hydrodynamic coefficients (added mass/inertia, radiation damping, load and displacement RAOs (response amplitude operator) are needed. To obtain the coefficients, Oracwave software 11.4 is used to obtain the hydrodynamic coefficients by incident regular waves from various directions. A wave period of 2.5–30.0 s and to 90° of wave heading is considered during the hydrodynamic coefficients analysis. In addition, it is assumed that the seawater density is 1025 kg/m^3 and water depth is 50 m. Also, added mass/inertia, radiation damping, load RAOs, and displacement RAOs under 90° are shown in Figures 4–7, respectively.

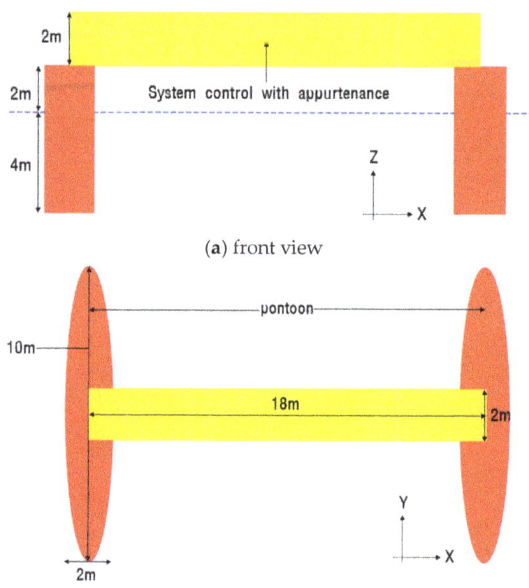

(**a**) front view

(**b**) top view

Figure 2. Floating dual vertical-axis tidal turbine system on extreme load (without turbine).

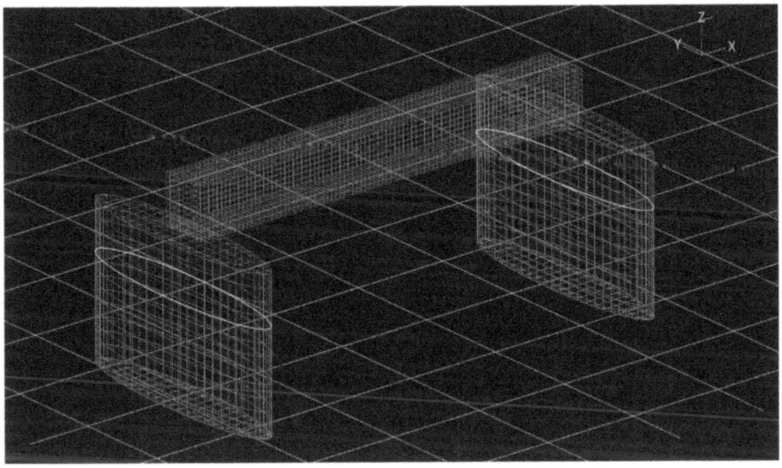

Figure 3. Mesh configuration for the floating dual vertical-axis tidal turbine system.

Table 1. Specification of floating dual vertical-axis tidal turbine system.

Name		**Unit**	**Value**
Pontoons		kg	60,000
Vertical current turbines		kg	500
System Control		kg	48,000
Appurtenance		kg	20,000
Total mass		kg	128,500
Center of gravity	X	m	0
	Y	m	0
	Z	m	−0.4454
Moment of inertia	Rx	$kg{\cdot}m^2$	3,738,050
	Ry	$kg{\cdot}m^2$	112,730
	Rz	$kg{\cdot}m^2$	1,076,700
Center of buoyancy	X	m	0
	Y	m	0
	Z	m	−2
Turbine power generation		**Unit**	**Value**
Turbine power		kW	50
Max thrust per turbine (Rated current)		kN	116

Table 2. Hydrostatic restoring stiffness.

Heave (kN)	**Roll (kNm/rad)**	**Pitch (kNm/rad)**
315.44	0	0
0	1.19	0
0	0	23,660

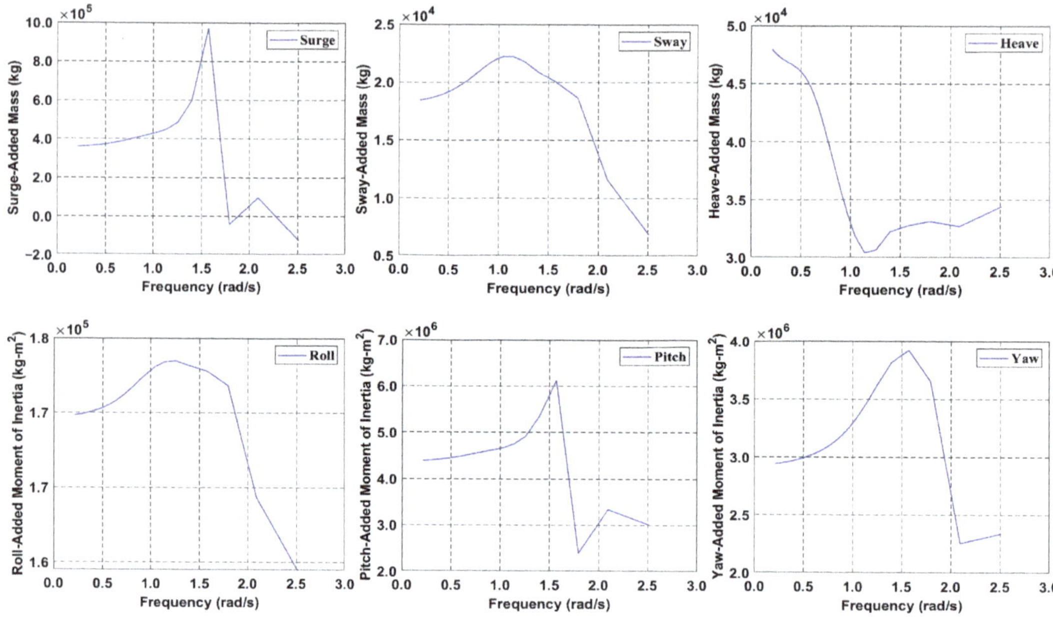

Figure 4. Added mass and moment.

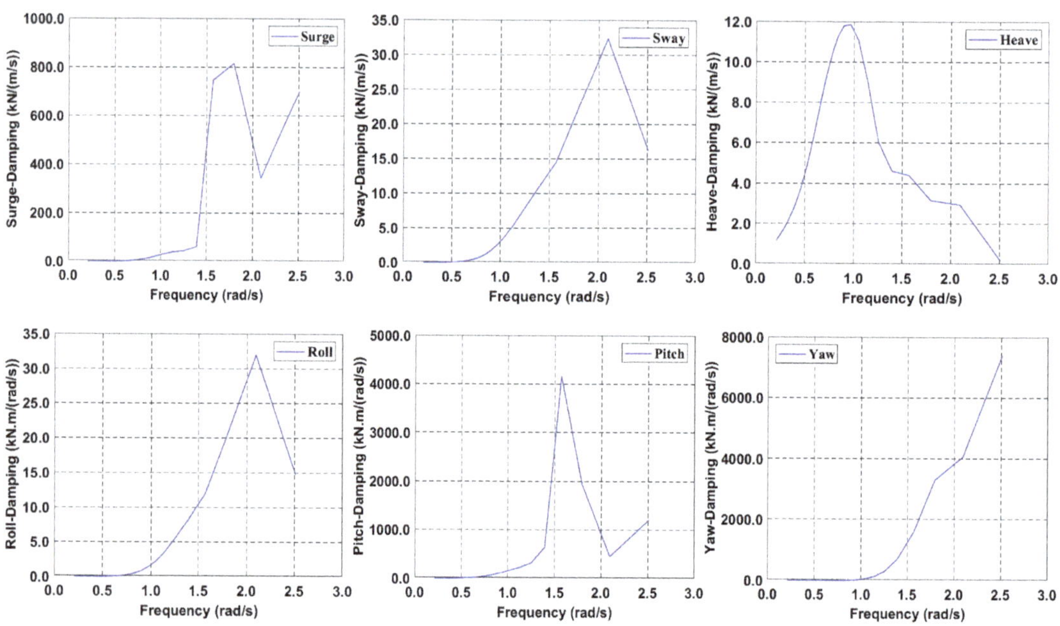

Figure 5. Radiation damping.

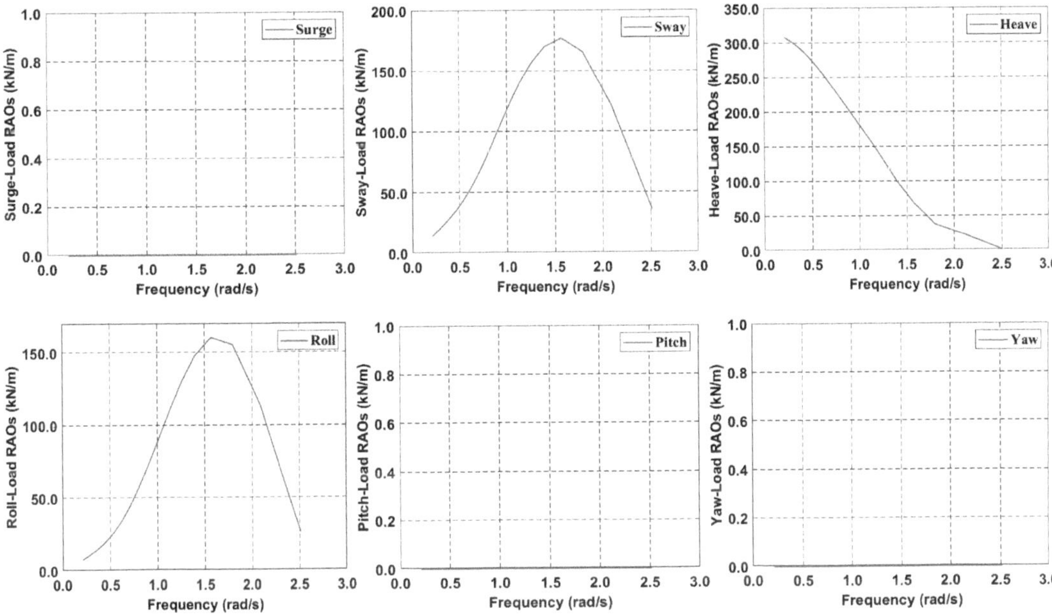

Figure 6. Load RAOs, 90°.

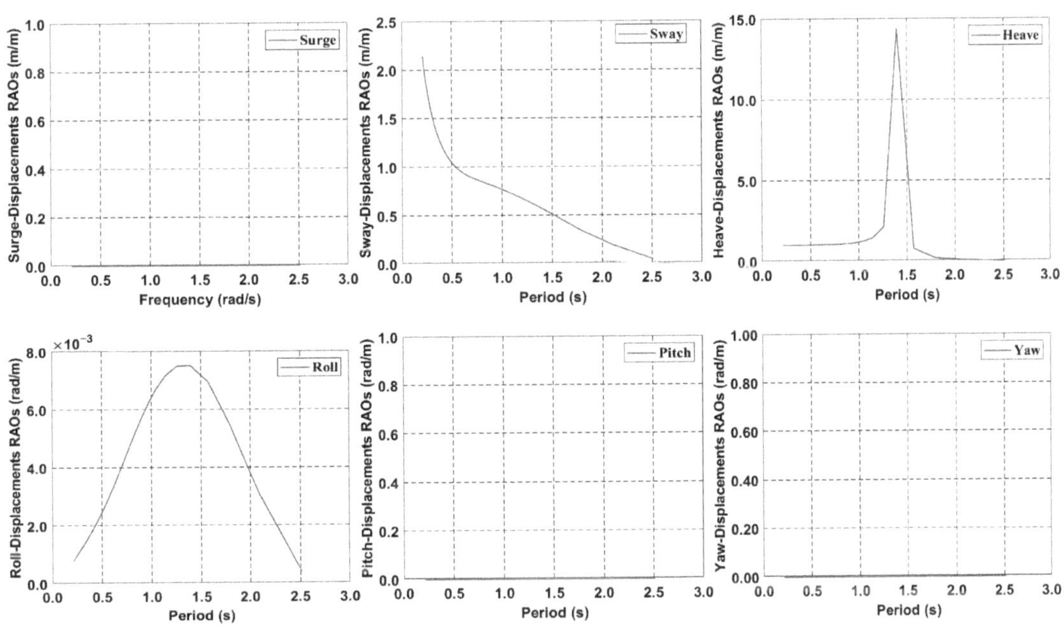

Figure 7. Displacement RAOs, 90°.

3.2. Mooring System

Figure 8 plotted the configuration of four mooring lines which were attached to enhance the station keeping capability of floating platform. The wave, current, and wind heading is also indicated in Figure 8. Details of the mooring properties are listed in Table 3 [11]. Also, the mooring system's fairlead and anchor location is tabulated in Table 4.

The fixed point is the location where there is a mooring line connected to the floating platform, whereas the anchor point is the location where there is a mooring line connected to the seabed. In addition, section length means how far each point mass is along the mooring lines based on the lumped mass method. Moreover, Table 5 listed the natural frequencies of the floating structure. Modal analysis in Orcaflex can be used to calculate the natural frequencies and mode shapes of the system. In addition, Figure 9 plotted the two different mooring types (taut and catenary) that are used in this numerical analysis as a design parameter. The taut mooring line length is 76 m and is a type that mooring line is kept tighter than loose. The mooring line's shape is close to a straight-line type rather than curve shape. It effectively distributes and reduces the vertical load using high elastic materials like synthetic fiber ropes (polyethylene, aramid fiber). Therefore, the anchor installation area (footprint) is smaller compared to catenary mooring systems. But, in shallow water, a taut leg system can be too stiff and increase line tension excessively.

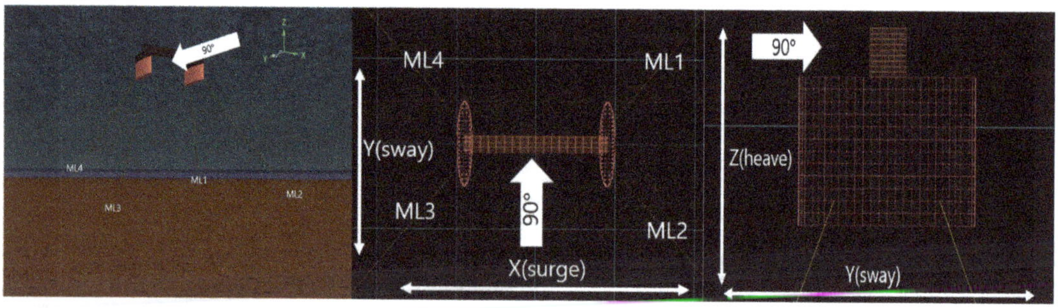

Figure 8. Mooring system configuration and environmental heading.

Table 3. Mooring properties.

Type	-	**6×19 Wire with Fiber Core**
Nominal diameter	m	0.15
Mass/unit length	kg/m	81.30
Displaced mass	kg/m	13.4
Axial Stiffness	MN	825.80
Bending stiffness	MN	0
MBL (Minimum Breaking Load)	kN	3835.50

Table 4. Mooring system's fairlead and anchor location.

Type	**Axis and Unit**	**ML1**	**ML2**	**ML3**	**ML4**
		(a) Taut			
Fixed Point	X	10	10	−10	−10
	Y	3	−3	−3	3
	Z	−3	−3	−3	−3
Anchor Point	X	50	50	−50	−50
	Y	50	−50	−50	50
	Z	−50	−50	−50	−50
Section Length	m	1	1	1	1
Number of Segment		76	76	76	76

Table 4. *Cont.*

Type	Axis and Unit	ML1	ML2	ML3	ML4
		(b) Catenary			
Fixed Point	X Y Z	10 3 −3	10 −3 −3	−10 −3 −3	−10 3 3
Anchor Point	X Y Z	75 75 −50	75 −75 −50	−75 −75 −50	−75 75 −50
Section Length	m	1	1	1	1
Number of Segment	-	115	115	115	115

Table 5. Natural frequencies.

Type Unit	1st (Surge Governed) Hz	2nd (Pitch Governed) Hz
	(a) Taut	
Value	0.12438	0.16179
	(b) Catenary	
Value	0.03406	0.04085

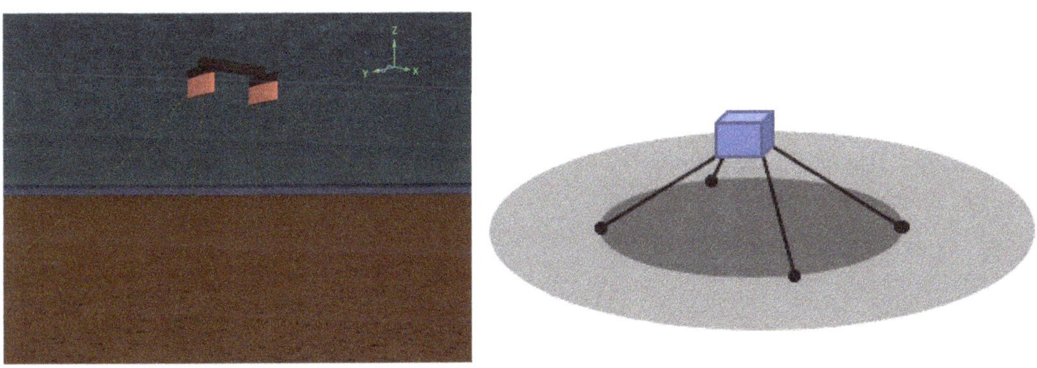

(**a**) taut

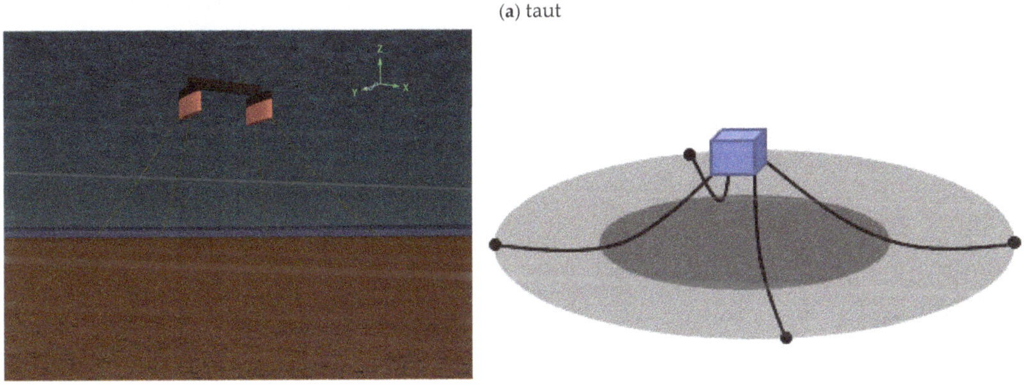

(**b**) catenary

Figure 9. Configuration of the two different mooring types.

However, in catenary type, the mooring line's shape is a catenary curve and partially touches the seabed. Thus, effective tension is not tight and when a horizontal load is received, the curve is tight and the restoring force acts [36]. As shown in Figure 9, the catenary's mooring line length is 115 m.

3.3. Describing of Environmental Loads

Environmental loads are listed in Table 6. The wave, current, and wind load are referenced by [11] and [18], respectively. Since environmental heading is 90°, the area that has received load in horizontal direction is only sway area. Also, this study presents an initial conceptual design of a floating dual vertical-axis tidal turbine system. Due to the nature of tidal energy generation, such systems are typically installed and operated in areas with relatively high current velocities, such as narrow straits or between islands. Given the system's geometrical configuration, it is assumed that the environmental load angle that yields the highest power generation efficiency is 90 degrees and the dynamic performance under parked conditions (non-operating) can be evaluated even in extreme environmental conditions. Therefore, additional analyses at other headings such as 0 and 45 degrees, are planned in future system design optimization stages.

Table 6. Environmental Conditions.

Type	Parameter	Unit	Extreme
Wave	Spectrum	-	JONSWAP
	Gamma	-	2.2
	Direction from X	deg	90
	Significant Wave Height (Hs)	m	8.7
	Peak Period (Tp)	s	17.2
Current	Surface Velocity (1/7 law is used)	m/s	1.0
	Direction from X	deg	90
Wind	Spectrum	-	API
	Mean Velocity	m/s	34
	Direction from X	deg	90

Figure 10 plotted wave elevation in time domain, frequency domain of wave, respectively. It can be seen that the peak frequency of wave is 0.05806 Hz. To confirm wave energy input correctly, it is important to match the wave-theoretical and wave-regenerated spectral density. It is well matched as shown in Figure 10. Also, JONSWAP (Joint North Sea Wave Project) spectrum is used as the wave model. The spectral density's fundamental frequency is controlled to match the theoretical and generated wave. It is set at 0.0012 Hz in this numerical analysis. Moreover, the JONSWAP spectrum can be defined as follows:

$$S(f) = \alpha g^2 (2\pi)^{-4} f^{-5} exp\left[-\frac{5}{4}\left(\frac{f_p}{f}\right)^4\right] \cdot \gamma^b \quad (17)$$

$$b = exp\left[-\frac{\left(f - f_p\right)^2}{2\sigma^2 f_p^2}\right] \quad (18)$$

where f, f_p are present frequency and wave peak frequency, respectively. α is spectrum energy factor. g is gravity acceleration. σ is spectral width factor. γ is peak enhancement factor.

Figure 11 plotted the wind and current profile at 90°. For wind, the API (American Petroleum Institute) spectrum is used as spectrum model. The API spectrum can be defined as follows:

$$S(f,z) = \frac{U_z^2 I_z^2}{f_p\left[1 + 1.5\left(\frac{f}{f_p}\right)\right]^{\frac{5}{3}}} \tag{19}$$

$$I_z = \frac{\sigma_u}{U_z} \tag{20}$$

where $\mathbf{U_z}$ is mean wind speed at z, $\mathbf{I_z}$ means turbulence intensity. In this study, it is set to 0.08 [37]. σ_u is standard deviation of wind speed. This value is 3.44 m/s. And f, f_p are present frequency and wind peak frequency, respectively.

Also, the current profile can be defined by the power law method and expressed as follows:

$$C = C_b + (C_f - C_b)\cdot\left[\frac{Z - Z_b}{Z_f - Z_b}\right]^{\frac{1}{p}} \tag{21}$$

where C is current speed at $\mathbf{Z}$ and C_f, and C_b are the current speeds at the surface and seabed, respectively. $\mathbf{1/P}$ is the power law exponent. In this numerical analysis, it is set to 1/7. It means that the current is gradually reduced from surface to seabed with 1/7 law. $\mathbf{Z_b}$ and $\mathbf{Z_f}$ are Z-coordinates of the surface and seabed, respectively.

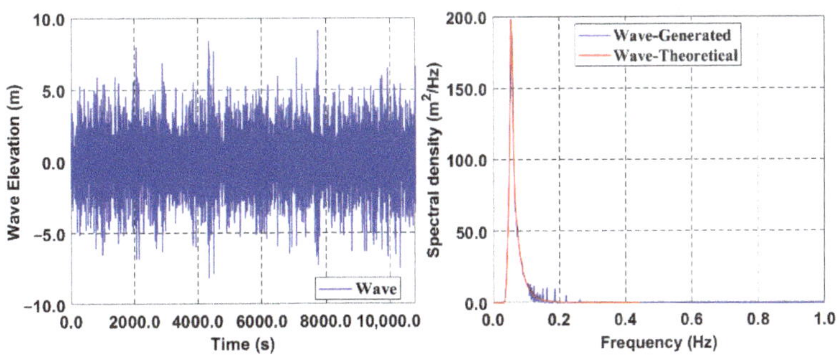

Figure 10. Wave elevation and spectrum, **90°**.

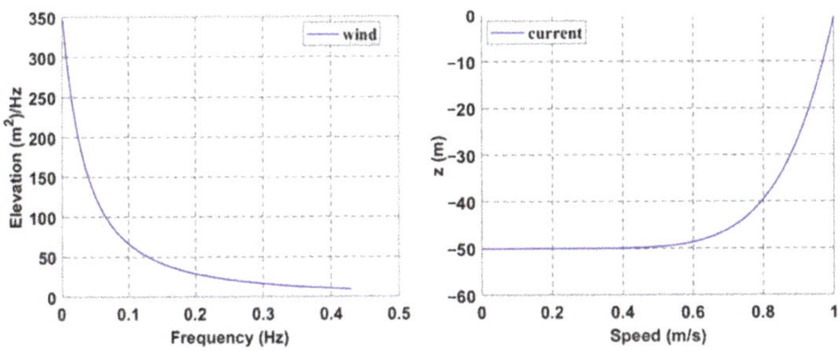

Figure 11. Wind and current profile, **90°**.

4. Results and Discussion

4.1. Taut vs. Catenary Mooring System Comparison

In this section, the results of the dynamic responses of the platform under two different mooring systems (taut and catenary) are explained. In Figure 12, (a) the dynamic response in time domain is plotted, whereas (b) presents the frequency domain results. The total simulation duration is set to 10,800 s, and taut is represented by the red line and catenary by the blue line, respectively. Since the environmental load heading is 90 degrees, surge, pitch, and yaw are converged to zero (no dynamic response). In sway and heave direction, it is observed that a large dynamic response is captured in catenary type rather than taut.

This is because the taut mooring lines are subjected to a larger effective tension than the catenary mooring lines. In this study, the maximum margin of rotational freedom is set to 40°. As can be seen from the roll results, it is confirmed that the angle does not exceed 40° in both mooring systems. Also, this can be confirmed in Figure 13. However, since fairlead point boundary condition is hinged, significant difference cannot be captured in roll direction between taut and catenary.

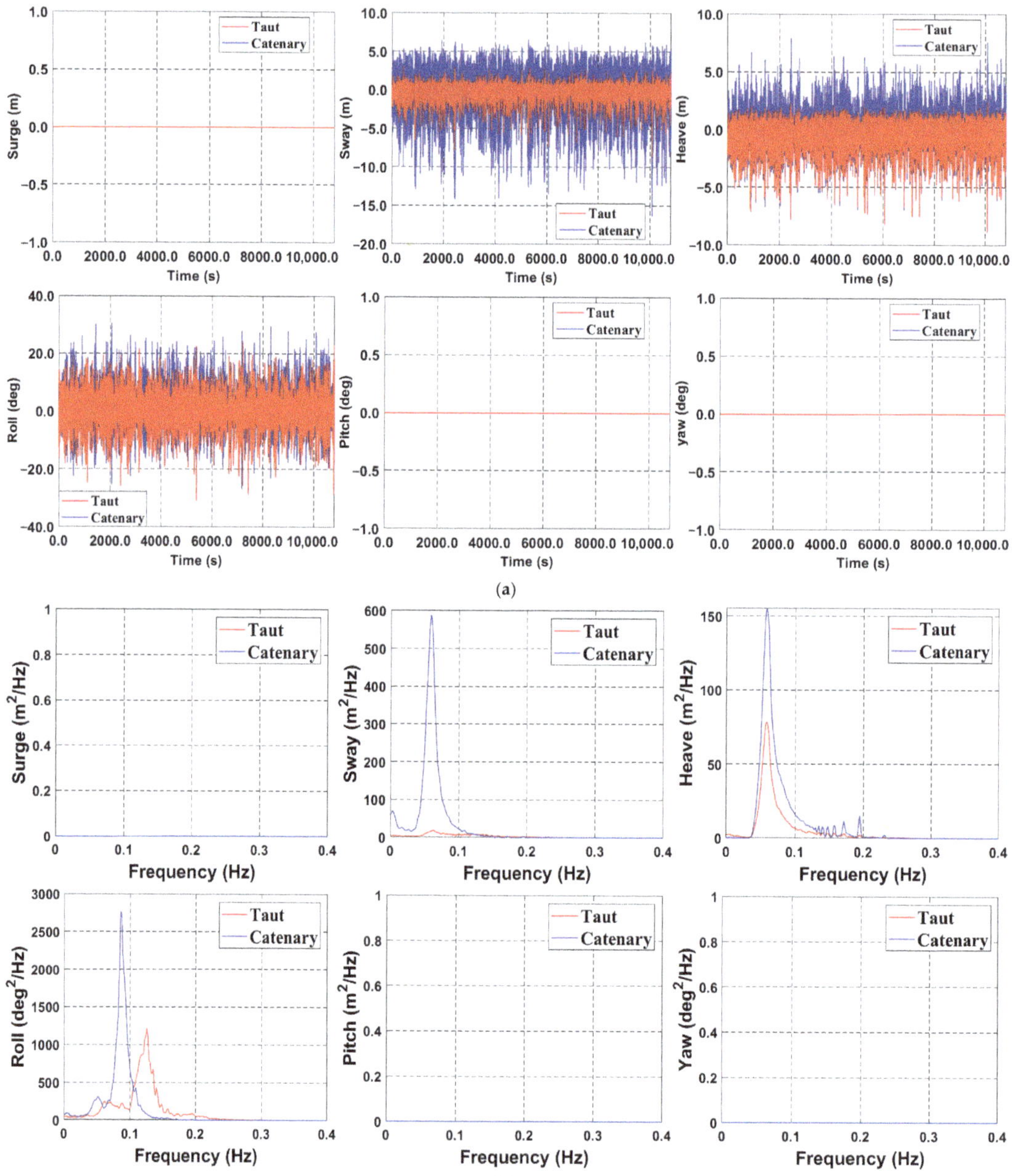

Figure 12. Floating platform motion, (**a**) time domain, (**b**) frequency domain, **90**°.

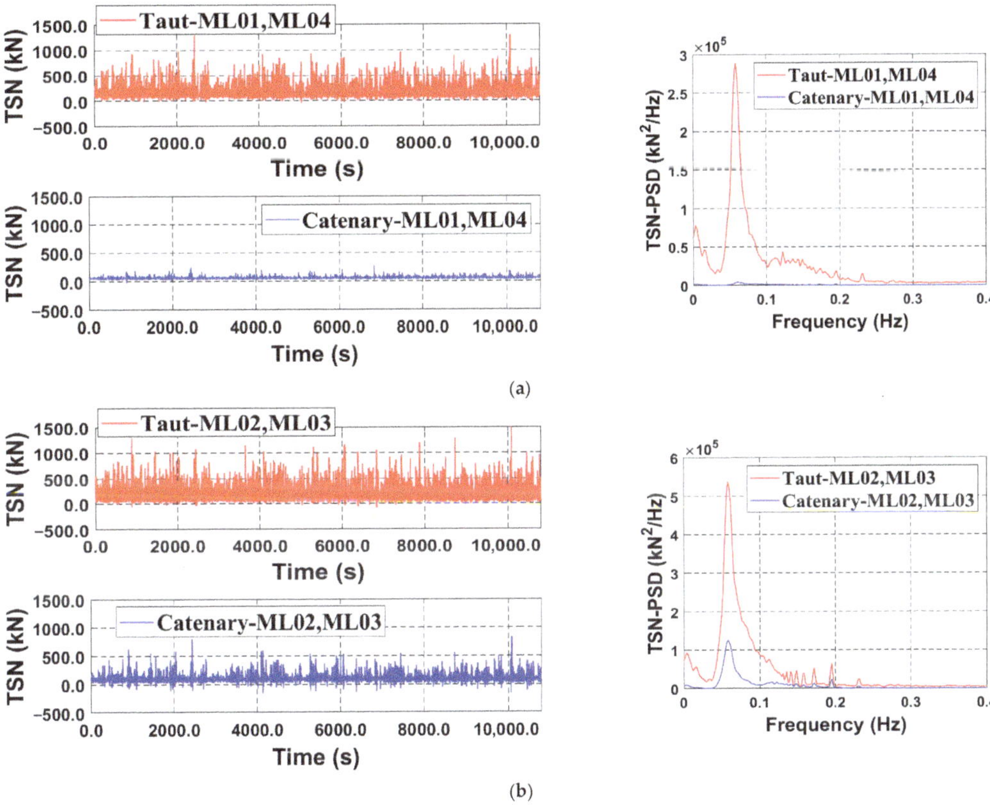

Figure 13. ML Effective Tension (TSN), (**a**) ML01&04, (**b**) ML02&03, **90°**.

In the frequency domain, peak frequencies around 0.058 Hz are observed in sway and heave. This matches the peak frequency of the wave spectrum which is plotted in Figure 10. Therefore, it can be confirmed that the influence of wave load is dominant in the sway and heave dynamic responses. However, in roll, unexpected peak frequencies are confirmed at 0.07332 Hz for catenary and 0.12438 Hz for taut, which are not same with the peak frequencies of wave load. This cannot be explained by the influence of wave load. Therefore, the natural frequencies of the system can be checked through the modal analysis of Orcaflex.

The results can be seen in Figure 14. The mode shapes are directed in the roll direction are mainly confirmed at 0.12438 Hz and 0.07332 Hz for taut and catenary, respectively. Therefore, it is confirmed that the unexpected peak frequencies of 0.12438 Hz and 0.07732 Hz in governed roll mode are more influenced by the resonance due to the natural frequencies.

Additionally, Figure 13 plotted the effective tension applied to each mooring line. Since the wave heading is 90°, the effective tensions of ML01 and ML04 are the same and ML02 and ML03 are also the same. As expected, the effective tensions applied to the taut system are confirmed to be greater than that of the catenary system. This is supported by the fact that the dynamic responses of taut in sway and heave are relatively small compared to those of the catenary system. As shown in Table 2, the MBL (minimum breaking load) of the mooring line used in this study is 3835.5 kN, so it can be confirmed that the applied mooring system can resist even in extreme environmental conditions. As a result of checking the PSD of the effective tension, the peak frequency of the effective tension can be confirmed at 0.058 Hz, which is the wave peak frequency. Also, it can be confirmed that the effective tension of the mooring line is also dominated by the wave load. Statistical characteristic mooring system type is listed in Appendix A.

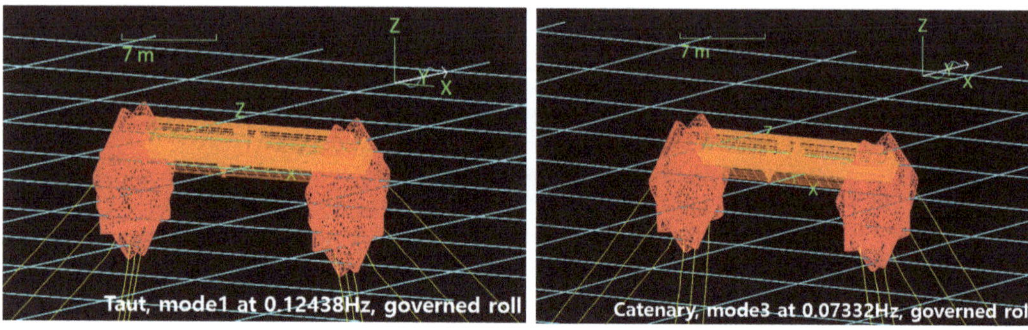

Figure 14. Natural frequencies and mode shape.

4.2. Mooring Failure

The failure of one mooring line's results will be discussed in this section. When designing a floating platform, mooring system failure scenarios when exposed to extreme environmental loads must be considered [11]. In the event of a mooring line failure case, ML02 will be disconnected intentionally during the simulation at 5400 s. Other conditions are the same as the previous Section 4.1.

Figure 15a,b plotted time-domain results before and after ML02 failure, respectively. In the figure, (a) presents the dynamic response of the floating platform before ML02 failure whereas (b) presents the dynamic behavior after ML02 failure. Figure 16a presents the dynamic response of the floating platform before ML02 failure in PSD whereas Figure 16b presents the dynamic response after ML failure in PSD. Likewise, in the intact case, before ML02 failure, surge, pitch, and yaw exhibit very little dynamic motion (almost close to zero), but relatively large dynamic responses are observed interestingly after the ML02 failure. This can be judged to be due to the imbalance of the effective tensions of the mooring lines and the application of the slow-varying drift load caused by the ML02 failure. As evidence for this, it can be confirmed that the dynamic variation in the yaw direction is the largest after ML02 failure rather than other directions. In sway, it can be confirmed that the stationkeeping capability decreases significance in both taut and catenary due to ML02 failure and the offset caused by ML02 failure cannot be recovered. However, no significant change is confirmed in heave. Furthermore, after the ML02 failure, unexpected peak frequencies occurred in all degrees of freedom except heave in the frequency domain. The slow-varying drift load due to the ML02 failure is caused by the unexpected peak frequency.

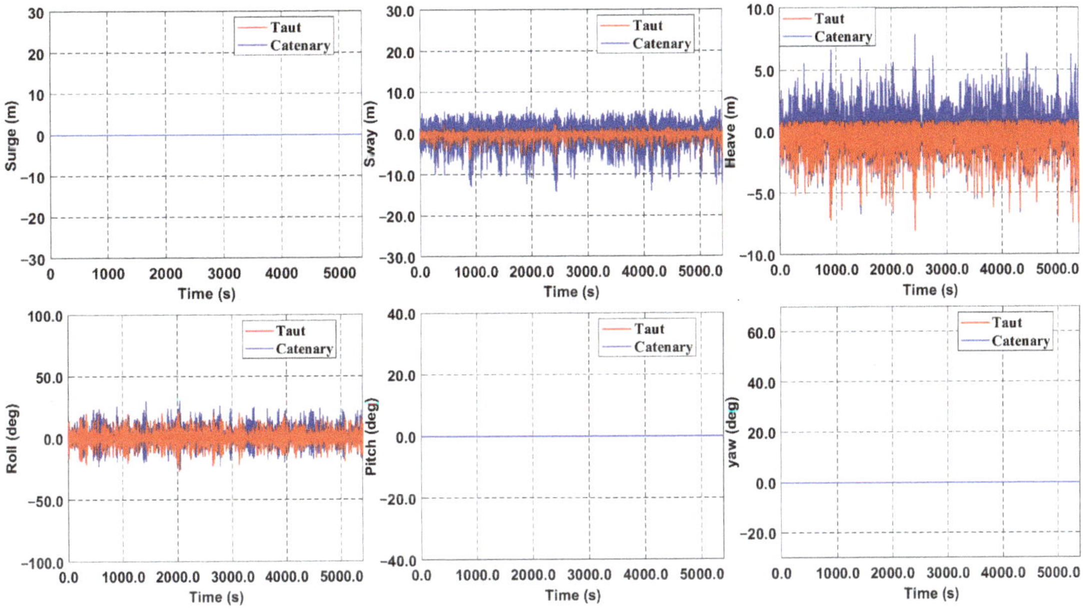

(a)

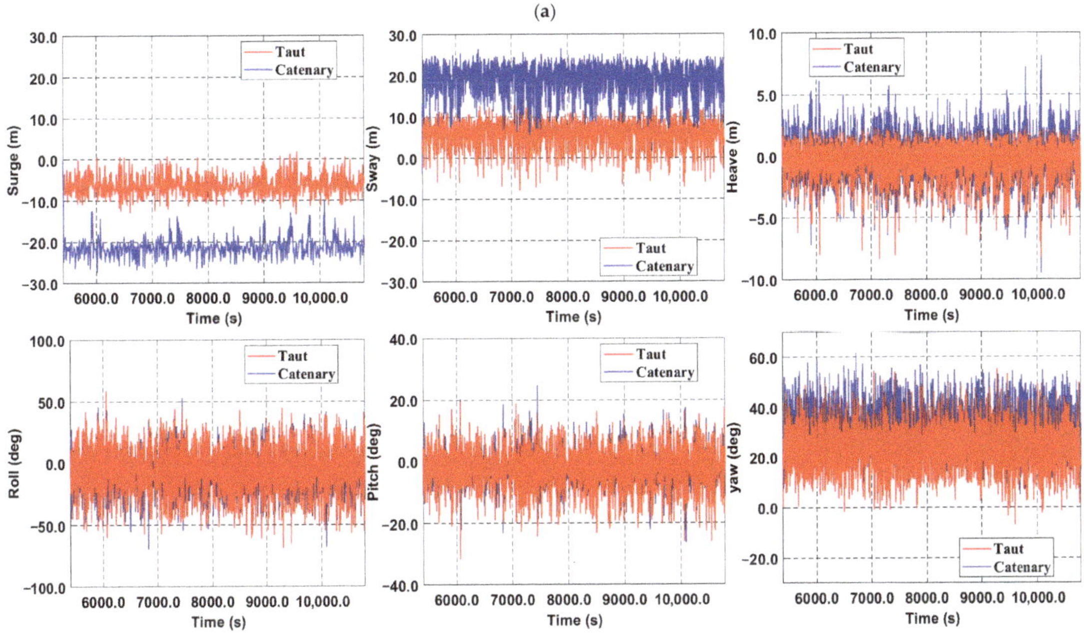

(b)

Figure 15. Floating platform motion time domain, (**a**) before failure, (**b**) after failure, 90°.

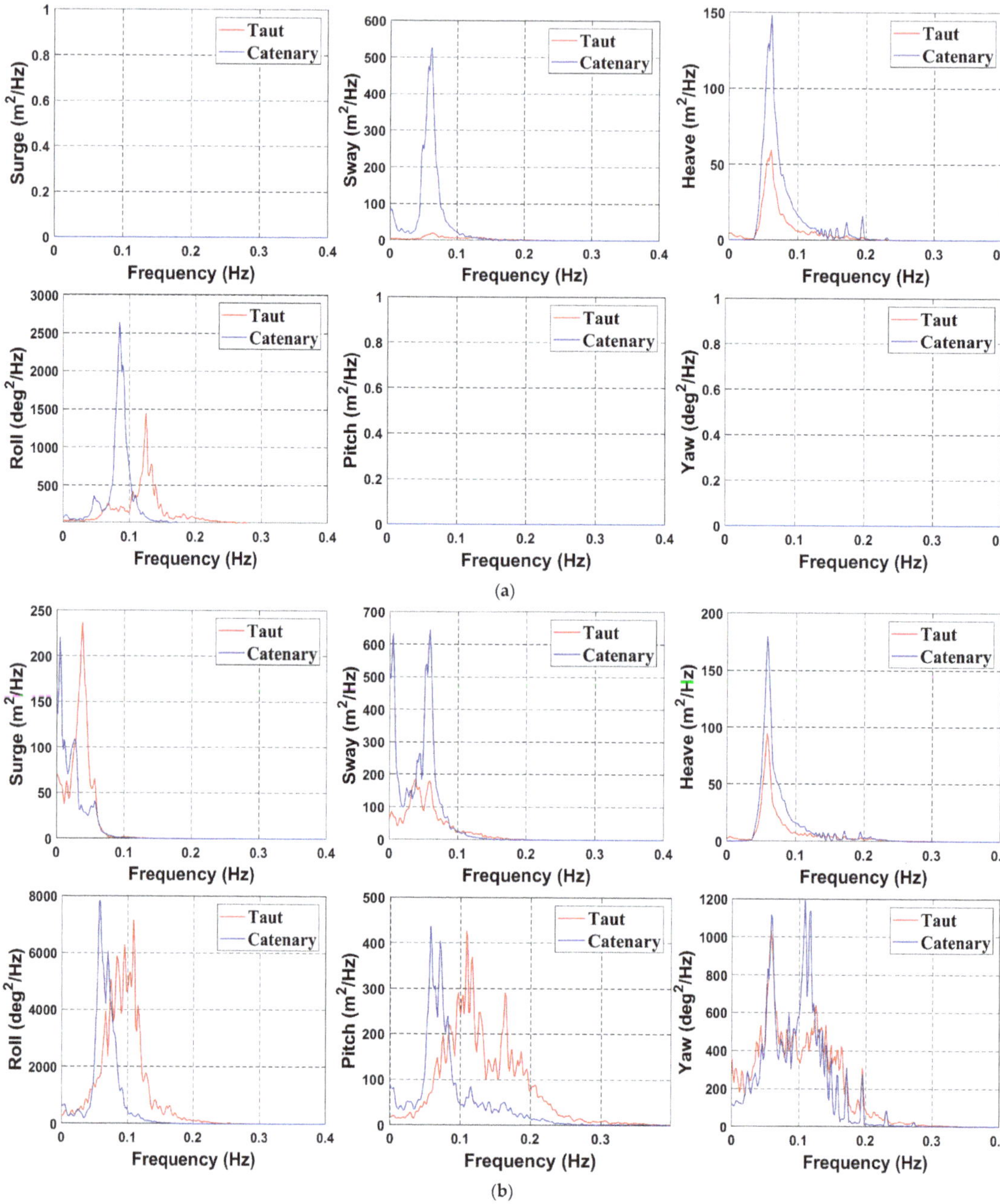

Figure 16. Floating platform motion frequency domain, (a) before failure, (b) after failure, 90°.

In addition, Figure 17 plotted the time and frequency domain of the effective tension before the ML02 failure. As discussed in the previous section, before the ML02 failure, the effective tension of ML01 and ML04 are the same and ML02 and ML03 are also the same. Also, Figure 18 plotted the time and frequency domain of the effective tension after the ML02 failure. As expected, each mooring line receives a different effective tension. It is confirmed that the maximum mooring load exceeded the minimum breaking load

(MBL) in both mooring systems following the failure. This phenomenon is attributed to the unbalanced distribution of effective tension, which became concentrated on the mooring line ML03. This is due to the slow-varying drift load and imbalance of the effective tension of mooring lines. Since ML02 is separated, it is not shown in Figure 18. For both the taut and catenary systems, it can be confirmed that ML01 and ML03 receive a larger effective tension than before the ML02 failure, whereas ML04 receives a relatively small effective tension. Also, it can be confirmed that the wave load is dominant in the frequency domain of the effective tension for both mooring systems. The statistical characteristics of mooring failure scenario is summarized in Appendix B.

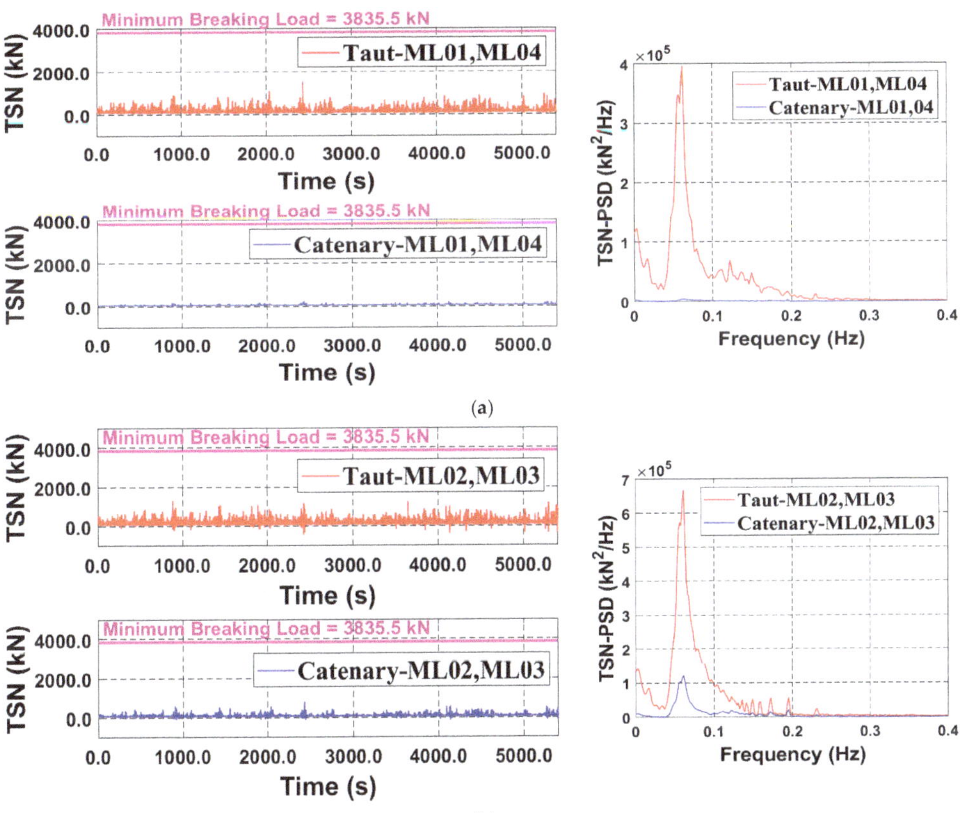

Figure 17. Before failure, effective tension (**a**) ML01 and ML04, (**b**) ML02 and ML03, **90**°.

4.3. Power Cable with Lazy-Wave Type

Usually, dynamic power cables are connected to an offshore renewable energy system to transmit electricity. They typically experience large load variations at the hang-off location and touch-down point due to the dynamic response of the floating platform [38]. To prevent damage caused by floater motion especially at the touch down point, several buoyancy modules can be attached to make a lazy-wave shape [39]. Thus, in this section, to prevent potential damage, the lazy-wave type power cable is applied here. Figure 19 plotted a configuration of two different mooring systems (taut and catenary) with a lazy-wave power cable in numerical analysis. The red and blue lines represent the Decline section and Buoyancy section, respectively. Additionally, dynamic power cable properties, which are used in this numerical analysis, are tabulated in Table 7 [40]. Also, Figure 20 shows the time and frequency domain of the effective tension applied to the dynamic power cable. The midpoint is set to an arclength 120 m along the dynamic power cable.

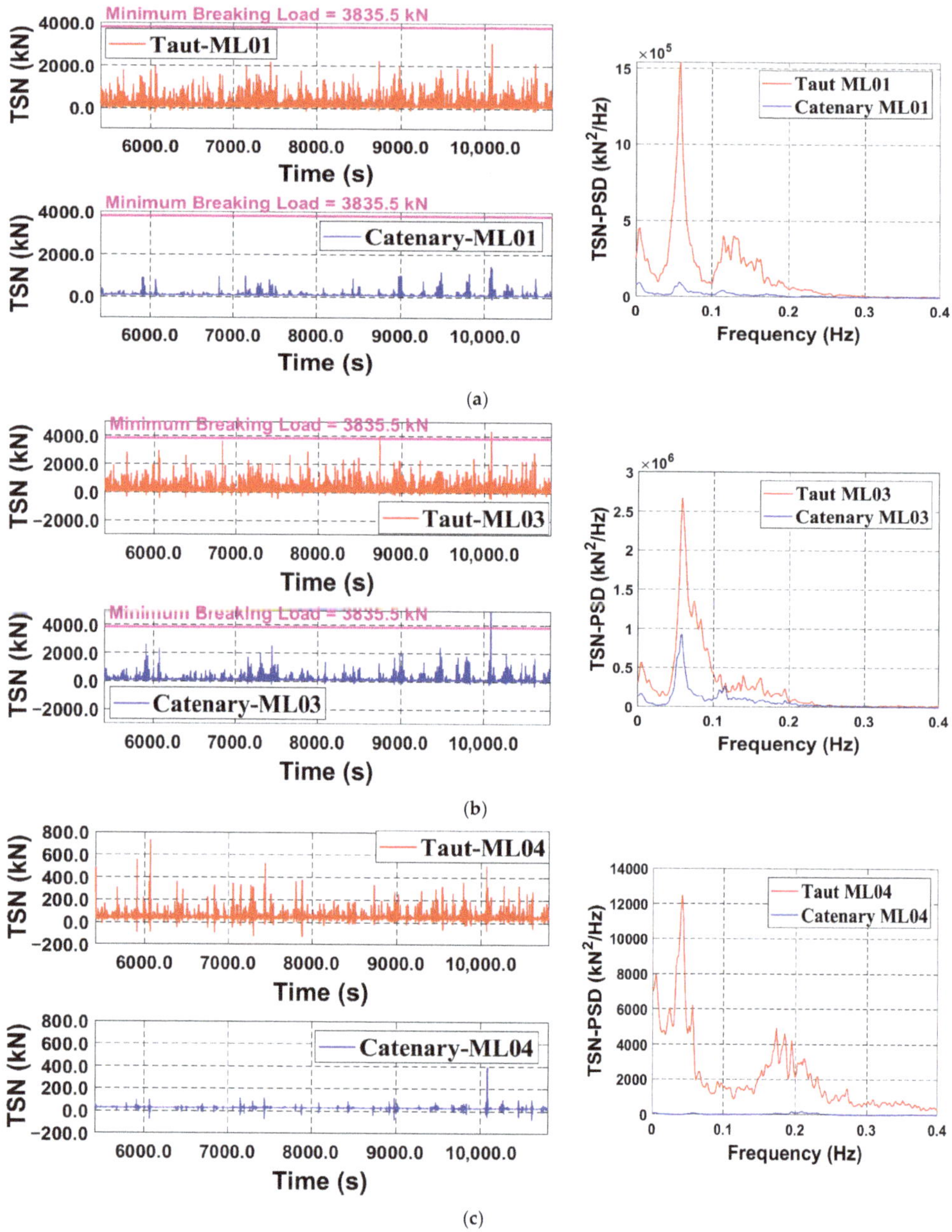

Figure 18. After failure, effective tension (**a**) ML01, (**b**) ML03, (**c**) ML04, **90°**.

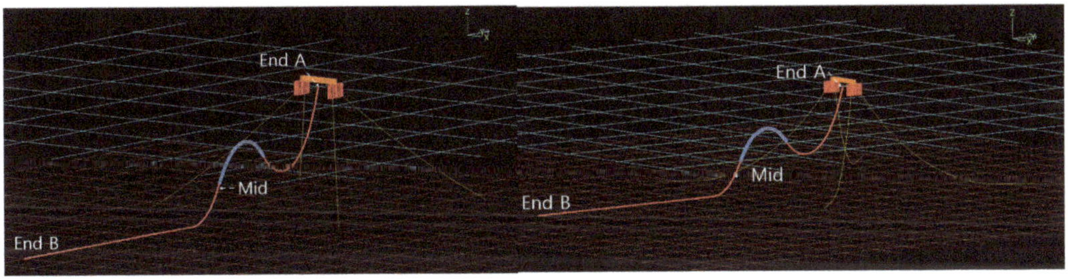

(a) Taut (b) catenary

Figure 19. Two different mooring systems with lazy-wave power cable.

Table 7. Dynamic power cable properties.

Property	Buoyancy Section	Decline Section
Submerged weight (N/m)	−463.83	516.16
Length (m)	40	200
Out diameter (m)	0.43	0.23
Axial stiffness (kN)	316.80×10^3	331.20×10^3

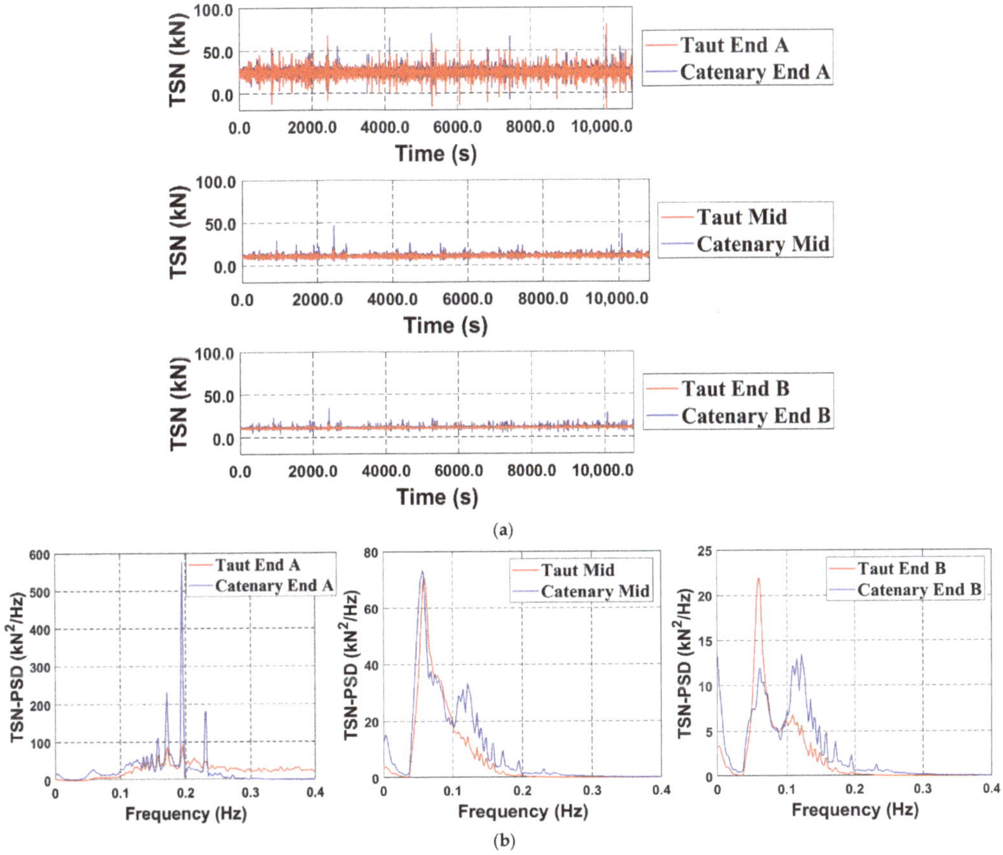

(a)

(b)

Figure 20. Effective tension of dynamic power cable (**a**) TSN (**b**) TSN-PSD.

As shown in Figure 20a, it can be confirmed that a larger effective tension is applied as it goes from End B (anchor) to End A (hang-off). Apart from the self-weight effect, it can be judged that a large effective tension is applied because End B is fixed to the seabed and far from the hang-off location, which received little dynamic response during the simulation, whereas End A is connected to the floating platform directly which was affected by a repetitive and dynamic response. In the frequency domain, the peak frequency of the midpoint is around 0.058 Hz, and it can be confirmed that the wave excitation load is dominated. However, an unexpected peak frequency of the catenary can be confirmed at End A and End B, respectively. In order to identify the cause of the unexpected peak, the natural frequency of the dynamic power cable should be computed using Orcaflex modal analysis. As shown in Figure 21, the natural frequencies can be captured at 0.13209 Hz and 0.18589 Hz, respectively. Thus, it can be confirmed that the peak frequency of the catenary at End A and B is more dominant due to resonance rather than the wave excitation load. To confirm the dynamic power cable integrity, it is assumed that allowable MBR of the dynamic power cable is 2.3 m and 4.3 m (=10 OD) in buoyancy and decline section, respectively. It can be confirmed that the minimum value of bend radius (MBR) during the simulation is 17 m in the taut system and 19 m in the catenary system. Thus, structure integrity of the dynamic power cable can be confirmed in initial design. Similarly, the statistical characteristics of dynamic power cable with the lazy-wave case is listed in Appendix C.

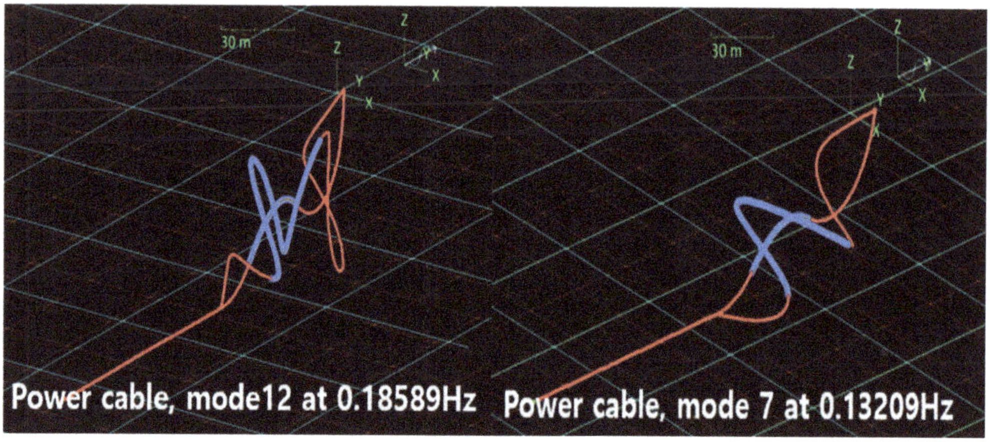

Figure 21. Natural frequencies and mode shapes of dynamic power cable.

5. Conclusions

Numerical analysis is performed to analyze the dynamic response of the floating dual vertical-axis tidal turbine platform under extreme environmental pressure. The floating platform is applied to the two different mooring systems, taut and catenary. JONSWAP and API are used as wave and wind spectra, respectively. The environmental heading is 90° and total simulation duration is 10,800 s. Also, the mooring system failure scenario is also checked. It assumes that ML02 is disconnected to the floating platform at 5400 s intentionally. Finally, the floating platform is equipped with dynamic power cable and the dynamic response of power cable under extreme environment conditions is also analyzed.

- In comparing the taut and catenary mooring systems with a 90° environmental heading, the catenary system shows larger sway responses due to lower effective tension in its mooring lines, while taut lines, with higher effective tension, show smaller heave responses. Roll responses are similar for both systems due to hinged BC. The sway and heave frequencies align with the wave spectrum (0.058 Hz), while roll exhibits unexpected peaks due to resonance, not wave loading. Taut mooring lines experience higher tension, which is reflected in the system's dynamic response and the peak frequency of effective tension, confirming wave load dominance.

- When a mooring line fails, the dynamic responses of surge, pitch, and yaw increase significantly, especially yaw. The slow-varying drift load on the floating platform further contributes to the peak frequencies of the dynamic responses. However, the effective tensions remain primarily influenced by the wave load.
- In dynamic power cable analysis, the effective tension increases from End B (anchor) to End A (hang-off) due to End A being connected to the floating platform, which moves dynamically, applying a larger tensile load. In the frequency domain, the catenary at End A peaks around 0.19 Hz, and at End B, it peaks around 0.13 Hz, aligning with the natural frequencies of the dynamic power cables. The peaks are influenced by resonance, and the wave load is confirmed to still dominate, as other frequencies coincide with the wave load's peak frequency.

6. Future Works

In this analysis, the moment induced by the lifting of the dual vertical-axis turbine is not considered, as it pertains to a non-operational condition. This additional moment could result in relatively high initial effective tensions in mooring lines ML01 and ML04. More accurate results may be obtained in future studies by incorporating this factor. Furthermore, response amplitude operators (RAOs) calculations that account for various mooring system configurations should be explored during the design optimization phase. Additionally, slackness in the mooring lines has been consistently observed, which may contribute to increased fatigue loading.

Author Contributions: Conceptualization, J.P.; methodology, Y.L., J.P. and W.C.C.; software, Y.L. and W.C.C., validation, Y.L., J.P. and W.C.C.; formal analysis, Y.L. and W.C.C.; investigation, Y.L., J.P. and W.C.C.; re-sources, Y.L., J.P. and W.C.C.; data curation, Y.L., J.P. and W.C.C.; writing—original draft preparation, Y.L.; writing—review and editing, J.P. and W.C.C.; visualization, Y.L. and W.C.C.; supervision, J.P. and W.C.C.; project administration, J.P. and W.C.C.; funding acquisition, J.P. and W.C.C. All authors have read and agreed to the published version of the manuscript.

Funding: This research is a part of the project titled "Development of Advanced Technology for Ocean Energy, and Harbor and Offshore Structures (PEA0321)", funded by the Korea Institute of Ocean Science and Technology (KIOST), Korea.

Data Availability Statement: Data are contained within the article.

Conflicts of Interest: The authors declare no conflicts of interest.

Appendix A

Appendix A is plotted with the summary results of dynamic response and effective tension.

Table A1. Mooring types effect, time domain, 90°.

Freedom Unit		**Surge m**	**Sway m**	**Heave m**	**Roll deg**	**Pitch deg**	**Yaw deg**
Taut	Mean	0	−0.449	−0.429	0.04	0	0
	STD	0	1.098	1.45	6.865	0	0
	Max	0	3.033	2.459	24.171	0	0
	Min	0	−8.624	−8.842	−30.921	0	0

Table A1. *Cont.*

Freedom Unit		Surge m	Sway m	Heave m	Roll deg	Pitch deg	Yaw deg
Catenary	Mean	0	−0.74	−0.034	2.313	0	0
	STD	0	3.616	2.057	7.640	0	0
	Max	0	6.532	7.923	30.487	0	0
	Min	0	−16.281	−8.225	−26.715	0	0

Table A2. Mooring types effect, effective tension, 90°.

Type Unit		Taut kN	Catenary kN
ML14	Mean	153.212	60.689
	STD	113.55	16.115
	Max	1289.896	274.215
	Min	−38.74	−26.887
ML23	Mean	190.421	97.183
	STD	145.46	61.827
	Max	1511.166	833.312
	Min	−79.261	−158.054

Appendix B

Table A3. Before ML2 failure, time domain, 90°.

Freedom Unit		Surge m	Sway m	Heave m	Roll deg	Pitch deg	Yaw deg
Taut	Mean	0	−0.464	−0.505	0.096	0	0
	STD	0	1.058	1.358	6.65	0	0
	Max	0	2.608	1.015	21.275	0	0
	Min	0	−7.482	−8.075	−26.707	0	0
Catenary	Mean	0	−0.717	−0.036	2.301	0	0
	STD	0	3.578	2.045	7.686	0	0
	Max	0	6.532	7.923	30.487	0	0
	Min	0	−14.162	−7.388	−25.314	0	0

Table A4. After ML2 failure, time domain, 90°.

Freedom Unit		Surge m	Sway m	Heave m	Roll deg	Pitch deg	Yaw deg
Taut	Mean	−6.136	5.468	−0.352	−6.033	−2.311	23.537
	STD	2.435	3.164	1.534	18.971	5.565	8.961
	Max	1.954	12.818	2.481	59.038	19.977	55.238
	Min	−13.357	−7.941	−8.355	−68.329	−31.484	−6.862
Catenary	Mean	−21.263	18.167	−0.299	−10.374	−3.328	30.962
	STD	2.185	4.746	2.051	14.653	4.33	8.597
	Max	0	26.675	8.156	53.188	24.909	61.782
	Min	−27.257	−3.612	−9.47	−69.247	−26.29	0

Table A5. Before mooring failure, TSN, 90°.

Type Unit		Taut kN	Catenary kN
ML14	Mean	174.082	60.574
	STD	123.188	16.219
	Max	1501.423	238.638
	Min	26.052	−26.887
ML23	Mean	197.677	96.865
	STD	154.26	61.498
	Max	1295.865	791.48
	Min	−442.319	−153.867

Table A6. After mooring failure, TSN, 90°.

Type Unit		Taut kN	Catenary kN
ML01	Mean	254.422	114.046
	STD	250.146	79.71
	Max	3098.351	1467.488
	Min	−207.825	−102.775
ML03	Mean	297.397	151.748
	STD	342.295	184.918
	Max	4380.641	4945.962
	Min	−449.497	−327.033
ML04	Mean	57.968	31.103
	STD	32.433	5.451
	Max	732.878	400.337
	Min	−127.078	−76.501

Appendix C

Table A7. Dynamic power cable with lazy wave, TSN, 90°.

Type Unit		Taut kN	Catenary kN
End A	Mean	24.584	25.385
	STD	3.949	3.488
	Max	81.51	70.868
	Min	−22.49	−7.729
Mid	Mean	10.9	11.438
	STD	1.721	2.013
	Max	21.095	46.807
	Min	4.991	3.868
End B	Mean	11.16	11.621
	STD	0.928	1.149
	Max	18.412	34.615
	Min	7.320	4.091

References

1. Edenhofer, O.; Jacobsen, J.B.; Díaz Anadón, L.; van Aalst, M.; Cartalis, C.; Dessai, S.; Vera, E.; Hertwich, E.G.; Lena, K.; Soussana, J.F.; et al. *Scientific Advice for the Determination of an EU-Wide 2040 Climate Target and a Greenhouse Gas Budget for 2030–2050*; European Scientific Advisory Board on Climate Change: Copenhagen, Denmark, 2023.
2. Shetty, C.; Priyam, A. A review on tidal energy technologies. *Mater. Today Proc.* **2022**, *56*, 2774–2779.
3. Sahu, A.; Yadav, N.; Sudhakar, K. Floating photovoltaic power plant: A review. *Renew. Sustain. Energy Rev.* **2016**, *66*, 815–824.
4. O'Rourke, F.; Boyle, F.; Reynolds, A. Tidal energy update 2009. In *Renewable Energy*; Routledge: Oxfordshire, UK, 2018; pp. Vol3_451–Vol3_476.
5. Chowdhury, M.S.; Rahman, K.S.; Selvanathan, V.; Nuthammachot, N.; Suklueng, M.; Mostafaeipour, A.; Amin, N.; Techato, K. Current trends and prospects of tidal energy technology. *Environ. Dev. Sustain.* **2021**, *23*, 8179–8194.
6. Denny, E. The economics of tidal energy. *Energy Policy* **2009**, *37*, 1914–1924.
7. Ashuri, A.; Medawaty, I. *Eco-Friendly Wastewater Treatment Technology in Tidal Area*; KnE Social Sciences: Dubai, United Arab Emirates, 2019; pp. 55–66.
8. Park, E.S.; Lee, T.S. The rebirth and eco-friendly energy production of an artificial lake: A case study on the tidal power in South Korea. *Energy Rep.* **2021**, *7*, 4681–4696.
9. Priyanka, D. Eco-Friendly Energy Resources. *Environ. Conserv. Chall. Threat. Conserv. Biodivers.* **2022**, *4*, 87–98.
10. Jo, C.H.; Hwang, S.J.; Park, H.J. *Mooring System Design for an Underwater Floating Tidal Current Power Device*; Inha University: Incheon, Republic of Korea, 2018.
11. Arini, N.R.; Thies, P.R.; Johanning, L.; Ransley, E.; Brown, S.; Xie, N.; Greaves, D. Feasibility Study of Mooring Lines Design for a Floating Tidal Turbine Platform Using Double Hull Structure. In Proceedings of the International Conference on Offshore Mechanics and Arctic Engineering, Glasgow, UK, 9–14 June 2019; Volume 58899, p. V010T09A007.

12. Song, J.; Kim, J.; Lee, J.; Kim, S.; Chung, W. Dynamic response of multiconnected floating solar panel systems with vertical cylinders. *J. Mar. Sci. Eng.* **2022**, *10*, 189.
13. Yang, R.Y.; Yu, S.H. A study on a floating solar energy system applied in an intertidal zone. *Energies* **2021**, *14*, 7789. [CrossRef]
14. Li, L.; Gao, Y.; Yuan, Z.; Day, S.; Hu, Z. Dynamic response and power production of a floating integrated wind, wave and tidal energy system. *Renew. Energy* **2018**, *116*, 412–422.
15. Muliawan, M.J.; Karimirad, M.; Moan, T. Dynamic response and power performance of a combined spar-type floating wind turbine and coaxial floating wave energy converter. *Renew. Energy* **2013**, *50*, 47–57.
16. Chung, W.C.; Kang, H.Y.; Kim, M.H. Multi-scale approach for chain-mooring OPB-induced failure considering time-varying interlink bending stiffness and fairlead condition. *Appl. Ocean Res.* **2020**, *98*, 102128.
17. Chung, W.C.; Kang, H.Y.; Kim, M.H.; Pestana, R.G. Riser Structural Health Monitoring with Numerical Sensors. In Proceedings of the ISOPE International Ocean and Polar Engineering Conference, Honolulu, HI, USA, 16–21 June 2019; p. ISOPE–I.
18. Chung, W.C.; Kim, M.; Jin, C. Real-time trace of riser profile and stress with numerical inclinometers. *Ocean Eng.* **2021**, *234*, 109292.
19. Chung, W.C.; Jin, C.; Kim, M. Dual-algorithm hybrid method for riser structural health monitoring using the fewest sensors. *J. Mar. Sci. Eng.* **2022**, *10*, 1994.
20. Guo, X.; Yang, J.; Lu, W.; Li, X. Dynamic responses of a floating tidal turbine with 6-DOF prescribed floater motions. *Ocean. Eng.* **2018**, *165*, 426–437.
21. Du, J.; Zhang, D.; Zhang, Y.; Xu, K.; Chang, A.; Zhao, S. Design and comparative analysis of alternative mooring systems for offshore floating photovoltaics arrays in ultra-shallow water with significant tidal range. *Ocean Eng.* **2024**, *302*, 117649.
22. Tang, Y.; Chen, X.; Huang, G.; Chen, M.; Zhang, K.; Jiang, Y.; Li, T.; Tao, T.; Yang, Y. Dynamic analysis of multi-module floating photovoltaic platforms with composite mooring system by considering tidal variation and platform configuration. *Ocean Eng.* **2024**, *312*, 119243.
23. Thomsen, J.B.; Bergua, R.; Jonkman, J.; Robertson, A.; Mendoza, N.; Brown, C.; Galinos, C.; Stiesdal, H. Modeling the TetraSpar floating offshore wind turbine foundation as a flexible structure in OrcaFlex and OpenFAST. *Energies* **2021**, *14*, 7866.
24. Sobhaniasl, M.; Petrini, F.; Karimirad, M.; Bontempi, F. Fatigue life assessment for power cables in floating offshore wind turbines. *Energies* **2020**, *13*, 3096. [CrossRef]
25. Taninoki, R.; Abe, K.; Sukegawa, T.; Azuma, D.; Nishikawa, M. Dynamic cable system for floating offshore wind power generation. *SEI Tech. Rev.* **2017**, *84*, 146.
26. Okpokparoro, S.; Sriramula, S. Reliability analysis of floating wind turbine dynamic cables under realistic environmental loads. *Ocean Eng.* **2023**, *278*, 114594.
27. Abrahamsen, A.B.; Faria, B.R.; Lund, R.S.; Verelst, D.R.; Roques, J.; Hochet, M.; Dykes, K. Reference dynamic power cable of floating offshore wind turbine for thermal and simple fatigue evaluation. In *Journal of Physics: Conference Series*; IOP Publishing: Bristol, UK, 2024; Volume 2767, p. 062036.
28. Yang, R.Y.; Chiang, W.C. Dynamic motion response of an oil tanker moored with a single buoy under different mooring system failure scenarios. *Ships Offshore Struct.* **2023**, *18*, 923–936.
29. Faltinsen, O. *Sea Loads on Ships and Offshore Structures*; Cambridge University Press: Cambridge, UK, 1993; Volume 1.
30. Salvesen, N.; Tuck, E.O.; Faltinsen, O. *Ship Motions and Sea Loads*; The National Academies of Sciences, Engineering, and Medicine: Washington, DC, USA, 1970.
31. He, J.H. A simple perturbation approach to Blasius equation. *Appl. Math. Comput.* **2003**, *140*, 217–222.
32. Newman, J.N. *Marine Hydrodynamics*; The MIT Press: Cambridge, MA, USA, 2018; p. 448.
33. González, M.; Uriarte, A.; Fontán, A.; Mader, J.; Gyssels, P. Marine dynamics. In *Elsevier Oceanography Series*; Elsevier: Amsterdam, The Netherlands, 2004; Volume 70, pp. 133–157.
34. Oricina. *Orcawave User Manual*; Orcina: Ulverston, UK, 2021.
35. Oricina. *Orcaflex User Manual*; Orcina: Ulverston, UK, 2024.
36. Zhang, S.X.; Tang, Y.G.; Liu, X.J. Experimental investigation of nonlinear dynamic tension in mooring lines. *J. Mar. Sci. Technol.* **2012**, *17*, 181–186.
37. Somoano, M.; Battistella, T.; Rodríguez-Luis, A.; Fernández-Ruano, S.; Guanche, R. Influence of turbulence models on the dynamic response of a semi-submersible floating offshore wind platform. *Ocean Eng.* **2021**, *237*, 109629.
38. Holcombe, A.; Hann, M.; Brown, S.; Cheng, S.; Rawlinson-Smith, R.; Nicholls-Lee, R.; Tosdevin, T.; Edwards, E.; Monk, K. Experimental–numerical model comparison of a dynamic power cable for a floating offshore wind turbine. *Ocean Eng.* **2025**, *321*, 120384.

39. Beier, D. Dynamic and Fatigue Analyses of Suspended Power Cables for Multiple Floating Offshore Wind Turbines. Master's Thesis, Universiti Islam Selangor, Kajang, Malaysia, 2023.
40. Zhang, D.; Zhao, B.; Zhu, K.; Jiang, H. Dynamic analysis of the subsea production system with lazy-wave risers attached to FPSO. *PLoS ONE* **2023**, *18*, e0291603.

Disclaimer/Publisher's Note: The statements, opinions and data contained in all publications are solely those of the individual author(s) and contributor(s) and not of MDPI and/or the editor(s). MDPI and/or the editor(s) disclaim responsibility for any injury to people or property resulting from any ideas, methods, instructions or products referred to in the content.

Journal of Marine Science and Engineering

MDPI

Article

The Time-Domain Design Stress Method for Fatigue Analysis of the Reactor Pressure Vessel in Floating Nuclear Power Plants

Jialong Yuan [1], Fuxuan Ma [1,2], Meng Zhang [1,2,3,*], Kai Shen [1] and Jinfeng Tang [1]

1 Yantai Research Institute, Harbin Engineering University, Yantai 264000, China; 2071800075@hrbeu.edu.cn (J.Y.); mafuxuan@hrbeu.edu.cn (F.M.); kaishen@hrbeu.edu.cn (K.S.); sanjin@hrbeu.edu.cn (J.T.)
2 College of Shipbuilding Engineering, Harbin Engineering University, Harbin 150001, China
3 College of Power and Energy Engineering, Harbin Engineering University, Harbin 150001, China
* Correspondence: zhangmeng@hrbeu.edu.cn

Abstract: Nuclear power technology has rapidly advanced with the growing global demand for clean energy. As one of the core components of nuclear power plants (NPPs), the design and lifespan evaluation of reactor pressure vessels (RPVs) are critically important. However, while fatigue analysis methods for RPVs in land-based NPPs are relatively well established, the application of these methods to floating nuclear power plants (FNPPs) faces great challenges. Existing analysis methods are difficult to directly apply, and no widely accepted fatigue analysis approach currently exists for this context due to the complex working conditions in marine environments. A time-domain design stress (TDDS) method is developed in this study for the fatigue analysis of RPVs in FNPPs. This method systematically analyzes the impacts of wave loads, internal pressure, and thermal effects on the fatigue life of RPVs by simplifying the wave environment into a time-domain model of roll and pitch motions and adopting the regular wave superposition techniques. This method further adjusts the initial phases of regular waves considering the uncertainty of various load combinations, and superimposes the stress components caused by regular waves with different initial phases, thermal loads, and pressure loads. Subsequently, stress history curves are analyzed using the rainflow counting method, and combined with the damage accumulation theory, the upper and lower limits of fatigue damage are obtained. The results demonstrate that compared to traditional methods in time-domain analysis, the proposed TDDS method provides greater accuracy in evaluating the fatigue life of RPVs in FNPPs, with the average error in fatigue damage values being only 0.033%. Furthermore, the TDDS method reduces analysis time by approximately 70%, which significantly improves computational performance. These findings underscore the reliability and effectiveness of this method in practical applications.

Keywords: floating nuclear power plant; reactor pressure vessel; fatigue analysis; design stress; uncertainty

1. Introduction

Nuclear energy has been widely adopted internationally and has become a key area in China's energy transition as a form of new energy. Floating nuclear power plants (FNPPs) have emerged as an efficient offshore energy solution, attracting increasing attention and recognition from various countries within the nuclear energy sector. FNPPs offer the advantage of flexible deployment, enabling the provision of stable power resources in diverse marine areas, including extreme environments [1]. This capability not only meets the power demands of offshore facilities and coastal cities but also supports multiple applications such as heating and seawater desalination. As a result, FNPPs are regarded as an ideal solution for offshore energy supply. However, their structural safety requires further in-depth investigation [2].

Nuclear pressure-bearing equipment is subjected not only to direct impacts such as high temperature, high pressure, fluid erosion, and neutron irradiation but also to indirect impacts from wave loads, wind loads, and current loads during the operation of FNPPs [3]. The combination of multiple loads significantly increases the stress range at structural hotspots, thereby greatly reducing the fatigue life of critical pressure-bearing equipment compared to the effects of a single load. The amplitude of structural responses varies accordingly, due to the different frequencies of various random loads. Furthermore, the potential high correlation between these loads further complicates the assessment of structural fatigue damage [4].

An analysis of the fatigue life and reliability of RPVs in FNPPs is of paramount importance. Traditional fatigue analysis methods for RPVs in land-based nuclear power plants (NPPs) cannot be directly applied to FNPPs; instead, fatigue analysis must account for the unique stress conditions imposed by the marine environment. However, although fatigue analysis methods have been extensively studied for land-based NPPs, there is still very little research on fatigue damage analysis for FNPPs. Therefore, fatigue analysis methods tailored to RPVs under marine environmental conditions are crucial to ensure the safe operation of FNPPs.

2. Background

As the stress response time history in the hull structure is difficult to obtain, the frequency domain spectral analysis method is typically used for fatigue assessment. In contrast, the typical transient method or fracture mechanics method is usually employed for RPV fatigue analysis since the stress-time history is relatively easy to obtain.

Spectral analysis is a commonly used method in hull fatigue analysis, particularly suited for assessing the fatigue damage of hull structures under random wave loads. The core concept of spectral analysis lies in treating external wave loads as a random process and utilizing wave spectra (such as the Pierson–Moskowitz spectrum or JONSWAP spectrum) to describe the distribution of wave energy across different frequencies. The structural response of the hull is represented through the dynamic characteristics of the structure, typically described by the power spectral density function (PSD) to characterize structural vibration responses in the frequency domain. The basic steps of spectral analysis can be summarized as follows: First, the PSD of the structural stress response is calculated, providing the foundation for subsequent fatigue analysis. Next, this PSD is linked to the rainflow stress range distribution to determine the number of cycles at different stress amplitudes. Finally, the selected S-N curve is used to reflect the material's fatigue properties, and combined with Miner's cumulative damage theory, the overall fatigue damage of the structure is computed [5]. Liu Yuchao [6] proposed a simplified spectral fatigue analysis method, which streamlines the calculation process by selecting key wave frequencies and determining the primary related frequencies. Additionally, a simplified approach for calculating local damage was introduced, effectively reducing computation time while ensuring the accuracy of the results. This method provides an efficient solution for the fatigue assessment and optimization design of hull structures. Yosri [7] employed the spectral fatigue analysis method to study the fatigue life of structural details on both sides of an oil tanker. Random environmental loads were derived in the frequency domain and mapped onto the hull's structural finite element model by using wave-fitting software based on 3D diffraction theory. Stress responses at the considered details were extracted from finite element analysis, and the Palmgren–Miner rule was applied to estimate cumulative damage. Makris Pavlos [8] combined the spectral analysis method with a probabilistic fatigue crack propagation model to quantify the uncertainty in fatigue life estimation of ship structures as caused by random wave loads and variations in material and geometric parameters. The research framework simulates the randomness of sea conditions and the sequences of sea states encountered by vessels that lead to fatigue accumulation. The framework enables reliability estimation over the entire lifecycle by constructing potential crack propagation path instances from a single realization of the effective random stress range and applying them to the number of cycles. Fan Wenliang [9] proposed a novel fatigue-damage spectral analysis method based on a nonlinear damage model by combining block cycle jumping techniques from the time-domain method with fatigue life spectral analysis. This approach not only improves computational efficiency but also more accurately reflects the nonlinear process of real fatigue damage.

The typical transient method is a commonly used approach in the fatigue analysis of RPVs specifically designed to evaluate fatigue under different operational transient conditions. The core idea of the typical transient method is to analyze the structure's response to instantaneous loads in the time domain, thereby assessing the fatigue damage of the structure. The basic steps are as follows: external random wave loads are numerically simulated and directly input into the structural dynamics model, and the transient response of the structure is calculated in the time domain. By integrating the equations of motion step by step, this method captures nonlinear effects, transient phenomena, and dynamic responses under complex sea conditions. Specifically, the typical transient method first simulates instantaneous mechanical behavior under the influence of waves or other external loads, then solves algorithms to obtain the structure's stress–strain time response history in the time domain, and finally calculates the fatigue damage of the structure using counting rules (such as the rainflow counting method). Junbo Jia [10] conducted a time-domain nonlinear dynamic analysis to study fatigue damage caused by wind, finding that the wind-induced fatigue calculation program based on time-domain nonlinear dynamics not only more accurately evaluates dynamic responses but also significantly reduces uncertainty while ensuring safety requirements. Tingsen Zheng [11] proposed a time-domain fatigue assessment method for the root bolts of floating offshore wind turbine (FOWT) blades. A finite element method was used to calculate the stress response of the bolts based on the time course of the environmental loads at the root of the blades, taking into account the combined effects of wind, waves, and currents. Stress ranges and corresponding cycles were determined using the rainflow counting method, and the fatigue life of the bolts was predicted using the S-N curve and Palmgren–Miner (PM) rule. However, the computational process of transient analysis is quite complex and requires long computation times. Ma Fuxuan [12] studied the fatigue issues of RPVs in FNPPs and proposed a quasi-time-domain method that combines principal component analysis (PCA) to simplify ocean loads. The wave stress spectrum was converted to time history, and fatigue stress was calculated after superimposing temperature and internal pressure. Numerical verification showed that the calculated results of this method deviated from the traditional time-domain method by no more than 24% while reducing computation time by 98.67%, providing an efficient and practical solution for FNPP RPV fatigue analysis.

Currently, research on the fatigue assessment of pressure vessels in marine environments is relatively limited, with the majority of studies focusing on fatigue evaluation methods based on fracture mechanics theory. Wang Xueping [13] used ANSYS APDL 2022 software to analyze the fatigue life and equivalent stress of internal cracks in marine engineering pressure vessels, exploring the effects of axial and circumferential cracks on fatigue life. He then compared the results of finite element simulations with fatigue life calculations based on fracture mechanics, drawing corresponding conclusions. Liu Tongqing [14] analyzed the stress conditions of sub-model cracks with different crack parameters for a certain type of marine SCR pressure vessel using finite element software. He combined fracture mechanics theory to assess crack propagation life and identified the crack parameters that significantly affect fatigue life through model comparison. Xu Shuai [15] performed fatigue reliability analysis of typical structural nodes on an LNG carrier using spectral analysis methods. A three-dimensional potential flow theory and the whole-ship finite element method were used to determine the fatigue hotspot stresses, and the distribution model of the short-term hotspot stress range was established based on spectral analysis. In this research, a least-squares fitting method based on equivalent fatigue damage was proposed to parameterize a long-term Weibull distribution model for fatigue loads from long-term prediction data. Zhang Hao [16] applied spectral analysis fatigue evaluation methods to marine nuclear-powered platforms, using finite element analysis to derive stress transfer functions and fatigue damage degrees, thus validating the reliability of the evaluation method. They used the principal stress in the 45° range of vertical welds as the reference stress in their fatigue assessment, from which they obtained stress transfer functions for various typical nodes.

Several other methods have been developed to improve the efficiency and effectiveness of fatigue damage assessment. Rasul et al. [17] developed internal ring-reinforced KT joints to minimize stress concentration factors (SCFs) and improve fatigue life. They proposed parametric equations for calculating SCFs using a finite element model and response surface methodology with eight input parameters and eleven outputs. The KT joint with optimized parameters showed the lowest SCF. The method was validated with finite element analysis, showing a less than 6% error, confirming its accuracy. Iqbal et al. [18] proposed empirical models to compute peak stress concentration factors (SCFs) for KT joints subjected to multiplanar bending, which is critical for offshore structures. Existing equations estimate SCFs only at the crown or saddle under uniplanar bending, but the peak SCF under simultaneous in-plane and out-of-plane bending occurs between these positions. The proposed model, developed using regression analysis and artificial neural networks (ANN) with 3716 ANSYS simulations, account for the relative magnitudes of the load components. The model was validated with data from the literature, showing an error of less than 1.5%, demonstrating its high accuracy in predicting peak SCFs.

In conclusion, the impact of the marine environment on the fatigue life of pressure vessels cannot be overlooked; however, there are significant differences between pressure vessels in marine environments and the RPVs of FNPPs. RPVs are relatively smaller in volume but are subjected to much higher internal pressures compared to conventional pressure vessels. This is due to the requirement for RPVs to operate stably under high-temperature and high-pressure conditions to sustain continuous nuclear reactions. Furthermore, RPVs face complex thermal transfer processes, involving significant heat generation within and heat exchange with external cooling media, imposing more stringent demands on materials in terms of thermal fatigue and thermal stress. This study proposes a novel fatigue evaluation method tailored to the unique operational conditions of FNPP RPVs to address these challenges. The commonly used design wave method in marine engineering is organically integrated with the fatigue strength analysis of RPVs for the first time. The proposed method systematically references ship design codes and ASME standards [19], incorporating a regular wave superposition technique to establish an analytical framework capable of addressing the coupled effects of loads in complex marine environments. This approach comprehensively considers wave loads, internal pressure, and thermal effects, significantly improving the accuracy and applicability of fatigue life evaluations. Consequently, this study outlines a new theoretical system for the fatigue analysis of FNPP RPVs.

3. Fatigue Analysis Methods

FNPPs operate in marine environments where their RPVs are subjected not only to wave loads induced by random sea states but also to internal pressure and thermal loads. The traditional fatigue analysis methods for land-based NPP RPVs are not directly applicable to such complex conditions. Consequently, it is necessary to explore and develop fatigue analysis methods tailored to marine environments. Currently, spectral analysis is commonly employed in ship fatigue analysis, while the fatigue analysis of RPVs primarily relies on typical transient methods and fracture mechanics approaches. Although these methods have demonstrated significant effectiveness in their respective fields, they face inherent limitations when applied to the unique operational scenarios of FNPPs.

3.1. Spectral Analysis

Structures are subjected to the combined effects of wind, wave, and current loads, among which wave-induced loads are the primary cause of fatigue failure in protective structures in the complex operating environment of FNPPs. Therefore, accurately assessing the fatigue strength of FNPP containment structures hinges on selecting an appropriate wave spectrum, calculating transfer functions, and evaluating fatigue damage. Spectral analysis is a widely used tool in the field of naval architecture and ocean engineering, particularly for investigating the relationships between loads and structural responses [8]. The theoretical foundation of this method lies in the linear system transformation of stochastic process theory. The FNPP is considered as a typical dynamic system, the wave process $\eta(t)$ serves as the input to the system, while the alternating stresses $X(t)$ induced by wave action represent the system's response. The relationship between input and response can be expressed using the following formula:

$$X(t) = L[\eta(t)] \tag{1}$$

where L is the operator for converting wave loads into alternating stress.

The system is classified as a linear system when L is a linear operator. In this case, the wave load calculations and structural response calculations are based on linear theory. The alternating stress obtained through the transformation is also a stationary random process when the wave load is considered a stationary random process. According to random process theory, the power spectral density functions of two stationary random processes are related as follows:

$$G_X(\omega) = \left|H(\omega)\right|^2 G_\eta(\omega) \tag{2}$$

where $G_X(\omega)$ represents the response spectrum of wave loads on the FNPP in irregular waves; $H(\omega)$ is the transfer function of the wave load; $G_\eta(\omega)$ represents the wave energy spectral density function; and ω represents the circular frequency.

The physical significance of the transfer function lies in its representation of the ratio between the amplitude of the response process and the amplitude of the input process when a linear dynamic system is subjected to simple harmonic excitation at an angular frequency ω [6]. The transfer function represents the ratio of the stress amplitude to the wave amplitude for the structure under a regular cosine wave excitation with an angular frequency ω, which is when the input process corresponds to waves and the response process corresponds to alternating stress. The power spectral density function of the structural response can be obtained by calculating the power spectral density function of the wave loads and the system's transfer function.

The next step is to calculate the short-term and long-term distributions of the stress range once the stress response spectrum of the structure is determined. The long-term state of ocean waves is typically considered to be composed of multiple short-term sea state sequences in naval architecture and ocean engineering. Each short-term sea state is described by wave characteristic parameters and the occurrence probability of that sea state. Waves are generally assumed to be a stationary Gaussian random process within each short-term sea state, and the alternating stress induced on ship structures by waves can also be regarded as a sequence formed by multiple short-term sea states [7].

The long-term distribution of the stress range can be obtained by combining the short-term distribution of the stress range for all sea states and directions in conjunction with the frequency of occurrence for each sea state and direction. This long-term distribution is typically piecewise continuous. The alternating stress process under a specific sea state and heading is often treated as a zero-mean, narrow-band stationary random process in practical applications.

The peaks of alternating stress follow a Rayleigh distribution for the short-term distribution of stress range according to random process theory. The probability density function is expressed as:

$$f_\sigma(\Delta\sigma) = \frac{\sigma}{m_0}\exp\left(-\frac{\sigma^2}{2m_0}\right) \tag{3}$$

where σ represents the stress peak and m_0 represents the 0-th moment of the power spectral density of the alternating stress process.

The long-term distribution of stress range S is typically represented by a two-parameter Weibull distribution, which effectively describes the variation of long-term stress. The probability density function and distribution function are as follows:

$$f_S(S) = \frac{\xi}{\alpha}\left(\frac{S}{\alpha}\right)^{\xi-1}\exp\left[-\left(\frac{S}{\alpha}\right)^{\xi}\right] \tag{4}$$

$$F_S(S) = 1 - \exp\left[-\left(\frac{S}{\alpha}\right)^{\xi}\right] \tag{5}$$

where α is the scale parameter and ξ is the shape parameter.

The corresponding fatigue accumulation damage can be calculated after obtaining the long-term distribution of the stress range. A commonly used method is to directly calculate the cumulative fatigue damage by combining the fatigue accumulation damage of each short-term distribution in practical calculations. Let the time spent by the considered hull structure in the i-th sea state and the j-th heading be T_{ij}, and let D_{ij} represent the cumulative damage during the time period T_{ij}. The considered long-term time period is T, and the corresponding long-term stress range distribution is composed of n_s sea states, with each sea state having a probability of p_i. The number of divided headings is n_H, and the probability of each heading is p_j. Thus, $T_{ij} = T \cdot p_i \cdot p_j$, and the total fatigue cumulative damage (D_T) during this period should be:

$$D_T = \sum_{i=1}^{n_s}\sum_{j=1}^{n_H} D_{ij} = \frac{T}{A}\Gamma\left(1+\frac{m}{2}\right)\sum_{i=1}^{n_s}\sum_{j=1}^{n_H} p_i p_j f_{0ij}\left(2\sqrt{2m_{0ij}}\right)^m \tag{6}$$

The actual fatigue life can be estimated based on the concept of linear cumulative damage after obtaining the cumulative fatigue damage (D_T) for the design service life, expressed as:

$$T_{life} = T_d \cdot \frac{1}{D} \tag{7}$$

Although spectral analysis is widely used in marine and offshore engineering, it still has limitations in the fatigue analysis of RPVs in FNPPs. Firstly, this method assumes a linear relationship between loads and structural responses; however, the load-response behavior of FNPPs is often nonlinear under complex environmental conditions, making it difficult to accurately describe using linear models. Secondly, the spectral analysis method assumes that wave loads are stationary random processes. In reality, the combined effects of wind, waves, and currents result in non-stationary load processes. Finally, the method neglects the influence of factors such as temperature and variations in material fatigue properties, which are critical in the fatigue analysis of FNPP RPVs. Therefore, the spectral analysis method is not suitable for the fatigue analysis of RPVs in FNPPs.

3.2. Typical Transient Method

The typical transient method is widely employed as the primary evaluation approach for the fatigue analysis of RPVs in land-based NPPs. RPVs undergo various transient conditions during operation, such as abrupt changes in temperature and pressure, which can potentially cause fatigue damage to the RPV [20]. Therefore, employing the typical transient method to analyze these conditions allows for effective prediction and assessment of an RPV's fatigue life, ensuring its structural integrity during long-term operation.

The fundamental principle of the typical transient method lies in simulating stress variations in pressure vessels under different transient conditions, calculating the stress amplitudes for each transient, and evaluating the impact of these stress cycles on fatigue life using the material's fatigue curve (S-N curve) [21]. The cumulative damage theory is further applied to predict the fatigue life over the entire service period based on these data. Transient changes in temperature and pressure induce internal stress fluctuations in the pressure vessel, which, when accumulated to a certain extent, may lead to fatigue failure. Therefore, the objective of transient fatigue analysis is to accurately calculate the stress cycles under each transient condition and assess their cumulative impact on fatigue life [22].

Finite element analysis is widely used to calculate the stress distribution under different transient conditions in the process of transient method analysis. The stress responses of various parts of the pressure vessel can be obtained by simulating the variations in temperature and pressure during transient conditions. Each transient condition causes fluctuations in the stress amplitude of the pressure vessel, which are typically characterized by the stress amplitude $S_{\max}$ [23]. The fatigue life of each transient condition can be calculated using the S-N curve, and the usage factor U_n for the transient can then be determined according to its number of cycles based on these stress amplitudes. This usage factor can be expressed by the following formula:

$$U_n = \frac{N_{actual}}{N_{allowable}} \tag{8}$$

where N_{actual} represents the actual number of cycles for the transient condition, and $N_{allowable}$ represents the maximum allowable number of cycles for the material under the corresponding stress. Multiple transient conditions are typically combined for evaluation in practical analysis. The effects of transient combinations must be considered to more accurately predict fatigue life. The usage factor for the combined transients can be calculated using the following formula:

$$U_{pq} = \frac{1}{N_p} + \frac{1}{N_q} \tag{9}$$

where N_p and N_q represent the fatigue lifecycle counts for transient conditions p and q, respectively. The overall usage factor of the pressure vessel can be obtained by performing cumulative analysis of each transient combination, which can then be used to determine whether the fatigue life of the equipment meets the design requirements.

The typical transient method is widely used in land-based NPPs to evaluate the fatigue life of pressure vessels. Its advantage lies in its ability to accurately simulate stress variations under different transient conditions and provide relatively accurate fatigue life predictions through cumulative damage theory. However, this method also has significant limitations. First, the typical transient method is primarily designed for transient conditions involving rapid temperature and pressure changes, whereas the operational environment of a FNPP is more complex, including the effects of waves, wind, currents, and other factors. This complexity leads to uncertainties and dynamic variations in the load process, which are difficult to model accurately using the transient method [24]. Secondly, the transient method typically assumes that each transient condition is independent, but in FNPPs, there may be strong coupling effects between different conditions. These coupled effects have a much more complex impact on the fatigue of the pressure vessel than simple transient combinations. Finally, the typical transient method does not account for the dynamic behavior and nonlinear responses of materials under extreme conditions, which makes it insufficient for the precise fatigue analysis required for FNPP RPVs. Therefore, the typical transient method is not suitable for the fatigue analysis of FNPP RPVs.

3.3. Fracture Mechanics Method

Fatigue assessments of marine pressure vessels primarily rely on methods based on fracture mechanics theory in the field of shipbuilding and ocean engineering. Fracture mechanics focuses on analyzing crack propagation within a structure and its impact on fatigue life, making it particularly suitable for structures with initial cracks or defects, such as marine pressure vessels. Fracture mechanics provides a more accurate assessment of the effect of crack propagation on the structural fatigue life compared to the traditional S-N curve method, especially in areas with significant stress concentration [25].

Crack initiation and propagation are key factors that determine the structural fatigue life in the fatigue analysis of pressure vessels. Typically, cracks begin at an initial size a_0 and gradually propagate under cyclic loading. The structure undergoes fracture failure when the crack reaches a critical size a_c. Therefore, the process of crack propagation from the initial crack size a_0 to the critical crack size a_c is the focus of fatigue life assessment [26]. The commonly used Paris' law in fracture mechanics describes the crack growth rate, and the formula is as follows:

$$\frac{da}{dN} = C(\Delta K)^m \tag{10}$$

where a is the crack length; N is the number of load cycles; ΔK is the range of the stress intensity factor; and C and m are material constants, determined through experiments.

The stress intensity factor K is a key parameter used to describe the stress field at the crack tip which depends on the crack length, load magnitude, and the structural geometry. The crack length gradually increases with the action of cyclic loads, which leads to the acceleration of the fatigue crack growth rate until the crack extends to the critical size, at which point the structure will fail. The stress intensity factor range ΔK is the main driving parameter for the crack growth rate, determined by the difference between the maximum stress intensity factor $K_{\max}$ and the minimum stress intensity factor $K_{\min}$, as follows:

$$\Delta K = K_{\max} - K_{\min} \tag{11}$$

ΔK will gradually increase as the crack propagates. Subsequently, the crack growth rate accelerates and eventually leads to the failure of the structure. The fatigue life assessment process for marine pressure vessels based on fracture mechanics includes several key steps: First, it is necessary to assume or detect the potential initial crack size a_0 in the pressure vessel. Next, the stress intensity factors $K_{\max}$ and $K_{\min}$ of the pressure vessel are calculated. Then, the crack growth rate $\frac{da}{dN}$ is calculated by using Paris' formula based on the stress intensity factor range ΔK. Finally, the crack growth process from the initial crack size a_0 to the critical crack size a_c can be predicted by integrating the crack growth rate formula, thereby determining the structural fatigue life [27].

The fracture mechanics method is suitable for the fatigue analysis of pressure vessels containing initial cracks as it can accurately assess the impact of crack propagation on fatigue life, the main advantage of which lies in its ability to directly analyze the initiation and propagation of cracks, providing a more precise prediction of fatigue failure in stress concentration areas compared to the traditional *S-N* curve method [28]. Fracture mechanics can provide precise fatigue life predictions at different stages of crack growth by using Paris' law.

However, the fracture mechanics approach focuses on analyzing crack extension and ignores the effects of factors such as temperature changes and changes in material fatigue properties on the structure. The behaviors of the pressure vessels in FNPP reactors under extreme operating conditions are more complex, and relying solely on fracture mechanics cannot fully account for the multi-factor coupling effects. Therefore, fracture mechanics has limitations in providing a comprehensive and accurate fatigue life assessment for the pressure vessels of FNPP reactors.

3.4. *Time-Domain Design Stress (TDDS) Method*

The methods proposed in the previous sections are not entirely applicable for the fatigue analysis of RPVs in FNPPs. The spectral analysis method commonly used for ship structural fatigue analysis provides detailed load analysis in the frequency domain but is primarily suited for onshore or fixed structures, making it inadequate to fully address the complex and variable load conditions of floating platforms [29]. While the typical transient method for RPVs has certain advantages in handling thermal–pressure coupled loads, it fails to account for wave loads and the dynamic characteristics of floating platforms. Similarly, the fracture mechanics method for marine pressure vessels effectively analyzes structural fracture behavior but is more focused on crack propagation, lacking comprehensiveness in fatigue analysis [30]. Therefore, a new method must be proposed to handle the fatigue issues of RPVs to address the complex environment of FNPPs.

This study proposes a new fatigue analysis method, namely the time-domain design stress (TDDS) method, for the fatigue analysis of RPVs in FNPPs. The method first converts irregular time-domain waves in sea states into regular time-domain waves, thereby simplifying the calculation of wave loads. Considering that the period of the wave load is quite different from that of the internal pressure–thermal load, it is difficult to determine the phase of the wave load corresponding to the internal pressure–thermal load at the time of moment 0. Therefore, the proposed method further adjusts the initial phase of the regular wave as it is known that the period of the regular wave is 360°, meaning that the initial phase is incremented from 0° to 360° in 30° steps with a total of 12 initial phases, i.e., 12 regular waves are generated and their periods are extended to the same length as the thermocompression loading time. The regular waves with different initial phases are superimposed on the stress components induced by the thermal pressure so that most of the possibilities of combining wave loads with internal pressure–thermal loads are obtained and multiple stress-time histories are calculated. These stress histories are then analyzed using the rainflow counting method and combined with cumulative damage theory to determine the upper and lower bounds of fatigue damage.

This study adopts the Design Wave Method to replace actual random wave loads on the hull for random waves. The design wave method simplifies the irregular waveforms of random sea states into regular waves by analyzing the energy distribution in the wave frequency spectrum, extracting the dominant frequency and significant wave height to construct an equivalent regular wave model. This method retains the primary dynamic characteristics of waves, reduces the computational complexity of load analysis, and facilitates superposition with other periodic loads, such as thermal–pressure loads. The specific steps are as follows:

First, a series of Response Amplitude Operators (RAOs) is calculated for wave loads under different wave heights and directions. These RAOs are then used to determine the wave direction and period parameters of the design wave. An appropriate wave spectrum is selected for the design sea area based on the short-term sea state information at the operational site of the FNPP, such as the *P–M* spectrum or the *JONSWAP* spectrum, and the wave energy spectral density function $S_w(\omega)$ is computed. The response spectrum of platform wave loads in irregular waves can be derived using spectral analysis once the wave load RAOs and the spectral density function are obtained, as follows:

$$S_R(\omega) = [RAO(\omega)]^2 \cdot S_w(\omega) \tag{12}$$

Then, the maximum possible value $R_{\max}$ of the load response corresponding to a specific short-term sea state is calculated as follows:

$$R_{\max} = \sqrt{m_0} \cdot \sqrt{2\ln(N)} \tag{13}$$

where m_0 represents the 0-th moment of the response spectrum; $m_n = \int_0^\infty \omega^n \cdot G_X(\omega) d\omega$; N represents the number of cycles during the duration, $N = D/T_a$; D represents the duration, typically taken as 3 h (10,800 s); and T_a represents the mean zero-crossing period of the response, $T_a = 2\pi\sqrt{m_0/m_2}$.

The wave amplitude A_D of the design wave system is determined as the maximum response value $R_{\max}$ from all short-term sea states divided by the maximum amplitude response of the load parameter RAO_C, expressed as:

$$A_D = \frac{R_{\max}}{RAO_C} LF \tag{14}$$

where LF is the load factor, typically ranging between 1.1 and 1.3.

Assuming that $\sigma_t(t)$ is the stress-time history under thermal–pressure load coupling, and $\sigma_p(t)$ is the stress-time history under wave load, the phase difference ϕ_i is defined and introduced into the stress calculation formula to consider the impact of multi-phase waves. Thus, the total stress-time history in the i-th combination can be expressed as:

$$\sigma_{total,i}(t) = \sigma_t(t) + \sigma_p(t + \phi_i) \tag{15}$$

where $\sigma_{total,i}(t)$ represents the total stress-time history in the i-th combination; ϕ_i represents the phase difference of the i-th combination; and n represents the number of phases.

Multiple different combined stresses can be generated in time domian. These stress-time histories can more accurately reflect the multiple loads acting on the pressure vessel in a FNPP, especially the complex coupling of thermal loads and wave loads. Thus, the total stress of the stress-time history can be obtained by summing the contributions from multiple phases, as follows:

$$\sigma_{total}(t) = \sum_{i=1}^{n} (\sigma_t(t) + \sigma_p(t + \phi_i)) \tag{16}$$

The rainflow counting method is used to extract representative stress cycles from complex stress-time histories for fatigue damage calculation. First, the peaks and valleys are identified from the stress-time history. By calculating the difference between adjacent peak and valley values, the stress amplitude A is obtained:

$$\Delta\sigma_i = \left|\sigma_{peak,i} - \sigma_{valley,i}\right| \tag{17}$$

Next, the closure condition of the rainflow counting method is used to determine when the stress cycle is complete. Once a complete cycle is determined, the stress amplitude and number of cycles are recorded. In this way, the entire stress-time history is decomposed into multiple independent stress cycles. The cumulative fatigue damage can be calculated using Miner's cumulative damage rule with the results derived from the rainflow counting process, as follows:

$$D = \sum_{i=1}^{n} \frac{n_i}{N_i} \tag{18}$$

where n_i is the number of cycles of the i-th stress cycle and N_i is the fatigue life of the material at the stress amplitude $\Delta\sigma_i$ for the n-th cycle.

The total fatigue damage of the structure can be obtained by summing the damage corresponding to all stress cycles. This method allows multiple fatigue damage results to be obtained, from which the upper and lower limits of fatigue damage can be determined. Compared to existing fatigue analysis methods, the TDDS method more comprehensively reflects the actual operating conditions of RPVs in FNPPs by considering the cumulative effect of multi-phase wave loading.

4. Structural Model and Loading Conditions

4.1. Structural Finite Element Model

The finite element model of the RPV is established using ANSYS APDL, as shown in Figure 1.

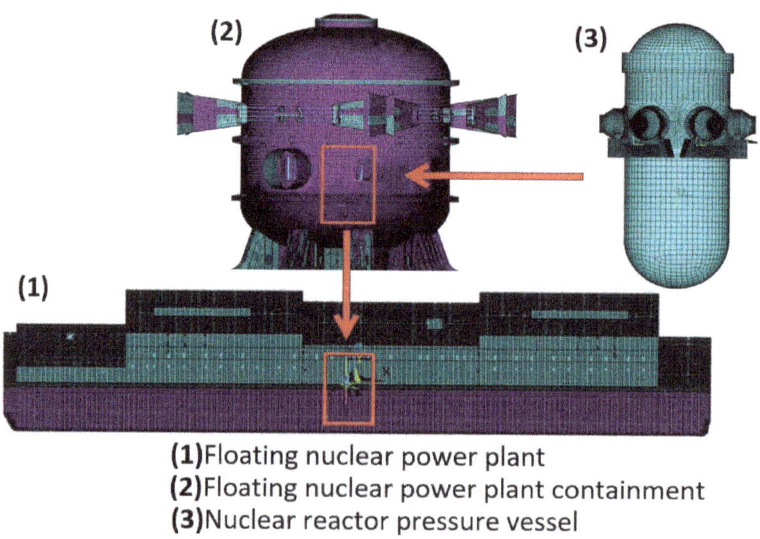

Figure 1. Schematic of a finite element model of a floating nuclear power station hull, floating nuclear power station containment, and RPV.

Pressure vessels are usually manufactured from high-strength, low-alloy steels, and 16MnD5 has been identified as the steel for nuclear power generation in accordance with the specification to ensure good mechanical properties and radiation resistance under extreme conditions [19]. Its main structure includes a cylindrical shell, top, and bottom heads. The shell bears the primary pressure and structural loads, with an inner lining that is resistant to corrosion and radiation. The top head features multiple openings for the installation of control rod drive mechanisms and instruments, while the bottom head is fixed to the base through a support structure. The side walls are equipped with pipeline interfaces for coolant and auxiliary systems. The finite element model of the nuclear reactor pressure vessel is shown in Figure 2.

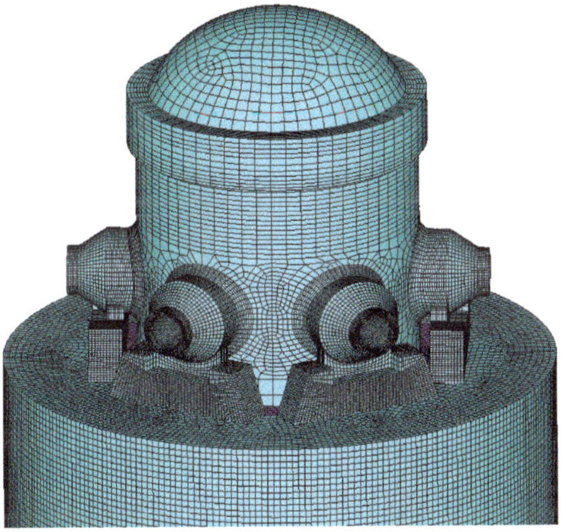

Figure 2. Schematic finite element model of a RPV with its base in containment.

Floating structures at sea are always in a state of dynamic equilibrium during operation; however, boundary conditions must be applied to avoid singularity in the stiffness matrix when performing structural finite element calculations. The inertial release method recommended by the standard [19] is used to balance the boundary conditions to constrain the hull in this study. The displacement boundary conditions for the inlet and outlet analysis model are as follows: First, the bottom of the model is constrained in the tangential direction to prevent rotation around the axis, ensuring that the model can freely expand in the radial direction. Second, the support pads are constrained in the vertical direction.

4.2. Thermal and Internal Pressure Load Conditions

The connection area between the inlet and outlet nozzles and the cylinder wall is a critical location for RPV fatigue analysis as these areas are subjected to complex loads. The selection of fatigue hotspots needs to combine the strength analysis results under hazardous conditions with the actual calculation process, from which points with larger stress responses are selected as fatigue hotspots.

Internal pressure and temperature stresses are calculated using the finite element analysis software ANSYS APDL 2022. The internal pressure under the most dangerous condition is applied to the inner surface of the RPV with a value of 15 MPa. The temperature, also under the most dangerous condition, is applied to the solid elements of the RPV, with a temperature of 343 °C.

When the most dangerous thermal and pressure condition is selected, the stress response of the structure is calculated. According to the relevant strength assessment requirements in the ASME code, the stress value at the evaluation point is chosen as the stress intensity at the assessment location. The results are shown in Figure 3. The stress contour map for the pressure vessel shell shows that the maximum stress occurs at the junction between the inlet/outlet nozzles and the cylinder wall, with a maximum stress of 286 MPa. Stress contour map of the pressure vessel shell is shown in Figure 4.

Figure 3. A RPV and its base in the containment shell finite element model cross-sectional view.

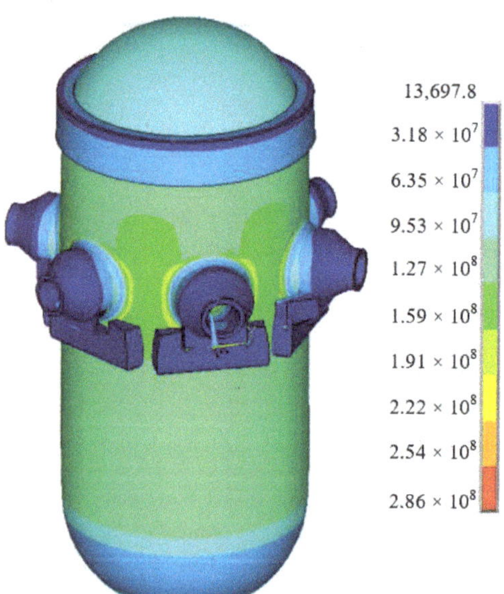

Figure 4. Stress contour map of the pressure vessel shell (Pa).

Key locations of high stress in the structure are identified based on the results of the finite element calculations of the overall model under typical working conditions combined with the structural characteristics of the reactor pressure vessel. The specific locations are listed in Table 1.

Table 1. Hotspot locations.

Fatigue Check Point Numbers	Location
1	Upper chamfer of the inner surface of the nozzle at the contact point with the inner wall
2	Lower chamfer of the inner surface of the nozzle at the contact point with the inner wall
3	Outer surface of the nozzle at the contact point with the nozzle support
4	Inner surface of the top cover at the contact point with the body

4.3. Marine Environmental Load Conditions

The first step is to determine the frequency and wave direction of the design wave to convert the irregular sea state into an equivalent regular wave, as explained in Section 3.4. The exported model file is subjected to hydrodynamic calculations using the three-dimensional potential flow software ANSYS AQWA. The hydrodynamic model of the structure is shown in Figure 5.

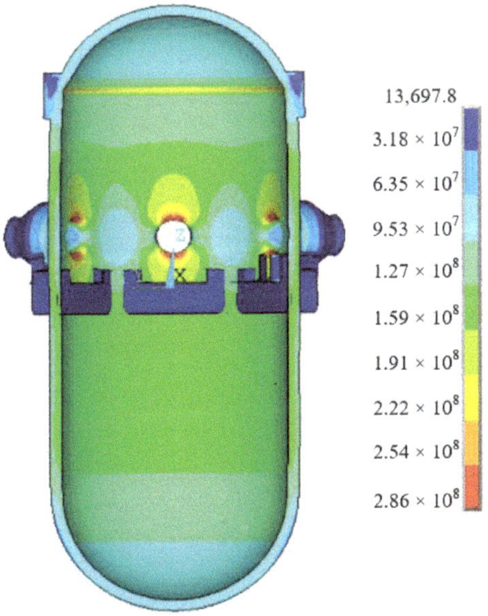

Figure 5. Schematic diagram of maximum stress under thermal–pressure loading (Pa).

The specific wave parameters are as follows: wave direction angles increase in 15° steps from head sea (0°) to following sea (180°), and the wave frequency range is from 0.2 to 1.8 rad/s. This frequency range and resolution can cover the spectral characteristics of the main waves in typical marine environments, with a step increment of 0.05 rad/s aiming to balance computational accuracy and efficiency. The frequency response function for the calculated load parameters was determined based on the primary load control parameters. The main control load parameter used was the acceleration of the floating nuclear power station (RX). Subsequently, a search was conducted across the wave-direction and wave-frequency ranges, and the wave direction and frequency corresponding to the maximum amplitude–frequency response were selected as the design wave's direction and frequency. The results indicated that the maximum amplitude–frequency response occurred at a wave direction of 90° and a wave frequency of 0.974 rad/s. For instance, when a sea state with an effective wave height of 5 m and a spectral peak period of 0.1 Hz is considered, the corresponding JONSWAP wave spectrum is shown in Figure 6.

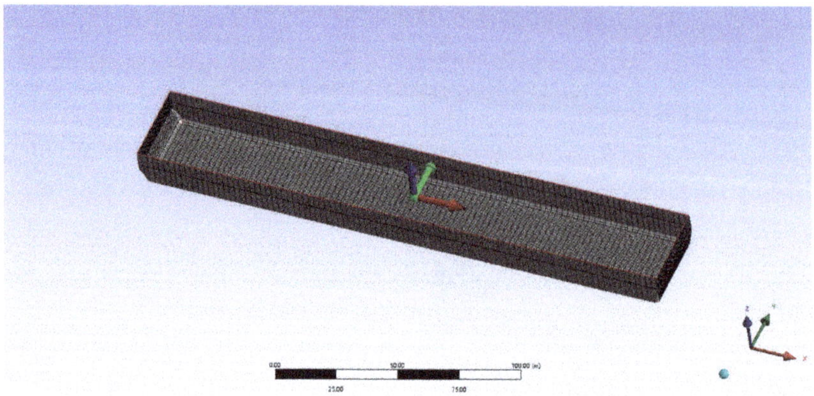

Figure 6. Hydrodynamic model of a floating nuclear power station.

The design wave parameters can be derived using the TDDS method outlined in Section 2 after completing the hydrodynamic analysis of the floating nuclear power station and obtaining the characteristic response load transfer function. The design wave parameters are shown in Table 2.

Table 2. Design wave parameters.

Characteristic Response Load	Wave Direction	Wave Frequency	Design Wave Height
Roll Acceleration	90	0.1553	0.21

The wave loads can be calculated and applied to the wet surface of the structural finite element model for solution after obtaining the design wave parameters. The effect of applying the wave loads is shown in Figure 7.

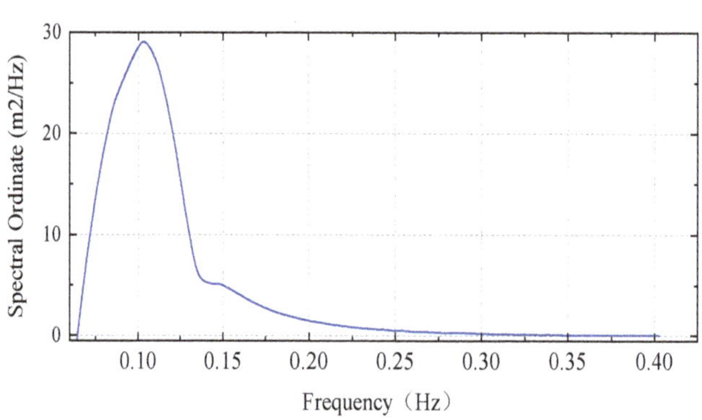

Figure 7. JONSWAP wave spectrum.

4.4. Stress State of the Structure Under Extreme Load Conditions

The temperature and internal pressure within the pressure vessel will rise rapidly to reach a steady operating condition when the reactor is in an open core state during operation. The stress state within the pressure vessel is affected by both internal pressure and temperature at this point. Meanwhile, the wave load from the marine environment is transmitted to the pressure vessel through the deformation and motion of the hull, further increasing the stress. The hull's movement, including pitch, roll, and yaw, generates dynamic loads within the pressure vessel. These dynamic loads may combine with the internal pressure and thermal loads, leading to stress concentrations.

Therefore, this section reports a strength analysis of the pressure vessel, selecting the most dangerous thermal pressure condition and the most critical wave load condition. From the RPV design conditions, it can be seen that 15 MPa is the maximum RPV internal pressure design operating condition and, similarly, 343 °C is the maximum temperature design operating condition. Both internal pressure, thermal loads, and wave loads were simultaneously applied in the finite element modeling; the effects of internal pressure (15 MPa) and temperature (343 °C) were combined with the wave loads computed under marine environmental conditions using the ANSYS APDL software, illustration of wave load application is shown in Figure 8. This approach allows for an accurate assessment of the stress distribution and maximum stress values of the structure under combined loading. The stress cloud diagram obtained from these calculations shows the stress state at various locations of the pressure vessel under combined loading. The stress cloud diagram of the RPV is shown in Figure 9.

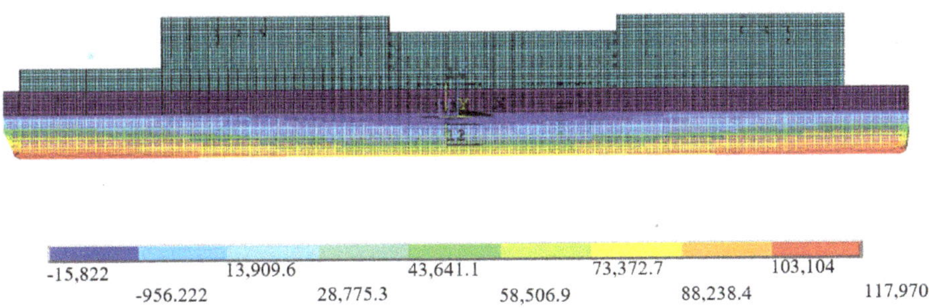

Figure 8. Illustration of wave load application.

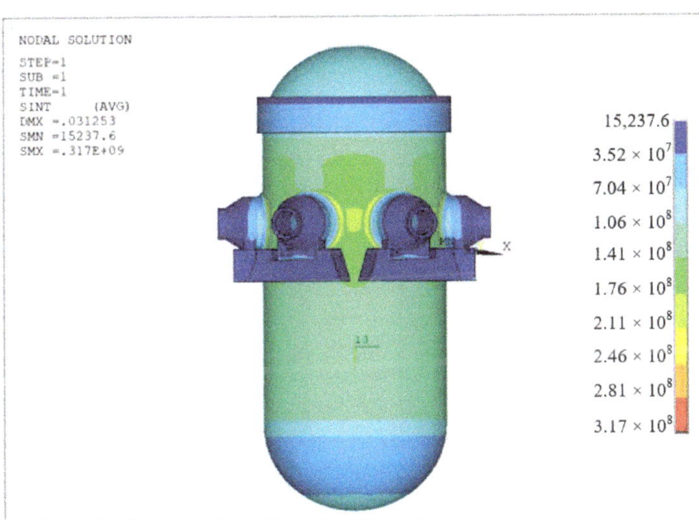

Figure 9. Stress cloud diagram of a RPV (Pa).

5. Fatigue Analysis Method Simulation Verification for RPVs in the Marine Environment

The pressure vessel of a FNPP is subjected to internal pressure from high-temperature, high-pressure coolant (such as water or helium) during operation. The coolant absorbs heat released from the reactor and circulates, producing constantly changing pressure loads, which induce cyclic stresses in the vessel walls and cause fatigue damage. Additionally, thermal expansion and contraction resulting from temperature fluctuations generate thermal loads, which exacerbate the thermal cyclic stresses in the material, potentially leading to fatigue failure [31].

Waves are the primary dynamic loads in the marine environment and have a significant impact on the pressure vessel structure of the FNPP. The wave loads are transmitted to the pressure vessel primarily through the deformation and motion of the hull, generating corresponding dynamic stresses and deformations [32]. Hull motion, particularly the inertial movements of the vessel, induces periodic loads inside the pressure vessel, becoming a major source of fatigue loading. The hull motion mainly involves heave, pitch, and roll motions under the influence of waves. The inertial forces generated by these movements deform and stress the internal structure of the pressure vessel, making them the primary source of fatigue loads in the marine environment [33].

Thermal compressive loads and wave loads are both important causes of the fatigue failure of reactor pressure vessels, and simulations are needed to validate the fatigue analysis method for reactor pressure vessels under marine environmental conditions. The design service life of FNPPs is forty years and, thus, the fatigue analysis for FNPP RPVs is performed with a duration of forty years. In this section, fatigue damage analysis is first conducted using the time-domain method recommended by the standards followed by the TDDS method, and a comparison is made between the results from both methods.

5.1. Time-Domain Fatigue Damage Analysis

In the time-domain analysis method, each sea state was modeled as a short-duration irregular wave lasting 30 h, and the results were weighted and averaged according to the wave-walk diagram to obtain the 30 h time-domain sea state. The stress response of the reactor pressure vessel was different in different short-term sea states. Table 3 shows the wave scatter diagrams in the Yellow and Bohai Seas, where the floating nuclear power plant is located [12]. Here, Tp represents the spectral peak period and Hs represents the significant wave height. The data in the table indicate the number of occurrences observed over 40 years, with each observation lasting 4 h.

The steps for calculating RPV fatigue damage using the time-domain method were as follows: Firstly, the transient calculations of wave loads for different sea states were carried out by using WORKBENCH 2022 software for 30 h, and the results were weighted according to the wave walk diagrams. The 30 h wave load data were then reasonably extended to 40 years, based on the cyclic nature of wave loads, in order to match the design life of 40 years in actual engineering. At the same time, for the thermal compression load, the corresponding stress components were also obtained based on the condition that the actual working condition lasts for forty years. Then, by superposing the forty-year wave load stress components with the forty-year thermopressure load stress components and calculating their stress intensities, the time-varying structural stress curves under the total loads were obtained, stress-time history curves for check points is shown in Figure 10 During this process, the irregular waves were chosen using the JONSWAP spectrum to map the time-domain results to the finite element model of APDL, and the statistical distribution of the hotspot stress-time course was obtained using the TDDS method.

Table 3. Wave scatter diagram.

Hs (m) \ Tp (s)	0–2	2–3	3–4	4–5	5–6	6–7	7–8	8–9	9–10	10–11	11–15	Total
0.0–0.5	2312	7647	14,374	1807	797	347	21	6	2	6	84	27,403
0.5–1.0			9086	33,155	5277	1462	173	4				49,157
1.0–1.5			231	8138	12,786	2744	51	10	1			23,961
1.5–2.0			6	143	4289	4957	297	3				9695
2.0–2.5				24	134	3443	1118	2				4721
2.5–3.0					25	542	1876	33				2476
3.0–3.5						14	886	303				1203
3.5–4.0							158	510	14			682
4.0–4.5							5	197	162			364
4.5–5.0								9	65	5		79
5.0–5.5									15	29		44
5.5–6.0									2	11		13
6.0–6.5										1	1	2
Total	2312	7647	23,267	43,267	23,308	13,509	4585	1077	261	52	85	11,980

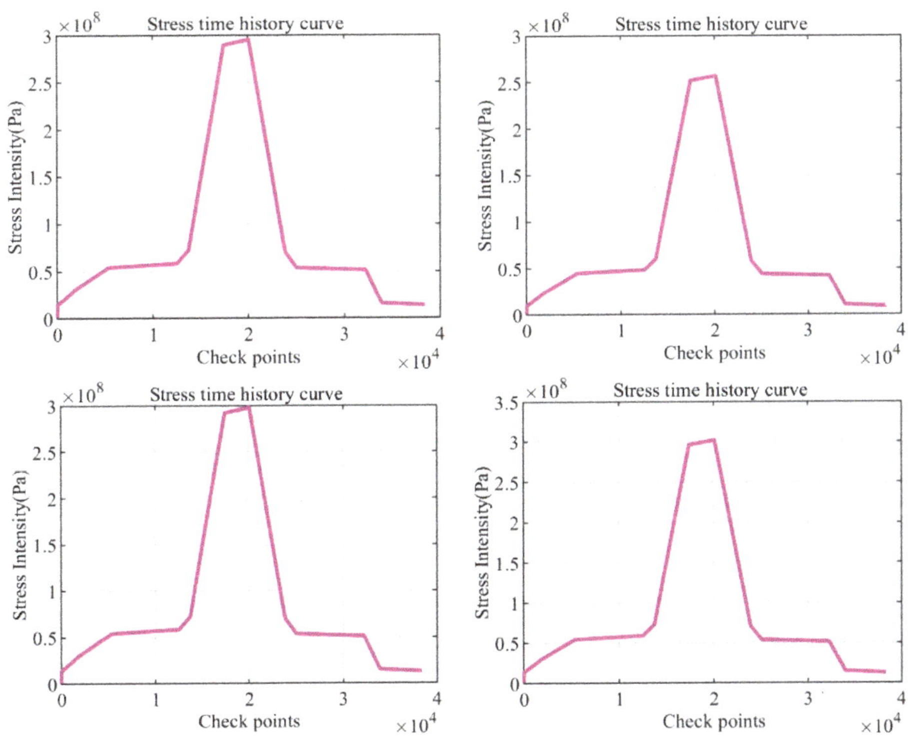

Figure 10. Stress-time history curves diagram for check points.

The stress-time history can be statistically processed into stress amplitudes through rainflow counting. The obtained amplitudes are then substituted into the S-N curve to obtain the corresponding number of cycles (N), which allows the calculation of the fatigue life at each stress amplitude. By applying Miner's linear damage accumulation theory, the damage values for each stress amplitude are summed to obtain the total damage value of the structure.

5.2. Fatigue Damage Analysis Using the TDDS Method

A RPV simultaneously experiences wave loads, internal pressure, and thermal loads during the operation of a FNPP. The combination of these loads is difficult to determine; therefore, a method based on a combination of thermo-compression coupled loads and wave loads was used in this study to assess the fatigue damage life of pressure vessels. The accumulated damage of the structure under different load combinations was analyzed.

Firstly, the six stress components under the thermal–pressure coupled load were extracted through finite element analysis. The complex wave loads are equivalent to regular waves to simplify the calculation. Twelve different initial phases were selected for combination with these loads on the basis of the thermal–pressure coupled loads. These initial phases represent the relative positions of the wave loads at different time points. A series of combined load stress components were obtained by combining the wave load at different phases with the six stress components of the thermal–pressure coupled loads at corresponding moments. This allows the calculation of the upper and lower limits of fatigue damage for FNPP pressure vessels. Next, these stress components were converted into stress intensity, and the stress amplitudes were obtained using the rainflow counting method. The final fatigue damage value range for the pressure vessel was determined in combination with the corresponding S-N curve. The results of the time domain method of damage are shown in Figure 11, the TDDS method damage range is shown in Figure 12.

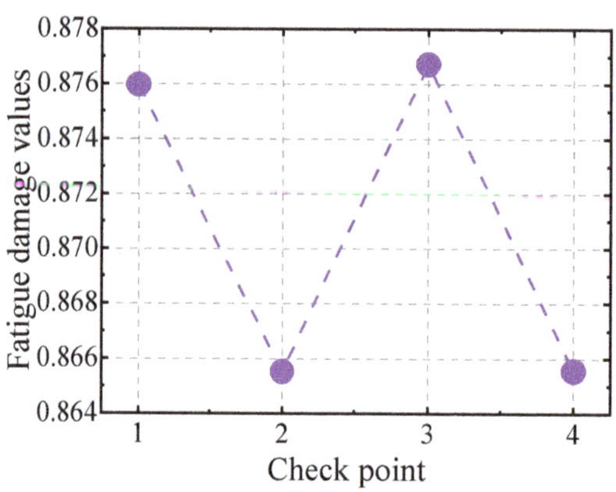

Figure 11. Schematic diagram of fatigue cumulative damage results using the time-domain method.

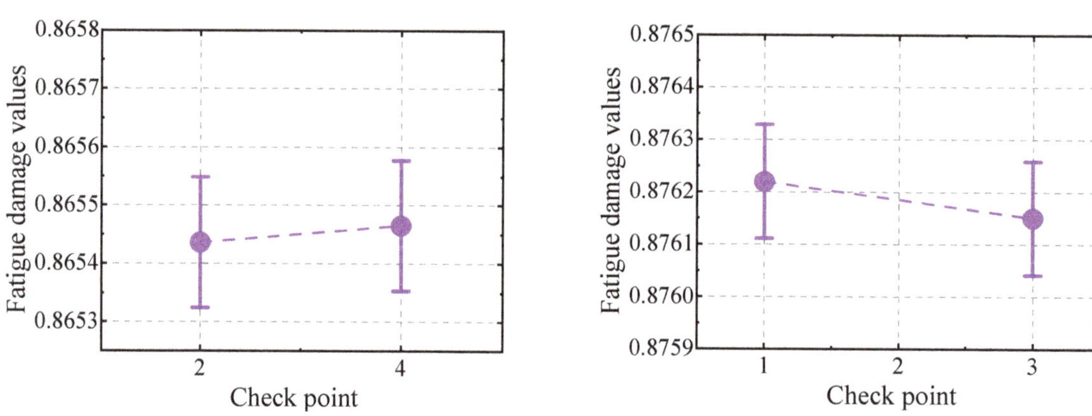

Figure 12. Schematic diagram of the fatigue cumulative damage range results at the check points using the proposed method.

From the results, it can be seen that the fatigue damage value error at Check point 1 is 0.021%, 0.019% at Check point 2, 0.074% at Check point 3, and 0.019% at Check point 4. In terms of time, the TDDS method saves 70% of the time compared to the traditional time-domain method.

The following data were derived by comparing the results of the average of the upper and lower bounds of the calibration point damage values derived from the TDDS method with those derived from the conventional time-domain method: At Check point 1, the fatigue damage value of the new method has an error of only 0.021% compared to the traditional time-domain method; at Check point 2, the error is 0.019%; at Check point 3, the error is 0.074%; and at Check point 4, the error is 0.019%. The TDDS method saves 70% of the analysis time when completing the analysis task compared to the traditional time-domain method.

6. Conclusions

A TDDS method was proposed in this study to evaluate the fatigue life of RPVs in FNPPs under combined conditions of wave loads, internal pressure, and thermal loads. Given the limitations of traditional spectral analysis, typical transient methods, and fracture mechanics methods in dealing with the complex working conditions of FNPPs, the new method is introduced. The specific conclusions are as follows:

1. The TDDS method was compared with the traditional time-domain method at four checkpoints. The fatigue damage value errors of the new method were 0.021%, 0.019%, 0.074%, and 0.019%, with an average error of only 0.033%. This demonstrates the method's exceptional accuracy in assessing fatigue life and its ability to accurately reflect the actual fatigue damage of FNPP RPVs.

2. In terms of analysis time, the TDDS method shows a significant advantage, saving approximately 70% of the time compared to the traditional time-domain method. By converting irregular waves in the time domain into regular waves and adjusting their initial phases, the method significantly reduces the time-domain calculation workload.

3. Numerical simulations and parameter analysis accurately identified key fatigue analysis locations, such as the nozzle–vessel wall connection. Stress concentrations were observed in areas like the chamfered upper and lower parts of the inner surface of the nozzle and internal wall contact, the external surface of the nozzle and nozzle support contact, and the inner surface of the top cover and body contact.

4. Future research will focus on extending the TDDS method to other fields, such as offshore engineering structures, floating platforms, and the fatigue analysis of critical equipment under complex loading conditions. The applicability of the regular wave superposition model in various environments and conditions will be further explored, with efforts to optimize model parameters and analysis processes for multi-field applications.

Author Contributions: J.Y.: conceptualization, methodology, software, formal analysis, writing—original draft, validation, data curation, and visualization; F.M.: investigation, and writing—review and editing; M.Z.: resources, project administration, writing—review and editing, supervision, and funding acquisition; K.S.: writing—review and editing; J.T.: data curation, and visualization. All authors have read and agreed to the published version of the manuscript.

Funding: This work was sponsored and supported by the Leading Innovation Project of China National Nuclear Corporation (KY90200220002).

Institutional Review Board Statement: Not applicable.

Informed Consent Statement: Not applicable.

Data Availability Statement: Data are contained within the article.

Conflicts of Interest: The authors declare no conflicts of interest.

References

1. Belyaev, V.M.; Bol'shukhin, M.A.; Pakhomov, A.N.; Khizbullin, A.M.; Lepekhin, A.N.; Polunichev, V.I.; Veshnyakov, K.B.; Sokolov, A.N.; Turusov, A.Y. The World's First Floating NPP: Origination and Direction of Future Development. *At. Energy* **2020**, *129*, 27–34. [CrossRef]
2. Xiang, G.; Fanfu, K.; Chenghua, Z.; Gang, Z. Research on Safety Criteria in Design of Floating Nuclear Power Plant. *hdlgc* **2021**, *42*, 244–249. [CrossRef]
3. Wang, Q.; Zhang, Y.; Zhang, H. The Development of Floating Nuclear Power Platforms: Special Marine Environmental Risks, Existing Regulatory Dilemmas, and Potential Solutions. *Sustainability* **2023**, *15*, 3022. [CrossRef]
4. Xiao, J. A Method for Hazard Analysis of a Floating Nuclear Power Plant Subjected to Ship Collision. *Ann. Nucl. Energy* **2023**, *190*, 109870. [CrossRef]
5. Yue, J. A Frequency-Time Domain Method for Ship Fatigue Damage Assessment. *Ocean Eng.* **2021**, *220*, 108154. [CrossRef]
6. Liu, Y.; Ren, H.; Feng, G.; Sun, S. Simplified Calculation Method for Spectral Fatigue Analysis of Hull Structure. *Ocean Eng.* **2022**, *243*, 110204. [CrossRef]
7. Yosri, A.; Leheta, H.; Saad-Eldeen, S.; Zayed, A. Accumulated Fatigue Damage Assessment of Side Structural Details in a Double Hull Tanker Based on Spectral Fatigue Analysis Approach. *Ocean Eng.* **2022**, *251*, 111069. [CrossRef]
8. Makris, P.; Silionis, N.; Anyfantis, K.N. Spectral Fatigue Analysis of Ship Structures Based on a Stochastic Crack Growth State Model. *Int. J. Fatigue* **2023**, *176*, 107878. [CrossRef]
9. Fan, W.; Li, Z.; Yang, X. A Spectral Method for Fatigue Analysis Based on Nonlinear Damage Model. *Int. J. Fatigue* **2024**, *182*, 108188. [CrossRef]
10. Jia, J. Wind Induced Dynamic Response and Fatigue Assessment—Nonlinear Time Domain versus Linear Stochastic Analysis. *Structures* **2021**, *34*, 4406–4422. [CrossRef]
11. Zheng, T. Time-Domain Fatigue Assessment for Blade Root Bolts of Floating Offshore Wind Turbine (FOWT). *Ocean Eng.* **2022**, *262*, 112201. [CrossRef]
12. Ma, F.; Li, H.; Zhang, M.; Qu, X. A Quasi Time-Domain Method for Fatigue Analysis of Reactor Pressure Vessels in Floating Nuclear Power Plants in Marine Environments. *J. Mar. Sci. Eng.* **2024**, *12*, 2085. [CrossRef]
13. Wang, X. Fatigue Life Analysis of Marine Engineering Pressure Vessels with Internal Cracks. Master's Thesis, Lanzhou University of Technology, Lanzhou, China, 2016.
14. Liu, T. Stress Analysis and Fatigue Simulation of SCR Pressure Vessel for Marine Diesel Engine. Master's Thesis, Jiangsu University of Science and Technology, Jiangsu, China, 2019.
15. Xu, S. Fatigue Reliability Analysis of LNG Carrier Based on Spectrum Analysis. Master's Thesis, Huazhong University of Science and Technology, Wuhan, China, 2012.
16. Zhang, H. Fatigue Life Assessment of Marine Nuclear Power Platform. *Ship Eng.* **2019**, *41*, 1–7+28. [CrossRef]
17. Rasul, A.; Karuppanan, S.; Perumal, V.; Ovinis, M.; Iqbal, M. Multi-Objective Optimization of Stress Concentration Factors for Fatigue Design of Internal Ring-Reinforced KT-Joints Undergoing Brace Axial Compression. *Civ. Eng. J.* **2024**, *10*, 1742–1764. [CrossRef]
18. Iqbal, M.; Karuppanan, S.; Perumal, V.; Ovinis, M.; Khan, A.; Faizan, M. Stress Concentration Factors in KT-Joints Subjected to Complex Bending Loads Using Artificial Neural Networks. *Civ. Eng. J.* **2024**, *10*, 1051–1068. [CrossRef]
19. The American Society of Mechanical Engineer (ASME). BPVC Section III: Rules for Construction of Nuclear Facility Components, Division 1, Subsection NB: Class 1 Components. 2023. Available online: https://www.asme.org/codes-standards/find-codes-standards/bpvc-section-iii-div-1-subsection-nb-class-1-components/2023/print-book (accessed on 22 October 2024).
20. Dai, X.; Yang, B. The Effect of Microstructure Evolution on Fatigue Property of SA508Gr.4N Steel for Nuclear Reactor Pressure Vessels. *J. Nucl. Mater.* **2023**, *584*, 154544. [CrossRef]
21. Dai, X.; Chen, Y.; Wang, P.; Zhang, L.; Yang, B.; Chen, L. Mechanical and Fatigue Properties of SA508-IV Steel Used for Nuclear Reactor Pressure Vessels. *J. Iron Steel Res. Int.* **2022**, *29*, 1312–1321. [CrossRef]
22. Lee, K.-H.; Kim, M.-C.; Lee, B.-S.; Wee, D.-M. Master Curve Characterization of the Fracture Toughness Behavior in SA508 Gr.4N Low Alloy Steels. *J. Nucl. Mater.* **2010**, *403*, 68–74. [CrossRef]
23. Niu, X.-P.; Zhu, S.-P.; He, J.-C.; Ai, Y.; Shi, K.; Zhang, L. Fatigue Reliability Design and Assessment of Reactor Pressure Vessel Structures: Concepts and Validation. *Int. J. Fatigue* **2021**, *153*, 106524. [CrossRef]
24. Bonadé, R.; Spätig, P.; Baluc, N. Fracture Toughness Properties in the Transition Region of the Eurofer97 Tempered Martensitic Steel. *J. Nucl. Mater.* **2007**, *367–370*, 581–586. [CrossRef]
25. Wu, X.; Kang, H.T. A Fracture Mechanics-Based Stress Approach for Fatigue Life Prediction of Welded Joints Considering Weld Profile Effect. *Theor. Appl. Fract. Mech.* **2023**, *123*, 103698. [CrossRef]
26. Wang, J.; Zeng, D.; Lu, L. Fatigue Life Evaluation for Ring-Welded Lap Joints of Stainless Steel Based on Fracture Mechanics Approach. *Theor. Appl. Fract. Mech.* **2023**, *127*, 104067. [CrossRef]

27. Esmaeili, F.; Rahmani, A.; Barzegar, S.; Afkar, A. Prediction of Fatigue Life for Multi-Spot Welded Joints with Different Arrangements Using Different Multiaxial Fatigue Criteria. *Mater. Des.* **2015**, *72*, 21–30. [CrossRef]
28. Liu, W.; Zhang, T. Multimedia Hashing and Networking. *IEEE Multimed.* **2016**, *23*, 75–79. [CrossRef]
29. Li, J.; Wang, D.; Zhang, M.; Liu, H.; Qu, X. Study of Stress Analysis Method for Floating Nuclear Power Plant Containment Under Combined Multiple Loads. In Proceedings of the 23rd Pacific Basin Nuclear Conference, PBNC 2022, Beijing & Chengdu, China, 1–4 November 2022; Springer Science and Business Media Deutschland GmbH: Beijing, China, 2023; Volume 283, pp. 800–811.
30. Ren, W.; Zhang, S.; Song, D.; Zhang, M.; Wang, W. The Dynamic Response Analysis Method of Steel Containment in Floating Nuclear Power Plant. In Proceedings of the 23rd Pacific Basin Nuclear Conference, PBNC 2022, Beijing & Chengdu, China, 1–4 November 2022; Springer Science and Business Media Deutschland GmbH: Beijing, China, 2023; Volume 283, pp. 764–775.
31. Ruan, X. Pressurized Thermal Shock Analysis of a Reactor Pressure Vessel for Optimizing the Maintenance Strategy: Effect of Asymmetric Reactor Cooling. *Nucl. Eng. Des.* **2021**, *373*, 111021. [CrossRef]
32. Dong, Y. Review on Uncertainties in Fatigue Loads and Fatigue Life of Ships and Offshore Structures. *Ocean Eng.* **2022**, *264*, 112514. [CrossRef]
33. Mendes, P.; Correia, J.A.F.O.; Mourão, A.; Pereira, R.; Fantuzzi, N.; De Jesus, A.; Calçada, R. Fatigue Assessments of a Jacket-Type Offshore Structure Based on Static and Dynamic Analyses. *Pract. Period. Struct. Des. Constr.* **2021**, *26*, 04020054. [CrossRef]

Disclaimer/Publisher's Note: The statements, opinions and data contained in all publications are solely those of the individual author(s) and contributor(s) and not of MDPI and/or the editor(s). MDPI and/or the editor(s) disclaim responsibility for any injury to people or property resulting from any ideas, methods, instructions or products referred to in the content.

Journal of
Marine Science and Engineering

MDPI

Article

Morphology and Effect of Load on Bridge Piers Impacted by Continuous Sea Ice

Li Gong [1,2], Yue Cui [1,*], Yunfei Du [1], Long Qin [1] and Xinyuan Zhao [1]

1 Department of Civil Engineering, Lanzhou Jiaotong University, Lanzhou 730070, China; gongl@mail.lzjtu.cn (L.G.); dyf980521@163.com (Y.D.); ql227790@163.com (L.Q.); 15294124865@163.com (X.Z.)
2 Gansu Normal University for Nationalities, Gannan Tibetan Autonomous Prefecture, Hezuo 747000, China
* Correspondence: cy17393182939@163.com

Abstract: In order to study the collision of sea ice on bridge piers of a sea-crossing bridge, this study establishes a finite element model of the impact of sea ice on bridge piers in aqueous media based on explicit dynamics analysis software and programming software using the arbitrary Lagrangian Eulerian (ALE) method. The results show that, when the sea-ice spacing is larger than the sea-ice edge length, the increase in sea-ice spacing leads to a decrease in the collision force and a significant increase in the probability of climbing and overturning. The increase in sea-ice mass significantly increases the impact force on the bridge abutment, and the peak value increases linearly with the increase in mass, and the sea-ice climbing and overturning phenomena are obvious. Different shapes of sea ice are obtained by cutting the sea-ice field with the two-dimensional Voronoi method, and the maximum impact force increases significantly with the increase in the average area. Irregularly shaped sea ice leads to a larger impact force and triggers the accumulation climbing phenomenon, which is verified by experiments, and the experimental values are in good agreement with the simulated values. In conclusion, this study reveals the significant effects of the spacing, mass, and shape of sea ice on the impact force of bridge piers, which provides an important reference for the design of bridge structures.

Keywords: numerical simulation; sea ice; ice load; impact of successive impacts; ice size

1. Introduction

In the project example, cross-sea bridge piers in the winter ice flow period are often hit by the sea ice swarm. The collision of the coupling force generated by the piers and collision avoidance facilities will cause varying degrees of damage and, in extreme cases, even cause the collapse of the bridge piers, the main beams falling off, and other phenomena, and the piers play a role in supporting the superstructure of the bridge. Once the failure occurs, it will cause the instability of the superstructure, which will seriously affect the normal navigation and transport on the sea surface and even bring great losses to people's lives and properties. Therefore, there is an urgent need to carry out the study of the loading pattern and effect of sea ice collision bridge piers in order to explore the impact of bridge piers in the collision effect of the law.

Due to differences in the strength, speed of movement, thickness, shape, and width of the structures, sea ice causes extrusion, creep, buckling, and splitting damage on the surface of upright structures. The different ways of sea-ice damage lead to differences in the forces exerted on the structures. Song et al. [1] regarded sea ice as a semi-infinite structure on an elastic foundation, regarded the buckling damage of the sea ice on the upright structure as a flat plate instability problem under the concentrated force, and gave the critical load of the sea ice under the action of piles and columns. The scaling experiments and field observations are the main parts of the simultaneous loading study, which can provide verification criteria for numerical simulations. Qu et al. [2] analyzed the creep compression force of low velocity sea ice with upright structure by means of experiment and field monitoring, and concluded that creep compression produces a small force relative to rapid compression damage. Lemström et al. [3] carried out modeling experiments on the process of interaction between ice and inclined structure in shallow water.

With the development of finite element software, numerical simulation has been favored by scholars from all walks of life. Metrikin et al. [4] verified that the errors between experimental and numerical results of the slewing and drag characteristics of a ship model–sea-ice area model in an ice pool were within acceptable limits; Gong et al. [5] simulated the dynamic response of an ice raft and a bridge abutment in an aqueous medium under different collision impact parameters; Zhu et al. [6] used an ice material-based model of sliding loads to investigate the joint action of ice indentation and sliding loads, ice fragmentation, and structural damage results under ice motion trajectory. An ice material model based on a concrete material model was used to simulate the mechanical impact of ice in the ANSYS/LS-DYNA program; Zhang et al. [7] simulated the collision process between flowing ice and a T-rigid bridge pier using LS-DYNA; and Wu et al. [8] proposed a complete bridge analysis model to study the dynamic response of the Bohai Sea-crossing bridge under the random ice loading of floating sea ice. Some scholars have also carried out a large number of studies in combination with the effects of sea traffic on sea ice loading, among which Xia et al. [9] considered the effects of drift ice and established a dynamic analysis model of drift ice loading on the coupled system of the shuttered bridges; Li et al. [10] investigated the excitation effect of the Kaoyong ice on the train. Wang et al. [11] led a team to use numerical simulation to study the dynamic analysis model of a bridge when the ice loading was applied as a static load applied at different locations of hull deformation and collision force, and the structural damage of the hull structure after interaction with ice. However, due to the complexity of ice morphology, most scholars have simplified the ice structure. In order to be closer to the real situation of sea ice morphology, a few scholars have simplified the ice morphology. Kawano et al. [12] found that Voronoi diagrams can be used to construct the geometry of ice crystals; Liu et al. [13] found that the geometry of ice crystals can be constructed by using Voronoi diagrams; Guo et al. [14] analyzed the collision damage of different sizes of drift ice on bridge abutments at different water depths. At present, the research of scholars at home and abroad is mostly experimental studies of different ice morphology and ice impact simulation in the form of ice loading, and the research on the impact of sea ice on sea-crossing bridges is mostly focused on the simulation of large ice and single small ice, while the research on the collision of bridge piers by considering the continuous sea ice in the aqueous medium is relatively little.

At present, scholars at home and abroad mainly focus on the simulation experimental studies for different ice morphologies, especially in the study of ice impact in the form of ice load applied to hydraulic buildings, and most of these studies focus on the simulation of the impact of a large ice cube or a single small ice cube. However, there is still an obvious gap in the research on continuous sea ice impacts on bridge piers in aqueous media, especially on factors such as continuous sea ice spacing, continuous ice mass, and continuous ice area. In order to fill this research gap, we used MATLAB (MATLAB-R2022A) and ANSYS/LS-DYNA (ANSYS-2022-R1/LS-DYNA-LSPREPOST4.3_X64) software to carry out detailed numerical simulations of the collision process between sea ice and bridge abutments based on a specific model of a sea ice area impacting a railway bridge abutment. This study not only contributes to an in-depth understanding of the loading pattern and effect of continuous sea ice on bridge piers, but also provides an important theoretical basis and technical reference for domestic and international cross-sea bridge abutment concrete in coping with the problem of sea-ice collision in order to enhance its safety and durability in extreme marine environments.

2. Continuous Sea-Ice Bridge Pier Impact Finite Element Mode

2.1. Fluid-Solid Coupling

Using the Lagrangian–Eulerian method, the sea ice and bridge abutments are defined as Lagrangian entities, the water and air media are defined as Eulerian entities, and the coupling of fluid and sea ice structures is achieved through the CONSTRAINED_LAGRANGE_IN_SOLID keyword in LS-DYNA (LS-DYNA-LSPREPOST4.3_X64) software to study the water–air coupling process in the ice collision area of the Yellow River mega-railway bridge of Yinchuan Airport in the course of change rule. The fluid–solid coupling model is shown in Figure 1, in which the water and air domains only cover the sea ice part to simplify the model for calculation.

Figure 1. Fluid-structure coupling mode.

2.2. Mesh Convergence Study

In order to study the influence of sea ice and abutment grid size on the numerical simulation results, two groups of ice grids, A_1 and A_2, and two groups of abutment grids, B_1 and B_2, were selected, and the X forward displacements obtained by numerical simulation under different grid sizes are shown in Table 1.

Table 1. Comparison of Peak equivalent stress and X-direction displacement peak at different mesh scales.

Mesh Serial Number	A_1	A_2	B_1	B_2	Experimental Value
Mesh size	$0.05 \times 0.05 \times 0.05$ m^3	$0.07 \times 0.07 \times 0.07$ m^3	$0.05 \times 0.05 \times 0.05$ m^3	$0.08 \times 0.08 \times 0.08$ m^3	
Peak equivalent stress ($\times 10^7$ Pa)	2.617 (4.0%)	2.928 (−7.5%)	2.517 (7.6%)	2.741 (−0.6%)	2.725
X-direction displacement peak ($\times 10^{-4}$ m)	9.749 (0.2%)	10.245 (−4.9%)	9.349 (4.3%)	9.812 (−0.5%)	9.766

From Table 1, the peak equivalent force of the ice grid with the experimental value error of 4.0% and −7.5%, respectively, the bridge abutment grid peak equivalent force with the experimental value error of 7.6% and −0.6%, respectively; the peak x–direction displacement of the ice grid with the experimental value error of 0.2% and −4.9%, respectively, the bridge abutment grid x–direction peak displacement with the experimental value error of 4.3% and −0.5%, respectively. In summary, the ice grid of A_1 ($0.05 \times 0.05 \times 0.05$ m^3) and the abutment grid of B_2 ($0.08 \times 0.08 \times 0.08$ m^3) are used in this paper for numerical calculation.

2.3. Selection of Model Material Parameters

The concrete continuous surface cap model CSCM developed by the Federal Highway Administration is used to simulate the dynamic performance and damage characteristics of bridge piers under the impact, and the keyword *MAT_CSCM_ CONCRETE is added to define it. Ice material parameters refer to the research results of Yang [15] and other research results; the 13th material *MAT_ISOTROPC_ELASTIC _FAILURE in the LS-DYNA material library is selected to simulate the characteristics of ice, and the model parameters are set as in Tables 2 and 3.

Table 2. Parameters of concrete material model.

LS-DYNA Model Parameter	Numerical Value	Unit
RO (mass density)	2500	$/(\mathrm{kg{\cdot}m^{-3}})$
RECOV (coefficient of recovery)	10	
FPC (free compressive strength without lateral limit)	29	MPa
DAGG (maximum aggregate size)	0.02	m

Table 3. Ice material parameters.

LS-DYNA Model Parameter	Numerical Value	Unit
RO (material density)	910	$/(\mathrm{kg{\cdot}m^{-3}})$
G (shear modulus)	2.2	GPa
SIGY (yield stress)	2.136	MPa
ETAN (plastic modulus)	4.26	GPa
BULK (bulk modulus)	5.26	GPa
EPF (Plastic strain at failure)	0.35	–
PRF (truncation stress)	−3.0	MPa

Referring to the setup of Gong [16] and others, the water and air media are described by means of an intrinsic model and an equation of state, using *MAT_NULL material to describe the water and air media, and the equations of state are realized by means of EOS-GRUNEISEN and EOS-LINER-POLYNOMIAL,

The Gruneisen equation of state is as follows:

$$P = \frac{\rho_0 C^2 \mu \left[1 + \left(1 - \frac{\gamma_0}{2}\right)\mu - \frac{\alpha}{2}\mu^2\right]}{\left[1 - (S_1 - 1)\mu - S_2 \frac{\mu^2}{\mu+1} - S_3 \frac{\mu^3}{(\mu+1)^2}\right]^2} + (\gamma_0 + \alpha\mu)E \tag{1}$$

where P is the pressure; $\mu = \rho/\rho_0 - 1$; $C, S_1, S_2, S_3, \gamma_0$ are the equation coefficients.

The linear polynomial equation of state is as follows:

$$P = C_0 + C_1\mu + C_2\mu^2 + C_3\mu^3 + \left(C_4 + C_5\mu + C_6\mu^2\right) \tag{2}$$

where P is the pressure, $\mu = \rho/\rho_0 - 1$; C_0, C_1, C_2, C_3, C_4, C_5, C_6 are polyno-mial equat-ion coefficients.

Parameters are set as in Table 4.

Table 4. Water medium state equation parameters.

LS-DYNA Model Parameter	Water Medium	Air Medium	Unit
Density RO	998	1.29	$/(\mathrm{kg{\cdot}m^{-3}})$
Cut-off pressure PC	-10×10^5	−10	Pa
Constant C	1647	0	
Constant S_1	1.92	0	
Constant S_2	−0.096	0	
Constant γ_0	0.35	0	
Coefficient C_4	0	0.4	

2.4. Ice-Abutment Impact Finite Element Modeling

2.4.1. Bridge Pier Model

As shown in Figure 2, a bridge pier in a sea-crossing bridge is selected, which is taken as 5.63 m in length, 1.95 m in width, 3.42 m in height, and 3.07 m in width at the bottom of the pier, with a circular collision avoidance body at the front and a diagonal inward 45° excavation depth of 0.36 m in the middle of the pier. The bridge pier was meshed finely by Hypermash software (Hypermash2022), imported into LS-DYNA modeling, and constrained by using the constraint keyword SET_NODE_LIST_(TITLE) in LS-DYNA to constrain all degrees of freedom of the base plate of the pier in all three directions, and the other surfaces were free surfaces. In the process of the impact of the sea ice layer on the bridge abutment, there will be an ice body damage erosion phenomenon. In order to ensure that in the ice body broken unit failure deletion, the ice and the bridge remain in contact with each other after the computation, the use of LS-DYNA in the contact keyword CONTACT_ERODING_ SURFACE_TO_SURFACE defines the erosion of the contact, the sea ice body for the slave surface (SSID), and the bridge abutment for the master surface (MSID). The sea ice body is the slave surface (SSID), the bridge abutment is the master surface (MSID), and the coefficient of static friction (FS) and the coefficient of kinetic friction (FD) are taken as the values of 0.15. In the finite element numerical simulation, the sea ice, bridge abutment, water medium, and air medium are all used in the single-point integration. Solid164 solid cells avoid the volume locking of the cells.

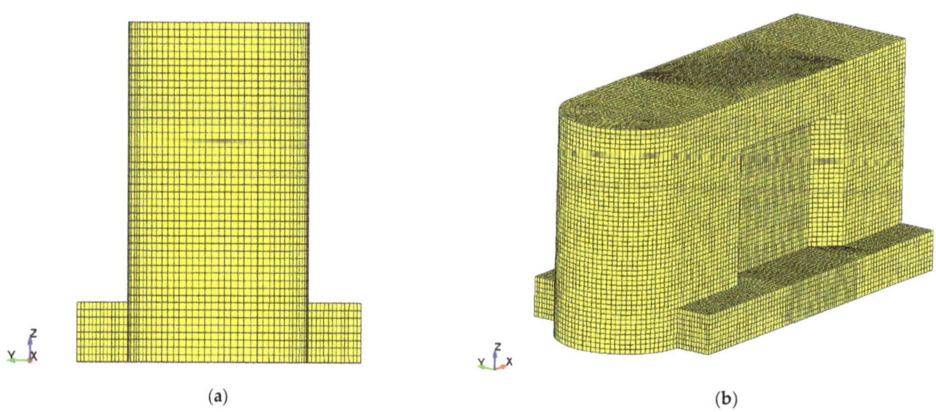

Figure 2. Mesh division of bridge pier: (**a**) main view (**b**) overall view.

2.4.2. Icebreaker Model

By statistically analyzing the existing data on sea ice bodies, a realistic distribution model of sea ice bodies is constructed based on the parameters of the size, shape, and density of the sea ice bodies. In this model, the main study is the influence factor of continuous sea ice collision, i.e., limiting the collision area of sea ice based on the actual observation of sea ice and previous research experience to select a sea ice body of 2 m × 0.4 m with a thickness of 0.1 m. Between ice and ice in the process of sea ice collision, an abutment will appear in the front row of ice in the impact of the abutment after the rebound and the rear row of ice collision, ice, and ice collision after the extrusion of the ice body crushing, to ensure that the sea ice continues to continue to collide with the bridge abutment, ice, and ice collision with the LS-DYNA in the one-sided contact. The keyword CONTACT_ ERODING_SINGLE _SURFACE defines the ice eroding contact with ice. The horizontal direction only considers the dragging force of the water flow on the sea ice; the initial velocity of the ice is chosen as the x–direction to provide power for the ice; the spacing between the sea ice body and the collision force between the shape of the sea ice and the abutment are considered; you can simulate the process of interaction between the sea ice and the abutment, and assess the degree of impact of the sea ice on the abutment, according to the real-time data measured by Hu [17], which are equal to the ice period from 2011 to 2020; the velocity of 1 m/s of the sea ice area is selected. Using discrete element analysis, the stability and safety of the bridge structure can be investigated under different ice-breaking conditions. By analyzing the movement trajectory of the sea ice body, the magnitude of the collision force, and other parameters, it is possible to assess the stress condition of the bridge piers and the potential risk of damage.

- Ice break-up spacing

Under the hydrodynamic air coupling mechanism, the collision of sea ice with the bridge abutment is not only the traditional collision problem between the abutment and ice, but also the collision problem between sea ice and overall ice. Therefore, not only must the impact of the speed of sea ice be considered, but the influence of the ship on the collision problem of the sea ice layer, sea ice spacing, and other parameter interactions must also be drawn on. In order to facilitate the calculation, the size of the calculation waters and air domain is reduced, the horizontal direction of the bridge ice collision spacing is shortened, and only the destructive effect of the impact force on the bridge abutment in a short time in the transverse direction and the subsequent impact of extrusion in a short time are considered. The high-speed frontal collision and extrusion are analyzed by finite element analysis. Considering the impact of continuous collision on the bridge abutment, i.e., the impact effect of sea ice with different spacing on the bridge abutment, according to the research of Wang [18] et al. (real-time data obtained through the UAV platform on 25 April 2021), the distribution density of the ice is 81.05% maximum, and the minimum is below 10%. A simple calculation yielded a spacing of sea ice between 0.15 m and 0.66 m. Therefore, four ice spacings of 0.2 m, 0.4 m, 0.6 m, and 0.8 m were selected for collision simulation, where 0.8 m is the value taken into account for extreme cases.

- Mass of sea ice cubes

Referring to the study of Yu [19] and others, the diameter of sea ice in the ice area usually ranges from 0.5 m to 8 m, and the corresponding area ranges from 0.8 m^2 to 200 m^2. The upper reaches of the Yellow River are mainly dominated by sea ice during the opening period, so sea ice areas with averages of 0.8 m^2, 1.2 m^2, 1.6 m^2, and 2 m^2 were selected for the study. By observing the flow ice in the section of Yinchuan Airport Yellow River Special Bridge, considering the ice condition of the Yellow River, the geographic location of Yinchuan Airport Special Bridge, and referring to the research of Chen [20] and others' research on flow ice, the thicknesses of 0.05 m, 0.1 m, 0.15 m, and 0.2 m of sea ice were selected for the study.

- Different shapes of sea ice

Combining different sea ice spacing conditions and different sea ice average area conditions, referring to the research of Zhu et al. [21], the two-dimensional Voronoi diagrams are generated by applying the MATLAB programming software referencing the exclusion method and the perturbation method, and the different shapes of the sea ice impact piers are investigated. The Voronoi diagrams are also referred to as the Dirichlet diagrams or the Tyson polygons, and the Voronoi diagram method is a geometric segmentation method that forms polygon boundaries by dividing the plane into a series of regions, where each region contains all points nearest to a point within that region. In the 2D Voronoi diagram of broken ice, each broken ice piece is considered a point, and the region around each broken ice piece is the corresponding Voronoi region of that broken ice piece. Considering these points as seeds, the exclusion and perturbation methods are introduced to control the degree of regularity, i.e., uniformity, of the distribution of the seed points.

Firstly, the number of seed points n and the ice area A are predetermined by introducing the geometric regularity parameter *delta* (with values ranging from 0 to 1) = d/d_{max}, and the maximum distance dmax for the repulsion method is calculated.

$$d_{max} = \sqrt{2A/\sqrt{3}n} \tag{3}$$

Then, the uniformly distributed initial seed points are generated, and the repulsion method is used to generate the Voronoi seed points. Finally, the perturbation method is used to adjust the position of the seed points, where the radius of the perturbation circle r is:

$$r = 0.1 \times d_{max} \tag{4}$$

The perturbation amount delatp is:

$$delta_p = (1 - delta) \times d_{max}/10 \tag{5}$$

The generation of two-dimensional fragmented ice areas with different geometrical regularities is shown in Figure 3.

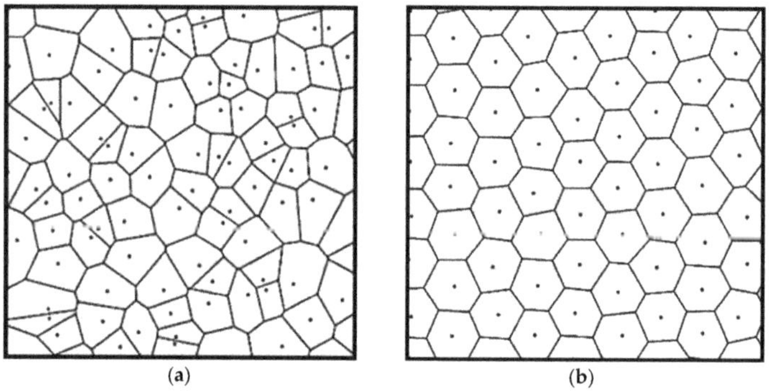

Figure 3. Two-dimensional Voronoi diagrams with different degrees of regularity: (**a**) *delta* = 0.1; (**b**) *delta* = 1.0.

In the case of selecting the same 2D (*delta* = 0.1) Voronoi diagram, different sizes of sea-ice area can be chosen by intercepting sea ice at different locations to achieve diversified effects. The specific steps are as follows:

1. Segment the Voronoi diagram: use the Voronoi segmentation algorithm to divide the 2D space into multiple non-overlapping regions, each representing the location of one sea ice cube.
2. Cutting regions: select regions at different positions and intercept them as needed. The location of the intercept can be determined by random selection, specifying the co-ordinate range of the region, or other methods.
3. Adjusting the size of crushed ice: For each intercepted region, adjust the size of the crushed ice as needed. Adjustments can be made based on the size of each region, a scale factor, or using a different scaling algorithm. Larger regions can generate larger crushed ice cubes, while smaller regions can generate smaller crushed ice cubes.
4. Generate shattered ice cubes: Based on the adjusted shattered ice cube size, randomly generate shattered ice cubes of corresponding size and shape in each intercepted region. A randomized algorithm can be used to generate the shapes of the sea ice cubes to ensure that there is a diversity of sea ice cubes within each region.

In this way, it is possible to select different locations of sea ice cubes on the same Voronoi diagram and size them according to the intercepted area to achieve the effect of different sea ice area sizes. This can increase the diversity and realism of the sea ice and fit the actual situation to study the overall effect of different spacing of different average areas on the bridge pier. As shown in Figure 4, sea ice with average areas of 1 m^2, 1.5 m^2, 2 m^2, and 2.5 m^2 were selected, respectively.

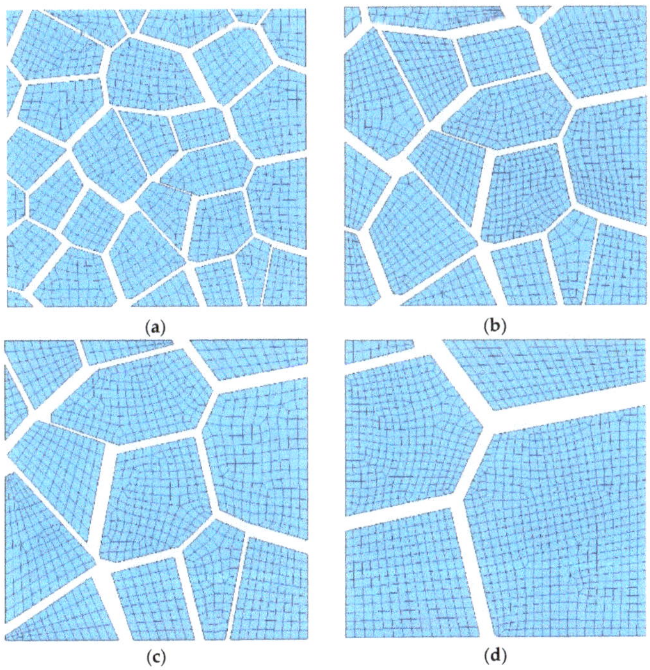

Figure 4. Average area of different sea ice blocks sea ice area: (**a**) A_0 = 1.0 m^2; (**b**) A_0 = 1.5 m^2; (**c**) A_0 = 2.0 m^2; (**d**) A_0 = 2.5 m^2.

3. Analysis of Parameters Affecting Ice-Pier Collision

3.1. The Impact of Different Spacing of Sea Ice on the Bridge Pier

Simulation results found that, along with the water cushion effect, the bridge pier by the impact area is mainly affected by the continuous collision of sea ice and the continuous extrusion effect. The continuous sea-ice collision railway bridge abutment is mainly divided into three stages: the first collision stage, the continuous collision stage, and the extrusion stage. Each stage is analyzed and illustrated by the equivalent stress diagram, and the impact force is calculated and derived by the keyword DATABASE_ASCLL_option. Four horizontally arranged sea-ice bodies of 0.1 m thickness, 2 m length, 0.4 m width, and 1 m/s speed were selected to collide with the bridge abutment, and the calculation time was increased with the spacing.

As shown in Figure 5, the sea-ice spacing of 0.2 m, the first row of sea ice collision with the bridge abutment rebound, and the rear row of ice collision rebound again, result in a continuous chain collision, i.e., the continuous collision stage (0.01 s~0.75 s), the sea ice spacing is small, the subsequent ice collision acceleration time is shorter, and continuous collision leads to the failure of the sea ice broken at the same time, accompanied by the emergence of the phenomenon of stress concentration; the stress is centered in the location of the collision contact. Subsequently, the chain collision of rows of ice is transmitted and diffused step by step, and finally concentrated in the ice–ice contact surface area. Subsequently, the extrusion phase (0.75 s~1 s) of the equivalent force of the bridge pier increases step by step with time, and it can be seen from the equivalent force cloud diagram that the extrusion stress is much larger than the impact stress. As shown in Figure 6, the sea-ice spacing is 0.4 m, the spacing is the same as the width of the sea ice, in the continuous collision stage (0.01 s~1.18 s) the number of multiple collisions is reduced, the second row of sea ice appeared along the y–direction climbing phenomenon, the sea ice did not appear obviously broken, the stress concentration phenomenon is mainly concentrated in the ice impact location, in extrusion stage (1.18 s~1.5 s) the piers equivalent effect force rising speed has slowed down, the sea ice climbing phenomenon is more obvious, and the covering phenomenon occurs between the ice blocks with the passage of time. As shown in Figure 7, the sea ice spacing 0.6 m, in the continuous collision stage (0.01 s~1.7 s) the sea ice climbing phenomenon increased significantly, and part of the ice overturned because the spacing of 0.6 m in the sea ice transverse size of 0.4 m gap is small and there is enough distance to make the sea ice overturn; the latter two rows of ice collide with the first two rows of ice when the speed did not slow down, and the first row of ice accelerated in the successive collision after a significant increase in the extrusion stage (1.7 s~2 s). With the sea ice climbing superimposed on the overall offset, lateral squeezing pressure increased significantly, and the effect of sea ice spacing at different angles can be considered subsequently. As shown in Figure 8, the sea ice spacing is 0.8 m, the continuous collision stage (0.01 s~2.4 s) due to the spacing in the sea ice lateral size is larger, the sea ice occurs after the flip, and the subsequent sea ice is covered together; the chain collision phenomenon is not obvious and the impact time span is larger. With the passage of time, the probability of extrusion overlap occurring in the extrusion stage (2.4 s~2.5 s) is significantly increased compared with other conditions.

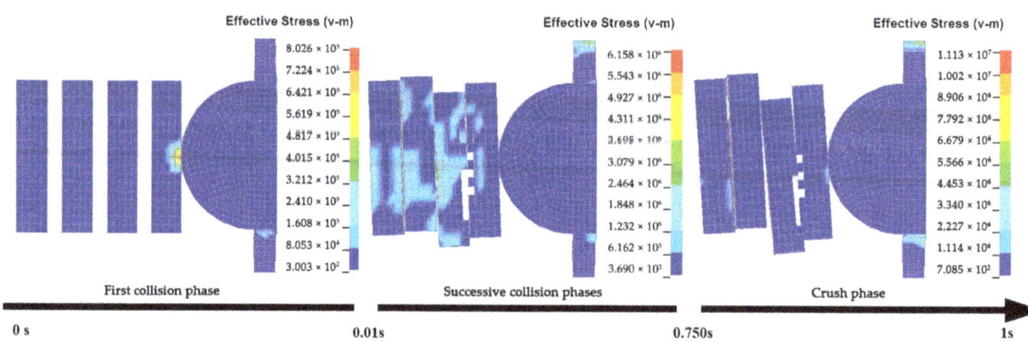

Figure 5. The maximum impact force of 0.2 m between sea ice corresponds to the equivalent stress cloud map.

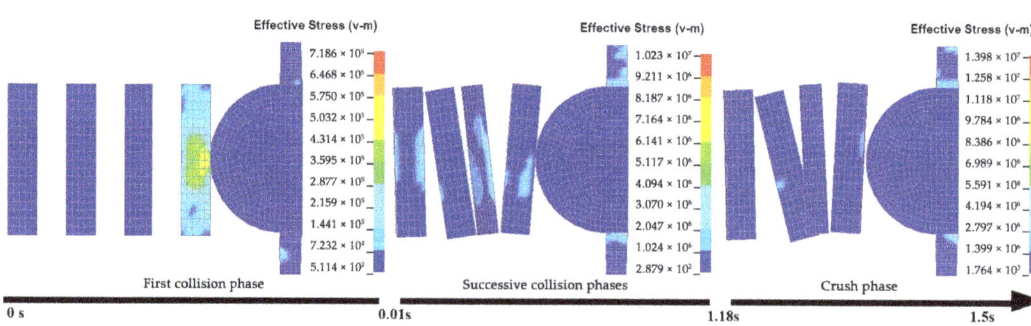

Figure 6. The maximum impact force of 0.4 m between sea ice corresponds to the equivalent stress cloud map.

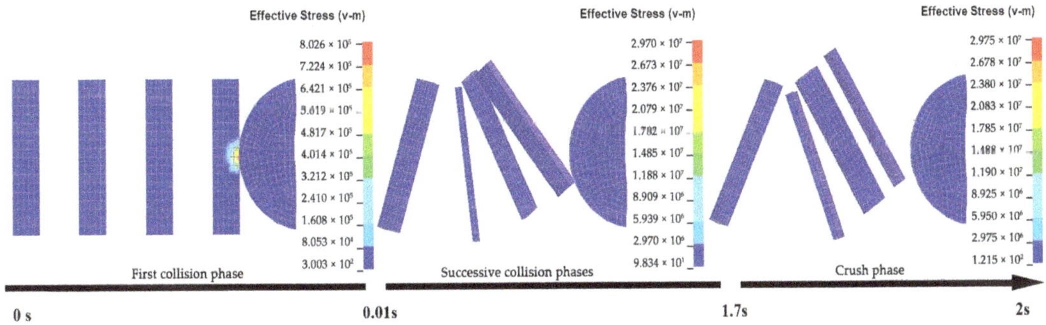

Figure 7. The maximum impact force of 0.6 m between sea ice corresponds to the equivalent stress cloud map.

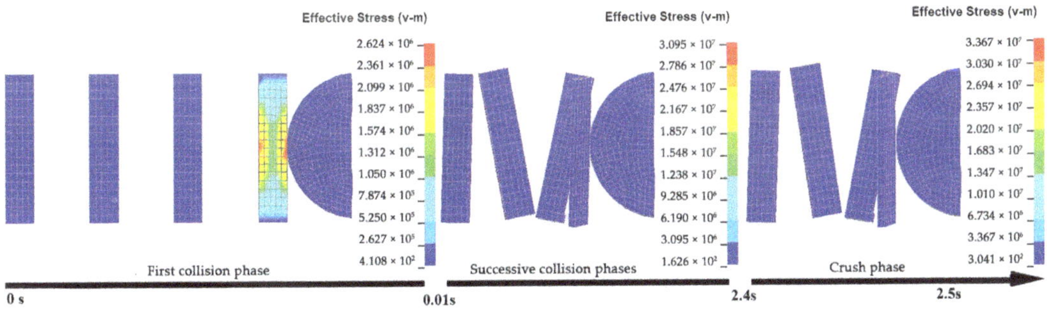

Figure 8. The maximum impact force of 0.8 m between sea ice corresponds to the equivalent stress cloud map.

Figure 9a shows the peak impact force, and sea-ice spacing of 0.2 m continuous collision more often; sea-ice spacing of 0.4 m secondary impact force peak obviously increased compared to the spacing of 0.2 m conditions; sea-ice spacing of 0.6 m in two rows of ice climbing impact the first row of ice, acceleration increased significantly, and the peak of the second collision is higher than the peak of the first collision; sea-ice spacing was 0.8 m after the row of ice speed decreased significantly on the second collision pier; the second impact on the bridge abutment is obviously reduced. Figure 9b is the equivalent force curve of the bridge abutment at the location of the collision, the sea ice's continuous impact on the bridge abutment when the equivalent force is obviously a peak phenomenon, the extrusion stage of the equivalent force slowly upward into the incremental trend, and with the passage of time the extrusion stress will be more than the impact stress to reach the maximum value, such as in Figure 9b 0.8 m spacing conditions. In summary, under different spacing, the ice sheet collided with the bridge abutment to form a continuous chain collision; the spacing gradually increased, resulting in a decrease in the collision force; the abutment equivalent force gradually increased, and with the passage of time the extrusion stage equivalent force is greater than the collision stage equivalent force. To sum up, the sea-ice continuous collision stage spacing was 0.6 m on the pier impact force maximum 60 kN, the impact is obvious, the extrusion stage of different spacing conditions of sea ice are extrusion obvious, and the stability of the bridge pier has a greater impact.

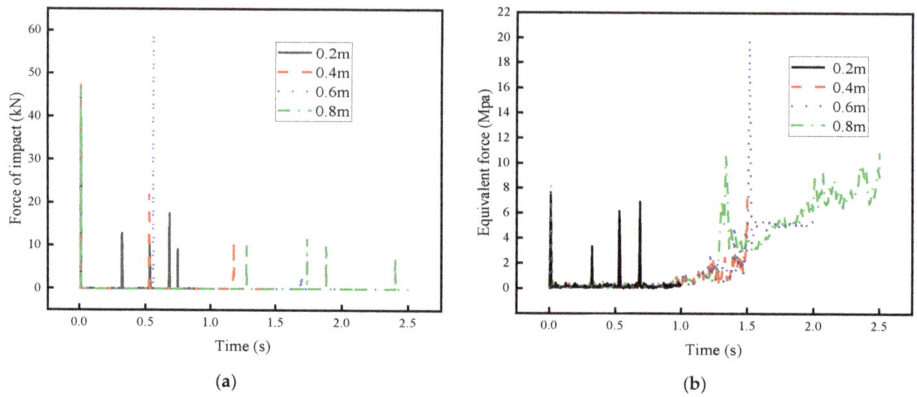

Figure 9. (**a**) Peak ice load diagram; (**b**) equivalent stress diagram.

3.2. Effect of Different Crushed Ice Masses on Bridge Abutments

Four horizontally arranged sea ice bodies of 2 m in length and 1 m/s velocity are selected to collide with the bridge pier. The sea ice thickness of 0.1 m was determined and the sea ice area changed by changing the sea ice width; the sea ice width of 0.4 m was changed by changing the sea-ice thickness of the integrated mass working conditions.

3.2.1. Area of Different Sea Ice Masses

Only the effect of the average area of different sea ice on the bridge abutment under the influence of different shapes was considered. As shown in Figure 10, the average area of different sea-ice rows was in collision with the abutment with the back row of ice and produced a continuous chain collision; sea ice began to break, accompanied by the emergence of stress concentration phenomenon, and the stress diffusion phenomenon with the increase in the average area tends to flatten out, while the maximum impact force occurs in the initial stage of collision in the subsequent stage of the extrusion collision. The abutment equivalent force is mainly distributed in the sea ice and abutments in contact with the side of the neighboring regions, namely the lower abutment connection. As the average area increases, the mass increases, the number of consecutive positive collisions decreases, the squeeze phase is advanced, and the equivalent force climb does not see an obvious pattern. As shown in Figure 11, the larger the average area, the greater the impact of continuous collision. The maximum impact force peak and the second largest impact force peak are linearly increasing, that is, under the same premise of the spacing, the acceleration of continuous collisions of sea ice will decrease step by step. The second largest impact force is manifested in the larger quality of the impact force, which shows that the effect of the mass of the sea ice impact on the bridge abutment is much larger than the impact of acceleration on the sea ice impact on the bridge abutment. The second largest impact force shows that the larger the mass, the larger the impact force. In summary, with the change of the shape of the sea ice, i.e., the average area of the sea ice increases, and the impact force on the bridge abutment increases step by step.

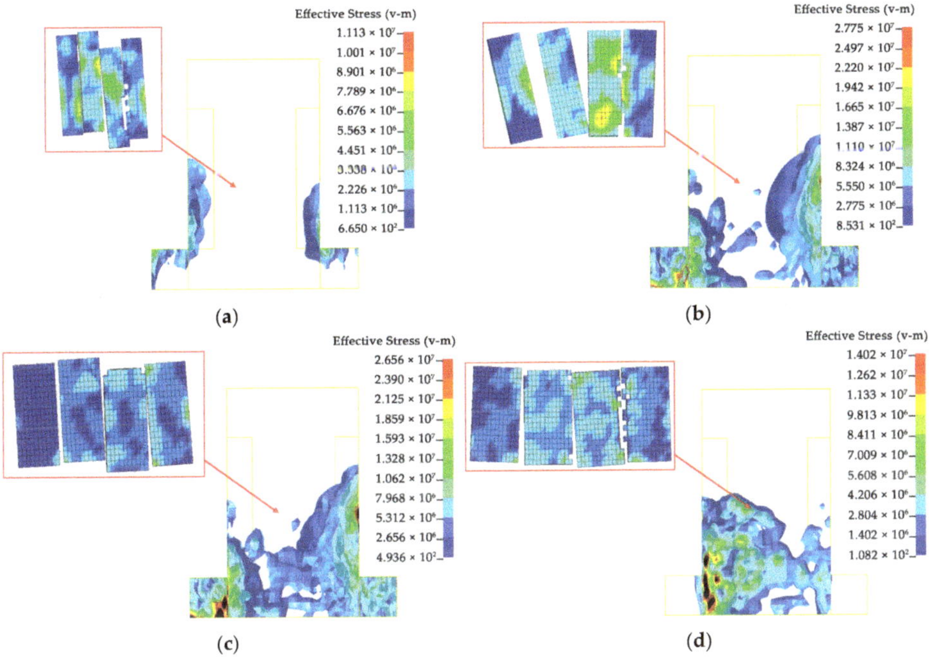

Figure 10. Maximum equivalent stress cloud map of different ice mass area: (**a**) $A_0 = 0.8$ m^2; (**b**) $A_0 = 1.2$ m^2; (**c**) $A_0 = 1.6$ m^2; (**d**) $A_0 = 2.0$ m^2.

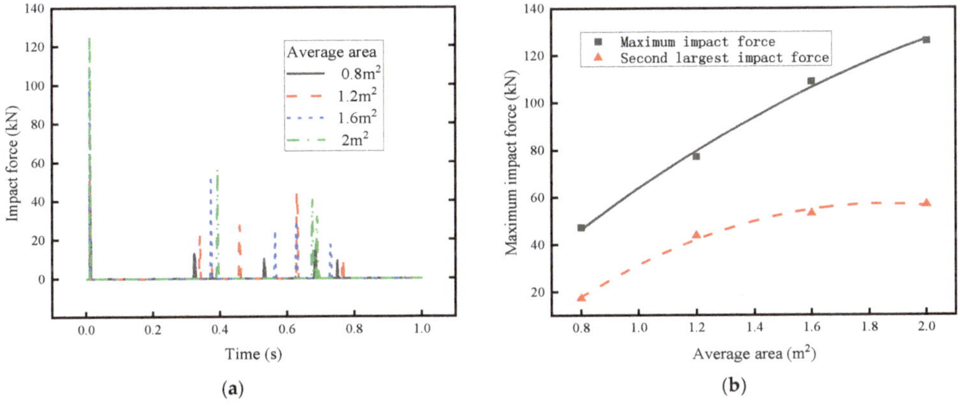

Figure 11. (**a**) Map of peak ice load with different sea ice area; (**b**) impact force curves for different ice cube areas.

3.2.2. Different Thicknesses of Sea Ice

Only the impact of different ice thickness on the bridge abutment under the influence of different shapes is considered. As shown in Figure 12, there are different sea-ice thickness ice rows in the collision with the bridge abutment with the back row of ice to produce a continuous chain collision. Sea ice was not obviously broken there is an abutment by the impact side of the stress concentration phenomenon, the equivalent force with the thickness increases step by step, the maximum impact force occurs in the initial collision stage, with different sea ice area conditions being the same as in the subsequent squeeze collision stage, the abutment equivalent force is mainly distributed in the sea ice, and the abutment contact is adjacent to the region of the side that is the lower abutment connection. The abutment contact is adjacent to the side of the region, i.e., the lower abutment connection. As the thickness increases, the mass increases, and the sea ice in many collisions occurs after the climb flip phenomenon, such as in Figure 12a which shows the thickness of 0.05 of the front side of the sea ice in the impact of the abutment after the climb. The back row of ice and the front side of the ice did not see the obvious extrusion phenomenon, such as in Figure 12d, which shows the thickness of 0.2 m of the sea ice layer last lateral flip extrusion abutment. With the increase in the thickness of the sea ice, the sea ice layer extrusion abutments became a more pronounced phenomenon. As shown in Figure 13, the greater the thickness, the greater the impact of continuous collision; the maximum impact force peak and the second largest impact force peak are linearly increasing, and the second largest impact force shows that the larger the mass, the larger the impact force. In summary, with the change of the shape of the sea ice, i.e., the thickness of the sea ice increases, the impact force on the bridge pier increases step by step.

The combination of ice mass, i.e., area and thickness, will determine the actual impact effect; with the increase in area and thickness, the impact force is linearly increasing. The large mass of ice with large area and thickness has higher kinetic energy, and the impact force on the abutment is larger, while the small mass of ice with small area and thickness is lower, but the frequent impact of ice may cause long-term fatigue on the abutment, and according to the change of the second largest impact force, we can judge that the impact of quality on the impact of ice on the abutment is much larger than the impact of acceleration on the impact of ice on the abutment. Overall, the impact of different masses of sea ice on bridge abutments will cause some damage to the bridge abutments, especially in the case of large masses, large areas, and thick ice. In order to protect the bridge structure, measures are usually taken to minimize this risk, such as setting up guards around the bridge or carrying out regular inspections and maintenance.

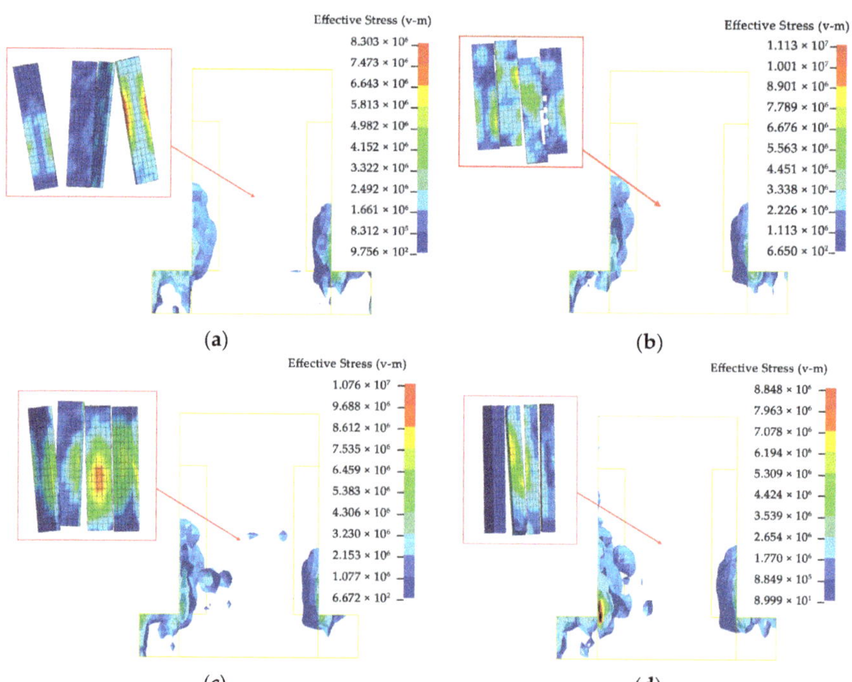

Figure 12. Maximum equivalent stress cloud map for different thickness of sea ice: (**a**) thickness = 0.05 m; (**b**) thickness = 0.1 m; (**c**) thickness = 0.15 m; (**d**) thickness = 0.20 m.

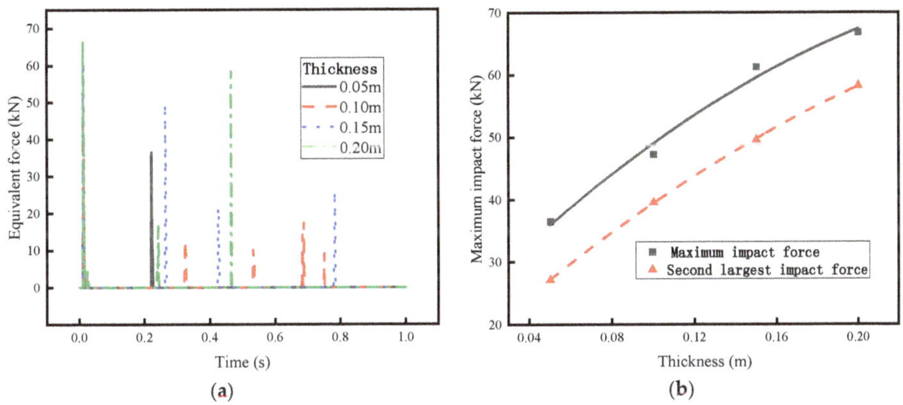

Figure 13. (**a**) Peak ice load map with different thickness of sea ice; (**b**) impact force curve of different ice thickness.

3.3. Impact of Different Shapes of Sea Ice on Bridge Abutments

Through the LS-DYNA keyword setting, the mass of sea ice, speed (1 m/s), density, density of sea-ice distribution (80%), and the design and strength of the bridge abutment are set to fixed values, and only the effect of the average area of different sea ice on the bridge abutment under the influence of different shapes is considered.

Figure 14 shows the average area of different sea-ice rows and piers after the collision of the back row of ice to produce a continuous chain collision, and sea-ice collision after breaking, accompanied by the emergence of the phenomenon of stress concentration; the stress diffusion phenomenon with the increase in the average area tends to flatten out, the maximum impact force of larger ice occurs in the first collision stage, followed by the sea ice appearing to be an obvious extension of the *y*–axis climbing phenomenon, and then into the extrusion stage. As shown in Figure 15a, in the process of collision between sea ice and bridge abutment, the equivalent force of bridge abutment is mainly distributed around the adjacent area of contact between sea ice and bridge abutment. And in the location of the collision contact between the sea ice and the abutment, the abutment is impacted by the region of the stress concentration phenomenon. As shown in Figure 15b, the average area of 1 m^2 of sea ice is due to the small average area, the mass of the sea ice, and the impact force are relatively small, and the impact on the abutment is small. It causes some impact and friction on the surface of the bridge abutment without causing serious damage. Shattered ice with an average area greater than 1.5 m^2: this size of shattered ice increases in size step by step, with higher mass and impact forces, and more pronounced impacts on the bridge abutments. Larger ice fragments may cause larger impact forces, resulting in damage to the surface of the bridge abutment. In addition, larger sea ice may also accumulate around the abutments, increasing the impact force of the water flow on the abutments, which can lead to the increased structural instability of the abutments. In summary, as the shape of the sea ice changes, i.e., the average area of the sea ice increases, the maximum impact force on the bridge abutment increases step by step.

It should be noted that the specific impact also depends on the velocity and density of the sea ice; if the velocity of the sea ice is higher or the density is larger, even if the average area is smaller, it may still cause a larger impact force on the bridge pier. Therefore, when assessing the impact of different average areas of sea ice on the bridge pier, it is necessary to comprehensively consider the velocity, density, and other factors.

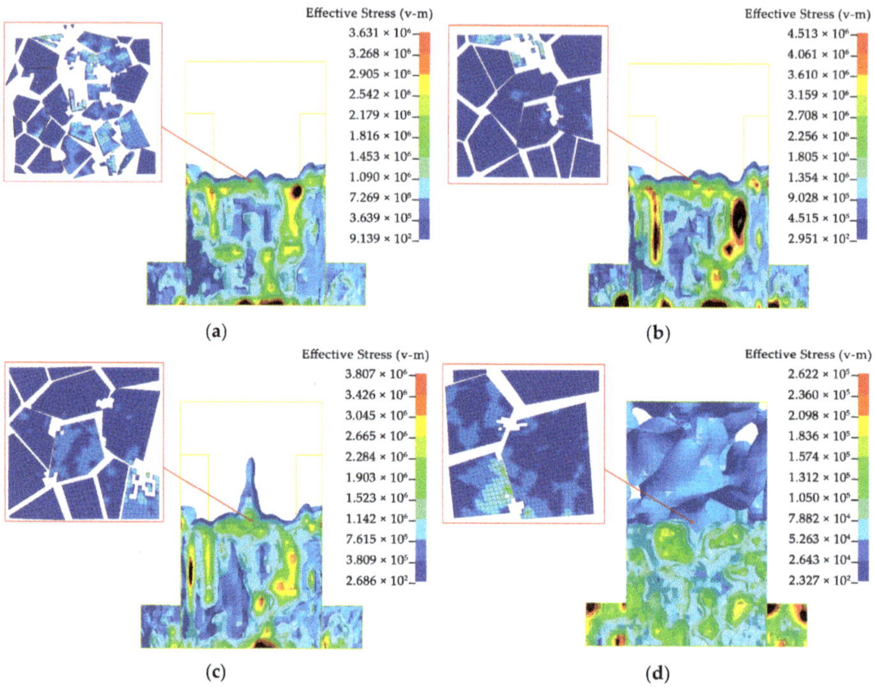

Figure 14. Equivalent stress nephogram of compressive pier with different shapes of crushed ice: (**a**) $A_0 = 1.0\ m^2$; (**b**) $A_0 = 1.5\ m^2$; (**c**) $A_0 = 2.0\ m^2$; (**d**) $A_0 = 2.5\ m^2$.

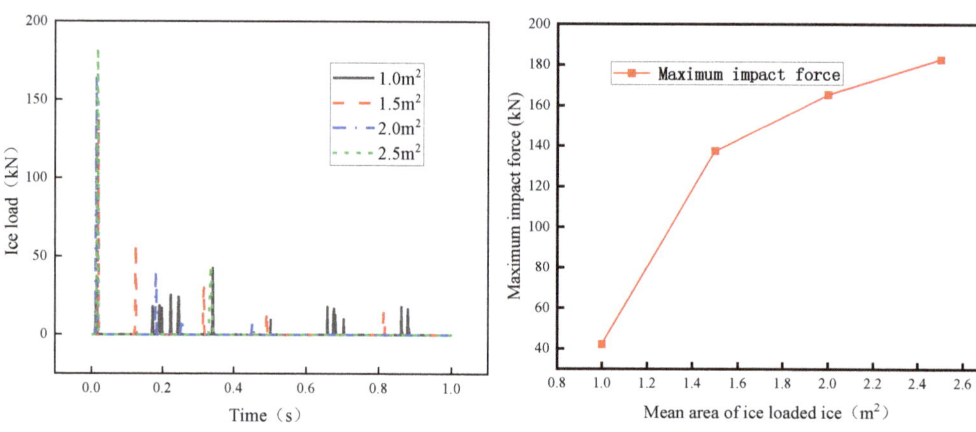

Figure 15. (**a**) Peak ice load of different shapes of sea ice; (**b**) maximum impact force curves of different shapes of sea ice.

4. Maximum Impact Force Verification Experiment

As shown in Figure 16, in order to verify the impact effect of small- and medium-sized drift ice on bridge piers, the experimental steps are as follows:

- Experimental preparation

The use of rectangular water tanks to simulate the water medium and the pier lining selection of C30 concrete slabs installed in the water tank to simulate the real bridge pier force situation.

- Equipment commissioning

Zeroing of all instruments to ensure the accuracy of the measurement of each instrument. Check whether each instrument works properly, including force sensors, speed sensors, and data recording equipment, to ensure the reliability of the data in the experimental process.

- Sea ice body preparation

Fix the crushed ice on the propeller to ensure the stability of the ice body. Adjust the position of the sea ice so that it is precisely aligned with the impact position of the bridge pier.

- Experimental operation

Start the power unit and set the motion parameters of the ice body according to the preset design ice speed and stroke so that the ice body accurately impacts the bridge pier.

- Data acquisition and analysis

Record the maximum impact force peak data of the ice body impacting the bridge pier. The experimental data were compared with the simulation results to analyze the effect of continuous collision of four different areas (0.8 m^2, 1.2 m^2, 1.6 m^2, and 2 m^2) of fragmented ice.

Sixteen groups of four sea ice bodies with widths of 0.4, 0.6, 0.8, and 1.0 m, length of 2 m, spacing of 0.2 m, and speed of 1 m/s were selected for experimental study. Through these steps, the experiment aims to evaluate the impact force of sea ice on bridge abutments at different areas in order to provide data support for actual bridge design and maintenance.

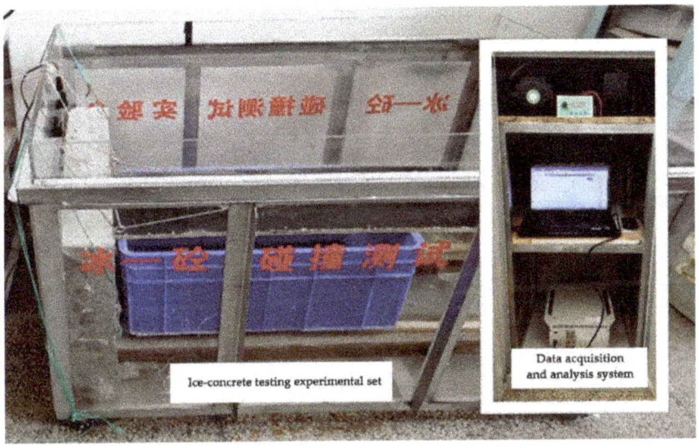

Figure 16. Ice–concrete collision test bench.

Figure 17a shows the magnitude of equivalent force when different average sea ice areas impact the bridge pier. It can be found that the irregularity of the peak equivalent force, the maximum equivalent force in the continuous collision stage, is under the condition of 1.6 m^2 of sea-ice area, and the maximum impact equivalent force does not occur in the first impact, which indicates that the maximum damage to the bridge pier by the sea ice under the effect of the continuous positive collision is brought about by the second impact, and there is no obvious regularity with the sea-ice area.

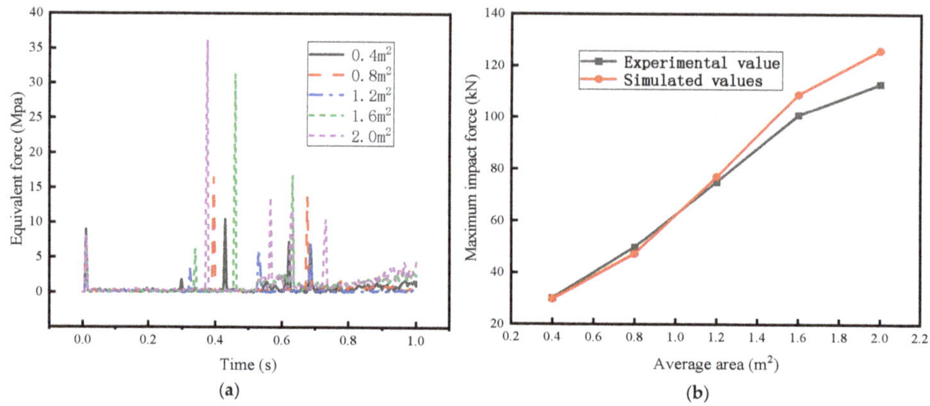

Figure 17. (**a**) Equivalent stress test data diagram; (**b**) comparison diagram between simulated data and experimental data.

According to the results of comparison between the simulated and experimental values under the four types of area conditions presented in Figure 17b, the results of the impact of different sea-ice areas show a linear trend in general, and with the increase in sea-ice area, the impact force shows an increasing trend, the simulated values of 0.4 m^2 and 0.8 m^2 sea ice are smaller than the experimental values, and according to the on-site observation, it is found that the larger the area of the sea ice, the more pronounced is the 'water cushion effect', and thus the smaller the area of sea ice the higher is the fit of the experimental values with the simulated values, but the error is still within the control. According to the field observation, the larger the sea ice area is, the more obvious the 'water cushion effect' is, and the larger the maximum impact force of the simulated value is, thus, the smaller the sea ice area is, the higher the fit between the experimental value and the simulated value is, and vice versa, thus the lower the fit is, but the error is still within the controllable range. It is concluded that under different average area conditions, the maximum impact force of sea ice hitting the contact area of bridge abutment shows a similar trend with the simulated value. The maximum error is 10.3%, the minimum error is 1.3%, and the average error is 6.38%. By carefully comparing the experimental data with the simulation results, good consistency was found between them, which further proves that the ice material model parameters used for numerical simulation, as well as the simulation results derived from the ALE algorithm, have a certain degree of reliability. These research results contribute to a more in-depth understanding of the effects of small- and medium-sized sea ice on bridge piers and provide important guidance for related engineering practices.

5. Conclusions

Continuous sea ice collision abutment dynamic response research, the application of finite element analysis software ANSYS/LS-DYNA and programming software MATLAB, based on different spacing of sea ice, different average area of sea ice, and sea ice average area conditions simulation of sea ice and railway bridge abutment impact role, through the simulation of the data generated by the analysis of the following conclusions:

(1) Consider the case of different spacing of sea ice: spacing 0.2 m, continuous chain collision is obvious; spacing 0.4 m, the peak value of the secondary collision increases significantly; spacing 0.6 m, sea ice climbing obviously, the secondary peak value is greater than the primary peak value; spacing 0.8 m, is the side of the length of the side of the 0.4 m twice, the subsequent ice extrusion overlap probability increases. Continuous positive collision ice spacing increases, the collision force decreases, and the probability of ice climbing and overturning increases. The stability of the bridge

abutment is most affected in the extrusion phase, and the impact of the subsequent extrusion increases step by step as the sea ice spacing increases and the collision force decreases, and the probability of sea ice climbing and overturning increases during the successive collision and collision phases. In the subsequent study, different bridge ice contact areas can be investigated under the influence of ice spacing.

(2) With the increase of sea ice mass, i.e., average area and thickness, the impact force on the bridge abutment increases step by step. In the initial collision stage, the peak impact force increases linearly with the increase of sea ice area and thickness, and the maximum impact force is 125.94 kN and 66.86 kN, respectively. In the crush collision stage, the distribution of the equivalent force of the bridge abutment is mainly on the side of the contact neighboring area. The climbing and overturning phenomenon of sea ice is more obvious with the increase of thickness; the stability of the bridge abutment is mainly in the first collision stage; there is a risk of impact instability; the second peak of the successive positive collision is higher; and the effect of mass on the impact of sea ice on the bridge abutment is more significant than the acceleration, so the subsequent research process needs to be set up for the stage of the collision of different shapes of collision avoidance bodies and study the advantages and disadvantages of the study.

(3) In the case of collision of different shapes of sea ice, the maximum value of impact force appears in the first collision stage as the average area of sea ice increases, and the subsequent sea ice shows an obvious climbing phenomenon and enters the extrusion stage. The average area of sea ice is larger than 1.5 m^2 with higher quality, and the maximum impact force is 182.8 kN, which is obvious to the bridge abutment. With the increase of the average area of sea ice, the maximum impact force on the bridge abutment increases step by step, and the experimental and simulated values of the maximum impact force of sea ice hitting the contact area of the bridge abutment show a similar trend. The larger sea ice in the irregular sea-ice layer causes a larger impact force, and the phenomenon of accumulation and climbing occurs, squeezing around the bridge abutment, increasing the impact force of the water flow on the bridge abutment, leading to an increase in the instability of the bridge abutment structure. In the subsequent study, only sea ice with an average area of 1.5 m^2 is studied, and different two-dimensional cutting surfaces are investigated and analyzed.

(4) Based on the design of the experimental program to study only the impact of sea ice area on the concrete structure of the bridge pier, under different area conditions, the maximum impact force of sea ice impact on the contact area of the bridge pier experimental values and simulated values show an increasing trend. The maximum error is 10.3%, the minimum error is 2.7%, and the average error is 7.65%. There is good consistency between the experimental value and the simulated value, which proves that the parameters of the ice material model used for numerical simulation and the simulation results of the area data based on the Voronoi cutting algorithm have a certain degree of reliability.

Author Contributions: L.G.: methodology, software, investigation, data curation, funding acquisition, writing—original draft, visualization. Y.C.: methodology, supervision, validation, resources, writing—review and editing. Y.D.: conceptualization, methodology, supervision, validation, resources, writing—review and editing. L.Q.: methodology, supervision, resources, writing—review and editing. X.Z.: methodology, software, supervision, writing—review and editing. All authors have read and agreed to the published version of the manuscript.

Funding: This study was the financially supported by the National Science Foundation (72261024, 51969011) of China; Gansu Provincial Science and Technology Programme Grant (23ZDFA002), Gansu Provincial Key Talent Project (2023RCXM27); Gansu Provincial Department of Education Higher Education Industry Support Programme Project (2022CYZC-32); Gansu Provincial Department of Transportation and Communication Science and Technology Project (2022-27).

Data Availability Statement: All data included in this study are available upon request by contacting the corresponding author.

Conflicts of Interest: The authors declare no conflicts of interest.

References

1. Song, W.D.; Ning, J.U. Critical force of ice on marine structures. *Glacial Permafr.* **2003**, *03*, 351–354.
2. Qu, Y. *Random Ice Load Analysis of Marine Structures Based on Field Experiments*; Dalian University of Technology: Dalian, China, 2006.
3. Lemström, I.; Polojärvi, A.; Tuhkuri, J. Modelscale tests on ice-structure interaction in shallow water: Global ice loads and the ice loading process. *Mar. Struct.* **2022**, *81*, 103106. [CrossRef]
4. Metrikin, I.; Kerkent, S.; Jochmann, P.; Løset, S. Experimental and numerical investigation of dynamic positioning in level ice. *J. Offshore Mech. Arct. Eng.* **2015**, *137*, 031501. [CrossRef]
5. Gong, L.; Dong, Z.Q.; Yang, T.T. Dynamic response analysis of ice-pier collision in Water. *J. Vib. Shock* **2024**, *43*, 72–82.
6. Zhu, L.; Cai, W.; Chen, M.; Tian, Y.; Bi, L. Experimental and numerical analyses of elastic-plastic responses of ship plates under ice floe impacts. *Ocean Eng.* **2020**, *218*, 108174. [CrossRef]
7. Zhang, J.M.; Yuan, Z.G.; Wu, Y.P. Analysis on the collision couple between drift ice and long-span prestressed concrete T-rigid frame bridge pier. In *Key Engineering Materials*; Trans Tech Publications Ltd.: Wollerau, Switzerland, 2015; Volume 648, pp. 17–24.
8. Wu, T.; Qiu, W. Dynamic analyses of pile-supported bridges including soil-structure interaction under stochastic ice loads. *Soil Dyn. Earthq. Eng.* **2020**, *128*, 105879. [CrossRef]
9. Xia, C.Y.; Lei, J.Q.; Zhang, N. Analysis of Dynamic Response and Train Operation Safety of high-speed Railway Bridge under Impact Load. *Eng. Mech.* **2012**, *29*, 101–107+120.
10. Li, P.; Li, Z.; Han, Z.; Zhu, S.; Zhai, W.; Lou, H. Running safety evaluation of high-speed train subject to the impact of floating ice collision on bridge piers. *Proc. Inst. Mech. Eng. Part F J. Rail Rapid Transit* **2022**, *236*, 220–233. [CrossRef]
11. Wang, B.; Yu, H.C.; Basu, R. Ship and ice collision modeling and strength evaluation of LNG ship structure. In Proceedings of the ASME 2008 27th International Conference on Offshore Mechanics and Arctic Engineering: American Society of Mechanical Engineers, Estoril, Portugal, 15–20 June 2008; pp. 911–918.
12. Kawano, Y.; Ohashi, T. Numerical study on c-axis orientations of sea ice surface grown under calm sea conditions using a particle method and Voronoi dynamics. *Cold Reg. Sci. Technol.* **2015**, *112*, 29–38. [CrossRef]
13. Liu, L.; Long, X.; Ji, S.Y. Discrete element method based on extended polyhedron and its ice load calculation on round pile. *Chin. J. Mech. Mech.* **2015**, *47*, 1046–1057.
14. Guo, Y.K.; Han, X.Y.; Meng, W.Y. Finite element analysis of pier safety under flow ice Collision. *J. North China Inst. Water Resour. Hydropower* **2013**, *34*, 16–18.
15. Yang, L.; Ma, J. Numerical Simulation analysis of collision between ship and offshore platform under ice medium. *China Offshore Platf.* **2008**, *23*, 29–33.
16. Gong, L.; Dong, Z.; Jin, C.; Jia, Z.; Yang, T. Flow–Solid Coupling Analysis of Ice–Concrete Collision Nonlinear Problems in the Yellow River Basin. *Water* **2023**, *15*, 643. [CrossRef]
17. Hu, Y.D.; Guo, W.N.; Zhang, Y. Analysis on the evolution of transit in the Ningzhou-Mongolia Reach of the Yellow River from 2011 to 2020. *Yellow River People* **2023**, *45*, 52–57.
18. Wang, E.L.; Hu, S.B.; Han, H.W. Study on flow density of Kaijiang River in Heilongjiang Province based on UAVS low altitude remote sensing and OTSU algorithm. *J. Hydraul. Eng.* **2019**, *53*, 68–77.
19. Yu, Y.W. *Numerical Simulation of Ship-Ice Collision in Ice Breaking Area*; Dalian University of Technology: Dalian, China, 2022.
20. Chen, Y.F. *Study on the Evolution Mechanism of ICE Transport and Its Application in Two-Phase Outwash Flow Channel*; China Institute of Water Resources and Hydropower Research: Beijing, China, 2019.
21. Zhu, H.R.; Ji, S.Y.; Liu, L. Structure and discrete element analysis of ice fragmentation region based on Voronoi cutting algorithm. *Chin. J. Comput. Mech.* **2019**, *36*, 454–463. (In Chinese)

Disclaimer/Publisher's Note: The statements, opinions and data contained in all publications are solely those of the individual author(s) and contributor(s) and not of MDPI and/or the editor(s). MDPI and/or the editor(s) disclaim responsibility for any injury to people or property resulting from any ideas, methods, instructions or products referred to in the content.

Journal of
Marine Science and Engineering

MDPI

Article

On the Static Stability and Seakeeping Performance of a Submerged Floating Tunnel Module in Wet Tow

Ikjae Lee [1,*], Chungkuk Jin [2,*], Sung-Jae Kim [3] and Moohyun Kim [1]

1 Department of Ocean Engineering, Texas A&M University, College Station, TX 77845, USA
2 Department of Ocean Engineering and Marine Sciences, Florida Institute of Technology, Melbourne, FL 32901, USA
3 Fisheries Engineering Division, National Institute of Fisheries Science, Busan 46083, Republic of Korea
* Correspondence: ijlee@tamu.edu (I.L.); cjin@fit.edu (C.J.)

Abstract: A case study is conducted for a submerged floating tunnel module (SFTM) in wet tow conditions. Inspired by the successful wet tow operations of spar platforms, a wet tow scenario is examined where a tunnel module, floating horizontally with a half-diameter draft, is towed by tugboats using towlines. To evaluate the static stability of the SFTM during wet tow, numerical static offset tests are performed at varying tow speeds to determine the equivalent system stiffness. These static offset tests consider surge, sway, roll, and yaw motions. Statistical analyses are subsequently performed based on the encounter-frequency approximation with varying equivalent stiffnesses. The most probable extreme motion analysis for 3 h under sea state 4 ($H_S = 2.44$ m and $T_P = 8.1$ s) shows that the beam sea condition causes the largest heave (0.6 m), and the stern sea (30 deg.) leads to the largest yaw response (0.85 deg.), which is likely to cause an instantaneous decrease in towing stability.

Keywords: submerged floating tunnel; wet tow; stability; seakeeping; encounter-frequency

1. Introduction

A submerged floating tunnel (SFT) is an innovative underwater structure designed for deep-water and long-distance crossings [1]. In recent years, the SFT system has been extensively studied (e.g., [2–12]). Most studies are in the preliminary stage and restricted to the operational phase. Research relevant to the installation and transportation phases is also critical; however, there has been only limited work on this topic. For example, Jin and Kim [6] have carried out some case studies on the installation and transportation stages of a submerged floating tunnel module (SFTM), but there are limited details and studies focused on transportation. Lee et al. [13] conducted a preliminary study on the static stability of an SFTM in the wet tow condition. It should be noted that SFTM refers to a segmented module that is later assembled with other modules to form a single SFT at the installation site. This study investigates horizontal wet tow scenarios for an SFTM using tugboats and towlines, motivated by successful wet tow operations of spar platforms. [14,15].

Since the static problem must be pre-solved before implementing dynamic problems, accurate modeling and thorough analysis of the static problem are important. On top of the well-defined static problem, dynamic analyses can be performed in both the time and frequency domains. Especially for mooring or towing problems, the initial stress effect, i.e., pre-tensions on the towed or moored structure, should be carefully evaluated.

In this study, numerical offset tests are performed for 4-degrees-of-freedom (DOF) SFTM motions that include surge, sway, roll, and yaw. These tests are conducted at varying

https://doi.org/10.3390/jmse13010077

tow speeds to obtain the equivalent system stiffness of the coupled SFTM system, allowing the evaluation of static stability. The equivalent stiffness can be determined through various methods, either analytically or numerically [16,17]. However, previous studies are mostly restricted to stationary moored structures tested for operation or installation scenarios. To the authors' knowledge, there has rarely been research on towing problems for SFTMs, where it is typical to perform coupled time-domain simulations without thorough static or steady-state analyses.

In the coupled floater-mooring analysis, two representative approaches are the lumped-mass and finite element rod theories [18–20]. However, in this study, we employed a continuum-based nonlinear finite element model [21,22]. A series of nonlinear static equations for the SFTM, towlines, and constraints are derived. Specifically, the total Lagrangian formulation for the towlines and rigid SFTM motions is fully coupled based on the method of Lagrange multipliers. A sequential procedure for the numerical static offset tests is proposed and performed for several wet tow conditions. Pre-tension effects due to the different tow speeds are first checked. Based on each tow speed's equilibrium state, offset tests are performed for the 4-DOF motions. Since the roll restoring moment becomes nearly zero when the geometric center coincides with the mass center, equivalent metacentric heights for the coupled wet tow system are obtained. The Munk moment effect, which is a destabilizing moment acting to turn the SFTM perpendicular to the incoming flow, is also investigated in the yaw offset test [23].

The paper consists of five sections. In Section 2, the mathematical modeling of the coupled SFTM system and the statistical analysis procedure are described. In Section 3, some configurations and parameters related to the SFTM's stability are discussed, and numerical results and key discussions of the offset tests are presented in Section 4, followed by conclusions in Section 5.

2. Mathematical Implementation

Based on a steadily moving coordinate system, a dynamic problem can be described as a static/steady problem. The hydrodynamic and aerodynamic loadings are evaluated with respect to the steadily moving reference frame, i.e., relative wind and current loadings are used. With reference to this frame, the motion of the tugs is assumed to occur at a constant speed, and thus they are modeled as fixed points. A sketch of the wet tow configuration is illustrated in Figure 1.

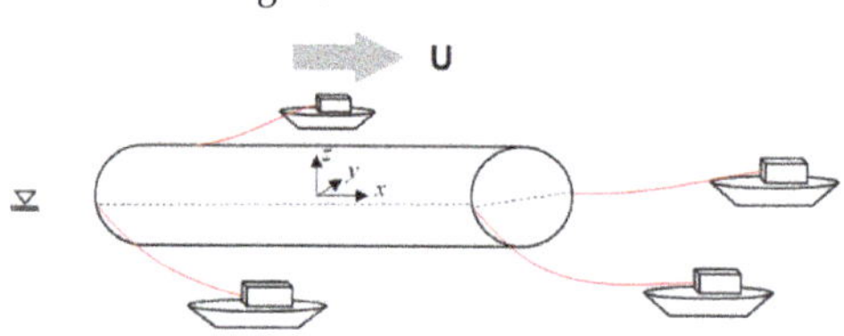

Figure 1. Wet tow configurations for the SFTM study with four tugboats and towing lines.

The SFTM motion is assumed to be a rigid body submerged with a half-diameter draft, and the towlines are modeled with a geometrically nonlinear beam finite element model. Static governing equations for the SFTM and towlines are separately derived and coupled using the method of Lagrange multipliers, which avoids direct calculations of stress resultants and fairlead modeling. The finite rotation effects of the towlines and the SFTM are considered based on Rodrigues' formula [22]. A coupled system of static equations is derived by taking the first variation of a constrained energy functional as

$$\delta W_{SFTM} + \delta W_{CABLE} + \delta \Pi = 0 \quad (1)$$

where the first and second terms on the left-hand side are the potential energies of the SFTM and the towlines, respectively, and the third term is a constraint of two position vectors at the fairlead positions. Each term will be discussed in the following sections.

2.1. Statics of the SFTM

The virtual work done by the rigid SFTM can be represented as

$$\delta W_{SFTM} = \delta \hat{\mathbf{q}}^{T}\, {}^{t+\Delta t}\hat{\mathbf{f}}_{SFTM}^{(i)} \tag{2}$$

where $\hat{\mathbf{q}}^{T} = [\xi_1, \xi_2, \xi_3, \theta_1, \theta_2, \theta_3]$, which is the 6-DOF displacement vector (surge, sway, heave, roll, pitch, and yaw) and $\hat{\mathbf{f}}_{SFTM}^{(i)} = \hat{\mathbf{f}}_{HS}^{(i)} + \hat{\mathbf{f}}_{D}^{(i)} + \hat{\mathbf{f}}_{I}^{(i)}$; here, subscripts HS, D, and I denote hydrostatic, drag, and inertia, respectively; and superscript (i) denotes the i-th iteration. Each term is given by

$${}^{t+\Delta t}\hat{\mathbf{f}}_{HS}^{(i)} = -\rho g\left[0, 0, A_{WP}\,{}^{t+\Delta t}\xi_3^{(i)}, \nabla \overline{GM}_T \sin {}^{t+\Delta t}\theta_1^{(i)}, \nabla \overline{GM}_L \sin {}^{t+\Delta t}\theta_2^{(i)}, 0\right]^T \tag{3}$$

$${}^{t+\Delta t}\hat{\mathbf{f}}_{D}^{(i)} = 0.5\rho \left\{ \begin{array}{c} C_{D_\tau} A_{f,\tau}\,{}^{t+\Delta t}\mathbf{U}_\tau^{(i)} \left|{}^{t+\Delta t}\mathbf{U}_\tau^{(i)}\right| + C_{D_n} A_{f,n}\,{}^{t+\Delta t}\mathbf{U}_n^{(i)} \left|{}^{t+\Delta t}\mathbf{U}_n^{(i)}\right| \\ \mathbf{r}^{(i)} \times \left(C_{D_\tau} A_{f,\tau}\,{}^{t+\Delta t}\mathbf{U}_\tau^{(i)} \left|{}^{t+\Delta t}\mathbf{U}_\tau^{(i)}\right| + C_{D_n} A_{f,n}\,{}^{t+\Delta t}\mathbf{U}_n^{(i)} \left|{}^{t+\Delta t}\mathbf{U}_n^{(i)}\right| \right) \end{array} \right\} \tag{4}$$

$${}^{t+\Delta t}\hat{\mathbf{f}}_{I}^{(i)} = \left[0, 0, 0, 0, 0, M_{MUNK}^{(i)}\right]^T \tag{5}$$

where A_{WP} is the waterplane area (assumed to be constant, for simplicity); ∇ is the displaced volume; $\overline{GM}_T$ and $\overline{GM}_L$ are the transverse and longitudinal metacentric heights, respectively; C_{D_τ} and C_{D_n} are the drag coefficients in tangential and normal directions; and $A_{f,\tau}$ and $A_{f,n}$ are the frontal areas of tangential and normal directions. Note that friction resistance is neglected because the drag coefficients employed here are measured in a fully submerged state and are generally greater than the surface-piercing body case. ${}^{t+\Delta t}\mathbf{U}_\tau$ and ${}^{t+\Delta t}\mathbf{U}_n$ are the tangential and normal directional components of incoming flow, i.e., ${}^{t+\Delta t}\mathbf{U}_\tau = \left(\mathbf{U} \cdot {}^{t+\Delta t}\boldsymbol{\tau}\right){}^{t+\Delta t}\boldsymbol{\tau}$ and ${}^{t+\Delta t}\mathbf{U}_n = \mathbf{U} - {}^{t+\Delta t}\mathbf{U}_\tau$ with $\mathbf{U} = -U\mathbf{i}$; here, U can be either relative wind speed in the air or tow speed in the water. M_{MUNK} is the so-called Munk moment, which is discussed later. All the forces at the i-th configuration can be linearized into

$${}_{0}^{t+\Delta t}\hat{\mathbf{f}}^{(i)} = {}_{0}^{t+\Delta t}\hat{\mathbf{f}}^{(i-1)} - {}_{0}^{t+\Delta t}\mathbf{K}^{(i-1)}\hat{\mathbf{q}}^{(i)} \tag{6}$$

where ${}_{0}^{t+\Delta t}\hat{\mathbf{f}}^{(i-1)}$ and ${}_{0}^{t+\Delta t}\mathbf{K}^{(i-1)}$ are the force vector and restoring stiffness matrix at $(i-1)th$ iteration, and the unknown displacement vector, $\hat{\mathbf{q}}^{(i)}$, will be calculated by solving a coupled algebraic equation. In the above, the left super- and subscripts denote deformed $(t+\Delta t)$ and undeformed (0) configurations, and it follows Bathe's notation [24].

2.2. Statics of Towlines

The virtual work done by towlines is given by

$$\delta W_{CABLE} = \delta \hat{\mathbf{u}}^{T}\, {}^{t+\Delta t}\hat{\mathbf{f}}_{CABLE}^{(i)} \tag{7}$$

which is decomposed into the internal and external works as $\hat{\mathbf{f}}_{CABLE}^{(i)} = \hat{\mathbf{f}}_{IN}^{(i)} + \hat{\mathbf{f}}_{EX}^{(i)}$. The external work is subdivided into $\hat{\mathbf{f}}_{EX}^{(i)} = \hat{\mathbf{f}}_{B}^{(i)} + \hat{\mathbf{f}}_{S}^{(i)} + \hat{\mathbf{f}}_{L}^{(i)}$, in which $\hat{\mathbf{f}}_{B}^{(i)}$ is the body force; $\hat{\mathbf{f}}_{S}^{(i)}$ is the surface force; and $\hat{\mathbf{f}}_{L}^{(i)}$ is the line force, respectively.

The internal force is derived based on the total Lagrangian formulation as

$$ {}^{t+\Delta t}\hat{\mathbf{f}}_{IN}^{(i)} = {}_0^{t+\Delta t}\hat{\mathbf{f}}_{IN}^{(i-1)} - {}_0^{t+\Delta t}\mathbf{K}_{IN}^{(i-1)}\hat{\mathbf{u}}^{(i)} \tag{8} $$

where

$$ {}_0^{t+\Delta t}\mathbf{K}_{IN}^{(i-1)}\hat{\mathbf{u}}^{(i)} = \left(\sum_{\updownarrow}\int_{{}^0V(\updownarrow)} {}_0^{t+\Delta t}\tilde{\mathbf{B}}_{L0}^{T}\tilde{\mathbf{C}}_0^{t+\Delta t}\tilde{\mathbf{B}}_L dV^{(\updownarrow)}\right)\hat{\mathbf{u}}^{(i)} + \left(\sum_{\updownarrow}\int_{{}^0V(\updownarrow)} {}_0^{t+\Delta t}\tilde{\mathbf{S}}_0^{t+\Delta t}\tilde{\mathbf{B}}_{NL} dV^{(\updownarrow)}\right)\hat{\mathbf{u}} \tag{9} $$

$$ {}_0^{t+\Delta t}\hat{\mathbf{f}}_{IN}^{(i-1)} = \left(\sum_{\updownarrow}\int_{{}^0V(\updownarrow)} {}_0\tilde{\mathbf{B}}_{L0}^{T\,t+\Delta t}\tilde{\mathbf{S}} dV^{(\updownarrow)}\right) \tag{10} $$

In the above, $\mathbf{B}_L$ and $\mathbf{B}_{NL}$ denote the linear and nonlinear strain-displacement compatibility 2nd rank tensors, where $\mathbf{B}_{NL}$ contains both the nonlinear part of the Green-Lagrange strain and finite rotations; ${}_0\mathbf{C}$ is the 4th rank constitutive tensor; ${}_0^{t+\Delta t}\mathbf{S}$ is the Second Piola–Kirchhoff stress tensor; $\updownarrow$ denotes the number of line elements; and the tildes above indicate the covariant and contravariant components, as it is convenient for curved geometry to be described in a natural coordinate system. A continuum-based geometrically nonlinear beam finite element model is employed for the present study, and more details can be found in Ref. [22].

On the other hand, the external works are performed by the cable's net buoyancy and hydrodynamic drag forces. Similar to Equation (8), all the external force terms are decomposed into restoring and force terms; thus, we can write

$$ {}^{t+\Delta t}\hat{\mathbf{f}}_{EX}^{(i)} = {}_0^{t+\Delta t}\hat{\mathbf{f}}_{EX}^{(i-1)} - {}_0^{t+\Delta t}\mathbf{K}_{EX}^{(i-1)}\hat{\mathbf{u}}^{(i)} \tag{11} $$

where the forcing vector and stiffness matrix can be obtained by linearizing the external forces from the abovementioned external loadings.

2.3. Coupled Static Equilibrium Equation and Iterative Solution

To couple prescribed two static equations, the method of Lagrange multipliers is introduced as

$$ \Pi = \sum_j \hat{\boldsymbol{\lambda}}_j^{T}\left({}^{t+\Delta t}\mathbf{x}_{CABLE,E_j}^{(i)} - {}^{t+\Delta t}\mathbf{x}_{SFT,E_j}^{(i)}\right) \tag{12} $$

where j denotes the order of towlines, and ${}^{t+\Delta t}\mathbf{x}_{CABLE,E_j}$ and ${}^{t+\Delta t}\mathbf{x}_{SFTM,E_j}$ are the end points of the towlines and the SFTM. Taking the first variation and linearization yields

$$ \delta\Pi = \delta\hat{\boldsymbol{\lambda}}^{T}\left({}_0^{t+\Delta t}\mathbf{K}_{\lambda-1}^{(i-1)}\hat{\mathbf{u}}^{(i)} + {}_0^{t+\Delta t}\mathbf{K}_{\lambda-2}^{(i-1)}\hat{\mathbf{q}}^{(i)} - {}_0^{t+\Delta t}\hat{\mathbf{f}}_{\lambda}^{(i-1)}\right) + \left(\delta\hat{\mathbf{u}}^{T}\,{}_0^{t+\Delta t}\mathbf{K}_{\lambda-1}^{(i-1)T} + \delta\hat{\mathbf{q}}^{T}\,{}_0^{t+\Delta t}\mathbf{K}_{\lambda-2}^{(i-1)T}\right)\hat{\boldsymbol{\lambda}}^{(i)} \tag{13} $$

which directly couples the two equations with kinematic conditions, and the stiffness matrices and forcing vectors are obtained by linearizing Equation (12). By combining the prescribed three equations (SFTM, towlines, and constraint) in a matrix form, we can construct a coupled algebraic equation:

$$\begin{bmatrix} {}^{t+\Delta t}_{0}\mathbf{K}^{(i-1)}_{CABLE} & \mathbf{0} & {}^{t+\Delta t}_{0}\mathbf{K}^{(i-1)T}_{\lambda-1} \\ \mathbf{0} & {}^{t+\Delta t}_{0}\mathbf{K}^{(i-1)}_{SFTM} & {}^{t+\Delta t}_{0}\mathbf{K}^{(i-1)T}_{\lambda-2} \\ {}^{t+\Delta t}_{0}\mathbf{K}^{(i-1)}_{\lambda-1} & {}^{t+\Delta t}_{0}\mathbf{K}^{(i-1)}_{\lambda-2} & \mathbf{0} \end{bmatrix} \begin{Bmatrix} \hat{\mathbf{u}}^{(i)} \\ \hat{\mathbf{q}}^{(i)} \\ \hat{\boldsymbol{\lambda}}^{(i)} \end{Bmatrix} = \begin{Bmatrix} {}^{t+\Delta t}_{0}\mathbf{f}^{(i-1)}_{CABLE} \\ {}^{t+\Delta t}_{0}\mathbf{f}^{(i-1)}_{SFTM} \\ {}^{t+\Delta t}_{0}\mathbf{f}^{(i-1)}_{\lambda} \end{Bmatrix} \tag{14}$$

Let ${}^{t+\Delta t}_{0}\mathbf{A}^{(i-1)}$, $\hat{\mathbf{y}}^{(i)}$, and ${}^{t+\Delta t}_{0}\mathbf{b}^{(i-1)}$ be coupled stiffness, unknown, and force vectors, respectively. Then, error tolerance, ε_d, of the iteration is calculated by

$$\frac{\left|\hat{\mathbf{y}}^{(i)}\right|}{\left|\mathbf{y}^{(i)}\right|} \leq \varepsilon_d, \text{ with } \mathbf{y}^{(i)} = \mathbf{y}^{(i-1)} + \hat{\mathbf{y}}^{(i)} \tag{15}$$

where the error tolerance is set at $\varepsilon_d = 10^{-4}$ for the present study.

2.4. Incremental Updates of Configuration with Finite Rotation Effects

As the iteration is repeated, the position vector is updated. As both the SFTM and the towlines contain rotational degrees of freedom, they need special attention due to geometric nonlinearity. The Rodrigues formula is employed in the present study to account for the finite rotation effects in updating the position vector and directors, ${}^{t+\Delta t}\mathbf{V}_k$:

$$\begin{aligned} {}^{t+\Delta t}\mathbf{x}^{(i)} &= {}^{t+\Delta t}\mathbf{x}^{(i)}_{C.R.} + \mathbf{Q}\left(\hat{\boldsymbol{\theta}}^{(i)}\right) \cdot \left({}^{t+\Delta t}\mathbf{x}^{(i-1)} - {}^{t+\Delta t}\mathbf{x}^{(i-1)}_{C.R.}\right) \text{ and} \\ {}^{t+\Delta t}\mathbf{V}^{(i)}_k &= \mathbf{Q}\left(\hat{\boldsymbol{\alpha}}^{(i)}\right) \cdot {}^{t+\Delta t}\mathbf{V}^{(i-1)}_k \quad (k = 1,2,3) \end{aligned} \tag{16}$$

where $\hat{\boldsymbol{\theta}}$ and $\hat{\boldsymbol{\alpha}}$ are rotational degrees of freedom for the SFTM and the towlines, respectively, and $\mathbf{x}_{C.R.}$ is the center of rotation. The Rodrigues formula is given by

$$\mathbf{Q}\left(\hat{\boldsymbol{\theta}}^{(i)}\right) = \mathbf{I} + \frac{\sin \hat{\theta}^{(i)}}{\hat{\theta}^{(i)}}\hat{\mathbf{Q}}\left(\hat{\boldsymbol{\theta}}^{(i)}\right) + \frac{1-\cos\hat{\theta}^{(i)}}{\left(\hat{\theta}^{(i)}\right)^2}\hat{\mathbf{Q}}\left(\hat{\boldsymbol{\theta}}^{(i)}\right)^2 \tag{17}$$

with $\hat{\mathbf{Q}}(\cdot)$ being a skew matrix of the cross product, and $\hat{\theta}^{(i)} = \left|\hat{\boldsymbol{\theta}}^{(i)}\right|$.

2.5. Stress Resultants

The Cauchy's stress tensor can be obtained as

$${}^{t+\Delta t}\boldsymbol{\sigma} = \det\left({}^{t+\Delta t}_{0}\mathbf{F}\right)^{-1} {}^{t+\Delta t}_{0}\mathbf{F} \cdot {}^{t+\Delta t}_{0}\mathbf{S} \cdot \left({}^{t+\Delta t}_{0}\mathbf{F}\right)^T \tag{18}$$

where ${}^{t+\Delta t}_{0}\mathbf{F}$ is the deformation gradient. Then, the stress resultants upon cross-sections along towlines can be obtained as

$${}^{t+\Delta t}\mathbf{F} = \int_{{}^{t+\Delta t}S} {}^{t+\Delta t}\boldsymbol{\sigma} \cdot {}^{t+\Delta t}\mathbf{n}\, dA \text{and } {}^{t+\Delta t}\mathbf{M} = \int_{{}^{t+\Delta t}S} \left({}^{t+\Delta t}\mathbf{x} - {}^{t+\Delta t}\mathbf{x}_{N.A.}\right) \times {}^{t+\Delta t}\boldsymbol{\sigma} \cdot {}^{t+\Delta t}\mathbf{n}\, dA \tag{19}$$

where ${}^{t+\Delta t}\mathbf{x}_{N.A.}$ denotes the position vectors of the cross-section at the neutral axis. Along the normal direction to the cable's sectional area, but not necessarily tangential to the cable's length direction due to the shear deformation, the force and the moment represent tensile force and torsional moment, respectively. Likewise, forces and moments normal and bi-normal to the outer surface of the cable represent shear forces and bending moments.

2.6. Motion Response of the SFTM

Based on the nonlinear static analysis, the linearized equivalent system stiffness from the hydrostatics and wet tow configuration can be obtained. Then, we further investigate the dynamic responses of the SFTM in the frequency domain. Steady-state analysis is implemented.

The hydrodynamic coefficients, i.e., added mass and radiation-damping matrices and wave-exciting load vector, can be obtained from 3D panel programs, e.g., WAMIT v7.4. The boundary integral equations for diffraction and radiation velocity potentials are solved as follows:

$$2\pi\varphi_D(\mathbf{x}) + \int_{\partial\Omega_f} \varphi_D(\boldsymbol{\xi})\frac{\partial G}{\partial n}(\mathbf{x};\, \boldsymbol{\xi})dS = 4\pi\varphi_I(\mathbf{x}),\ \mathbf{x} \in \partial\Omega_f \tag{20}$$

$$2\pi\varphi_j(\mathbf{x}) + \int_{\partial\Omega_f} \varphi_j(\boldsymbol{\xi})\frac{\partial G}{\partial n}(\mathbf{x};\, \boldsymbol{\xi})dS = -i\omega\int_{\partial\Omega_f} G(\mathbf{x};\, \boldsymbol{\xi})n_j(\boldsymbol{\xi})dS,\ j = 1,2,\ldots,6 \tag{21}$$

where velocity potential $\phi = \mathrm{Re}\{\varphi e^{-i\omega t}\}$, with ω being wave frequency. φ_I is the incident wave potential, φ_D is the diffraction wave potential, and φ_j is the radiation potential for j-th DOF. G is the free-surface Green function of either finite or infinite water depths [25,26]. $\partial G/\partial n = \nabla_{\xi} G \cdot \mathbf{n}$ with $\nabla_{\xi} = (\partial/\partial\xi, \partial/\partial\eta, \partial/\partial\zeta)$ and normal vector $\mathbf{n} = (n_1, n_2, n_3)$ defined over the wet surface of the SFTM, $\partial\Omega_f$. $\mathbf{x}$ and $\boldsymbol{\xi}$ denote position vectors of field and source points, respectively; when $\mathbf{x}$ coincides with $\boldsymbol{\xi}$ over the wet surface, solid angle 2π comes out as Cauchy's principal value, as shown in Equations (20) and (21). It should be noted that the body boundary condition neglects the effects of uniform flow and its disturbance, as we assume a very low wet tow speed. Additional integral term over the lid panels at mean water level (MWL) might be considered in case of removing irregular frequencies. Moreover, normal vector component n_j is defined as follows:

$$n_j = \begin{cases} n_j, & j = 1,2,3 \\ (\mathbf{x} \times \mathbf{n}) \cdot \mathbf{e}_{j-3}, & j = 4,5,6 \end{cases} \tag{22}$$

where $\mathbf{x}$ is the position vector measured from the center of the SFTM.

Once we obtain velocity potentials by solving Equations (20) and (21), added mass, radiation-damping coefficients, and wave-exciting loads can be obtained by numerically integrating φ_j and φ_D over each module's wet surface:

$$a_{ij} + \frac{i}{\omega}b_{ij} = i\omega\rho\int_{\partial\Omega_f} \varphi_j n_i dS \text{ with } i,\, j = 1,2,\ldots,6 \tag{23}$$

$$f_i = i\omega\rho\int_{\partial\Omega_f} \varphi_D n_i dS \text{ with } i = 1,2,\ldots,N \tag{24}$$

where ρ denotes sea water density. The added mass, radiation-damping matrices ($\mathbf{A}$ and $\mathbf{B}_R$), and the wave-exciting load vector ($\mathbf{F}_{EX}$) are of sizes 6×6 and 6×1, respectively.

The hydrostatic stiffness matrix is given by:

$$\mathbf{K}_{HS} = \rho g \begin{bmatrix} 0 & 0 & 0 & 0 & 0 & 0 \\ 0 & 0 & 0 & 0 & 0 & 0 \\ 0 & 0 & A_{WP} & A_{WP}y_F & -A_{WP}y_F & 0 \\ 0 & 0 & A_{WP}y_F & L_{22} + Vz_B - \gamma z_G & -L_{12} & -Vx_B + \gamma x_G \\ 0 & 0 & -A_{WP}x_F & -L_{12} & L_{11} + Vz_B - \gamma z_G & -Vy_B + \gamma y_G \\ 0 & 0 & 0 & 0 & 0 & 0 \end{bmatrix} \quad (25)$$

where A_{WP} is the waterplane area, V is the displaced volume, x_F and x_F are the centers of floatation in the waterplane along each direction of coordinates, and (x_G, y_G, z_G) are centers of gravity measured from the SFTM's center. The second moment of waterplane area is defined by:

$$L_{\alpha\beta} = \int_{\Gamma_{WP}} x_\alpha x_\beta dS \text{ with } x_1 = x \text{ and } x_2 = y(\alpha, \beta = 1,2\Big) \quad (26)$$

Considering the pre-calculated varying coupled system stiffness, $\mathbf{K}_T$, the motion response amplitude operators (RAOs) can be calculated as follows:

$$\mathbf{RAO}(\omega_e) = \frac{\mathbf{F}_{EX}(\omega_e)}{-\omega_e^2(\mathbf{M}+\mathbf{A}(\omega_e)) - i\omega_e(\mathbf{B}_V+\mathbf{B}_R(\omega_e)) + (\mathbf{K}_{HS}+\mathbf{K}_T)} \quad \text{with } \omega_e = \omega - kU\cos\beta \quad (27)$$

where $\mathbf{M}$ and $\mathbf{A}$ are the inertia and added inertia matrices; $\mathbf{B}_V$ and $\mathbf{B}_R$ are, respectively, the linearized viscous and wave radiation-damping matrices; $\mathbf{K}_{HS}$ is the hydrostatic stiffness matrix; $\mathbf{K}_T$ is the equivalent stiffness from towlines; $\mathbf{F}_{EX}$ is the wave-exciting load vector of the component given in Equation (24); and ω_e is an encounter frequency determined by tow speed U and wave-heading angle β. The denominator stands for inverse matrix.

Assuming a low wet tow speed, the added inertia, wave radiation-damping, and wave-exciting loads are evaluated in the zero-speed form. Any changes due to the presence of uniform flow are neglected, which may influence the frequency-dependent quantities due to the different free-surface boundary conditions and hydrodynamic pressure formula. Taking into account low wet tow speeds, such variation is assumed to be negligible. For moderate forward speeds, double-body or uniform flow approximations can be made. The forward speed effect on the hydrodynamic coefficients and loads, based on the uniform flow approximation, can be easily accounted for by manipulating the zero-speed components, which involve the m-term correction and encounter frequency conversion, as detailed in [27].

2.7. Statistical Analysis

The JONSWAP wave energy spectrum is given by [28]:

$$S_\zeta(\omega) = (1 - 0.287\ln\gamma)(5/16)H_S^2\omega_p^4\omega^{-5}e^{-b\omega_p^4/\omega^4}\gamma^a \text{ with } a = e^{-(\omega-\omega_p)^2/2\omega_p^2\sigma^2} \quad (28)$$

where ω_p is peak frequency; H_S is significant wave height; γ is the peak enhancement factor; $b = 1.25$; and σ is spectral width parameter, $\sigma = 0.07$ if $\omega \leq \omega_p$, otherwise $\sigma = 0.09$.

The wave energy spectra for different tow speeds are generated based on the encounter frequency conversion:

$$S_\zeta(\omega_e) = S_\zeta(\omega)\left(\frac{C_g}{C_g - U\cos\beta}\right) \text{ with } \omega_e = \omega - kU\cos\beta \quad (29)$$

where the group velocity $C_g = \frac{1}{2}C_p\left(1 + \frac{2kh}{\sinh 2kh}\right)$ with the phase velocity $C_p = \sqrt{\frac{g}{k}\tanh kh}$.

Now, the motion spectra and their statistics can be estimated based on the input wave spectrum and RAOs. In the linear time-invariant system, the motion spectrum can be obtained as follows:

$$S_{\xi_i}(\omega_e) = S_\zeta(\omega_e)|RAO_i(\omega_e)|^2 \tag{30}$$

Subsequently, in a typical statistical analysis with the assumptions of Gaussian and narrow-banded Rayleigh distributions, the wave elevation and motion response are estimated as follows:

- $\sigma = \sqrt{m_0}$: Standard deviation
- $\eta_E = \sigma_\eta\sqrt{2\ln(10800/T_2)}$: The most probable extreme amplitude for 3 h,

where η is either ζ(wave) or ξ_i (motion amplitude for i-mode); $m_i = \int_0^\infty (\omega_e)^i S(\omega_e)d\omega_e$; and mean wave period $T_2 = 2\pi\sqrt{m_0/m_2}$.

3. Considerations in the Stability Analysis of the SFTM

3.1. Roll Stability

As mentioned earlier, the SFTM is a circular cylinder horizontally afloat with a half-diameter draft. Thus, pressure loadings that act normally on the outer surface of the SFTM are directed toward the geometric center. Considering that the center of rotation is located near the geometric center, it can be easily expected that the resulting hydrostatic pressure-induced moment is zero. Accordingly, the transverse metacentric height ($\overline{GM_T}$) of the SFTM becomes zero unless the center of gravity is non-zero, as

$$\overline{GM_T} = \frac{I_{yy}^{WP}}{\nabla} + z_B - z_G = \frac{L(2R)^3/12}{\pi R^2 L/2} + \left(-\frac{4R}{3\pi}\right) - z_G = -z_G \tag{31}$$

where ∇ is the displaced volume of the SFTM; I_{yy}^{WP} is the second moment of waterplane area; and z_B and z_G are the centers of buoyancy and gravity, respectively, as illustrated in Figure 2a. As presented in the above equation, the metacentric height reduces to a vertical coordinate of the center of gravity in a negative sign. In this problem, an additional component involved in the roll stability is the towline's stiffness, which can result in a more important role in the metacentric height and resulting role stability.

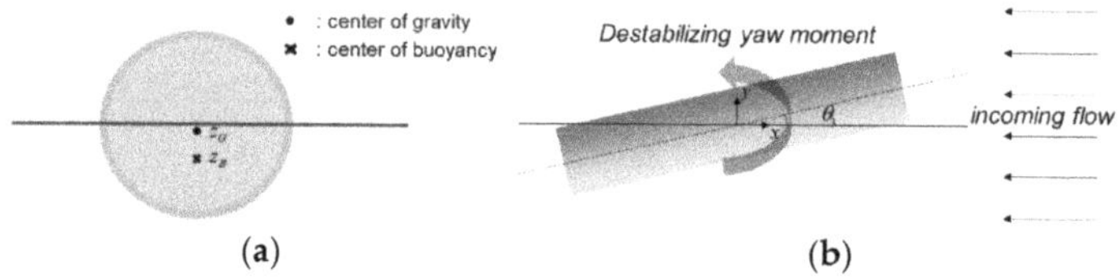

Figure 2. Sketches of the SFTM. (**a**) SFTM cross-section: center of gravity and buoyancy (frontal view). (**b**) Destabilizing yaw moment (Munk moment) in the presence of incoming flow (top-view).

3.2. Destabilizing Moment (Munk Moment)

The Munk moment is a destabilizing moment and acts to turn the body perpendicular to the incoming flow (see Figure 2b). To calculate the Munk moment, we calculate the double-body velocity potential, which is the zero-frequency limit of the frequency-dependent velocity potential. The Munk moment is calculated by:

$$M_6 = U^2(A_{22} - A_{11})\cos\theta_3\sin\theta_3 \tag{32}$$

where A_{11} and A_{22} are added mass for surge and sway at zero-frequency obtained from the commercial program WAMIT [29], for which a water depth of 70 m is assumed. The generated hydrodynamic mesh is illustrated in Figure 3. The quarter number of meshes is 492, which is fine enough considering the dimension and wavelength for a typical frequency range of wave energy spectra. The nondimensionalized added masses by water density are $A_{11}/\rho = 3.22 \times 10^3$ and $A_{22}/\rho = 3.06 \times 10^4$. Since the Munk moment is proportional to $A_{22} - A_{11}$, a body with a high aspect ratio may be subject to a larger Munk moment. We assumed in the present study that the Munk moment in air is proportional to the square of the tow speed, not the relative wind speed, as they are marginal. We will investigate the effects of the Munk moment in the yaw offset test in the next section.

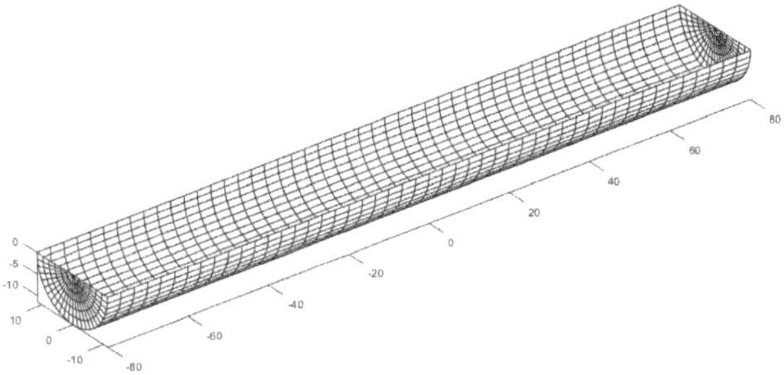

Figure 3. Hydrodynamic mesh of the SFTM generated for WAMIT v7.4 simulation.

4. Numerical Results and Discussions

4.1. Static Offset Test in Wet Tow Conditions

A sequential procedure is summarized in the flow chart in Figure 4. As we are dealing with a geometrically nonlinear problem, we split the offset test procedure into multiple sub-steps, and at each sub-step, incremental loads are imposed on the SFTM to generate load-displacement curves. The same procedure is repeated for different DOFs, tow speeds, and line arrangements; here, the different line arrangements refer to different angles, ψ, of the towlines at the fairleads of the SFTM, as shown in Figure 5. As shown, four tugboats are arranged in parallel, towing the SFTM with evenly distributed towline tensions. It should be ensured that the tugboats, under lateral loads from towlines caused by instantaneous SFTM motion, can maintain their dynamic stability. In addition, overstressing or slack of the towlines should be avoided by controlling the system. The front two tugboats provide propulsion and steer the towing direction, while the two rear tugboats control the orientation of the system's rear end [30]. The towlines are connected to the tugboats 3 m above the MWL. The range of input parameters and wet tow conditions used in the calculations is given in Table 1. Also, the principal dimensions of the SFTM and towlines are given in Table 2. Because the current towlines are modeled using a continuum-based 3D beam finite element approach, the solution is likely to diverge when the bending stiffness is too small. To avoid such a numerical barrier, the towline diameter is increased to 0.24 m while maintaining the axial stiffness (EA) of a steel wire rope employed in a static towing analysis performed by MARIN (Maritime Research Institute Netherlands), as in ref. [31].

Table 1. Wet tow conditions.

Parameters	Value
Tow speeds (m/s)	0, 1, 2, 3, 4
Wind speed (10 m above MWL) (m/s)	6
Angles (ψ) at fairleads (degree)	35, 40, 45
Water depth (m)	70

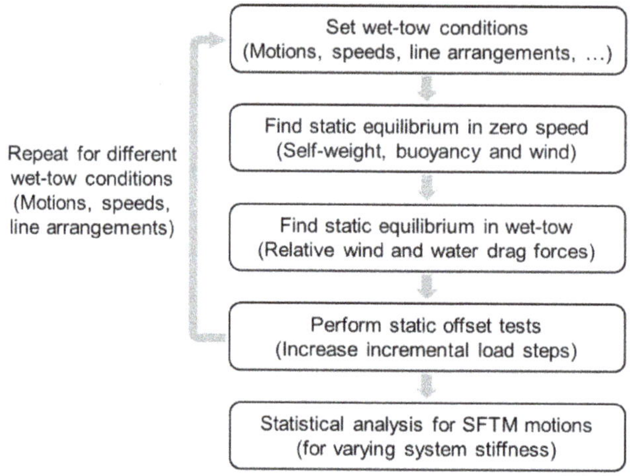

Figure 4. Flow chart of the overall procedure.

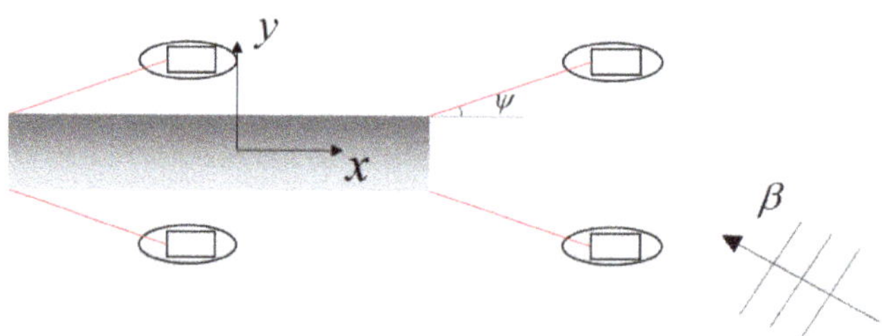

Figure 5. A wet tow configuration: birds-eye view.

Table 2. Principal dimensions of the SFTM and towlines.

	Specifications	Value
SFTM	Length (m)	160
	Diameter (m)	23
	Draft (m)	11.5
	Transverse metacentric height $(\overline{GM_T})$ (m)	0.1
	Drag coefficients in air $\left(C_{Da,\tau}^{SFTM}/C_{Da,n}^{SFTM}\right)$	0.82/0.4
	Drag coefficients in water $\left(C_{Dw,\tau}^{SFTM}/C_{Dw,n}^{SFTM}\right)$	0.82/0.4
Towlines	Density (kg/m^3)	1500
	Unstretched length (m)	60
	Axial stiffness (kN)	1.5×10^5
	Constraint at SFTM	Hinged
	Constraint at tugboats	Clamped (with fixed ψ)
	Drag coefficients in air $\left(C_{Da,\tau}^{CABLE}/C_{Da,n}^{CABLE}\right)$	0.0/0.0 (neglected)
	Drag coefficients in water $\left(C_{Dw,\tau}^{CABLE}/C_{Dw,n}^{CABLE}\right)$	0.01/1.2

4.2. Pre-Tensions

Pre-tensions were calculated for different tow speeds and towline angles, as shown in Figure 6, as they are determined by towing resistance that requires the bollard pull of tugboats. For validation purposes, an equivalent coupled towline–SFTM model was developed in the commercial software, OrcaFlex v11.4, from which the pre-tension for

$\psi = 45°$ is generated and given in Figure 6. First of all, the comparison between OrcaFlex v11.4 and the in-house program shows matching results at different tow speeds. The pre-tensions were measured at positions very close to the fairleads. It can be seen that the towline angles affect the pre-tensions, although they are connected as hinged-type constraints. Additionally, it is found that these pre-tensions increase quadratically with the increase in tow speed. This can be easily explained by noting that the drag force is proportional to the square of the incoming flow speed. Since pre-tensions have a positive impact on the system's restoring force, we restrict them to be $\psi = 45°$, ensuring higher pre-tensions. It should be noted that aerodynamic drag is considered to have a wind speed of 6 m/s at 10 m above MWL, but it is at a negligible level, i.e., the relative order of $O(10^{-3})$, to the hydrodynamic drag. Further, mean wave drift load is neglected in the calculation of pre-tension and load-displacement curves, as it is sea state-dependent.

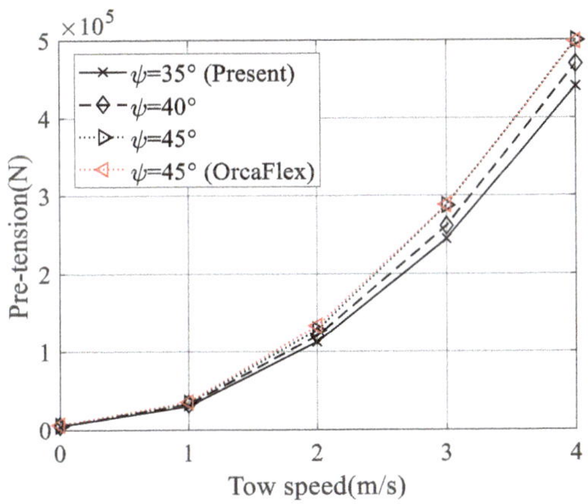

Figure 6. Variation of pre-tensions for different towline angles and tow speeds.

4.3. Load-Displacement Curves

At the static equilibrium of each tow speed, static offset tests are subsequently performed by imposing incremental load steps to generate nonlinear load-displacement curves. The corresponding offset DOF and surge DOF are enabled, whereas all the other DOFs are disabled. Snapshots of 4-DOF SFTM motions at $U = 4$ m/s are shown in Figure 7, and load-displacement curves are shown in Figures 8–10. Again, the comparison between OrcaFlex v11.4 and the in-house program is considered. The results show that at low displacements, they are well matched, although discrepancies are observed at higher displacements owing to the modeling differences between the lumped-mass method in OrcaFlex v11.4 and the continuum-based nonlinear finite element model in the present study. In the case of surge (See Figure 8a), the equivalent stiffness is non-symmetric with respect to the equilibrium position, as the wet tow configuration is also non-symmetric in surge. When positive loads exceed the towing drag of the SFTM, all the towlines will be instantaneously in slack and experience a significant reduction in stiffness. On the other hand, as shown in Figure 8b showing a sway load-displacement curve, due to symmetry, displacement-force curves are in symmetry as well. Also, rapid decreases in stiffness are observed, i.e., the slope of the curves decreases as displacement increases, due to the slack of the towlines located on the opposite side of the applied loads (See Figure 7b).

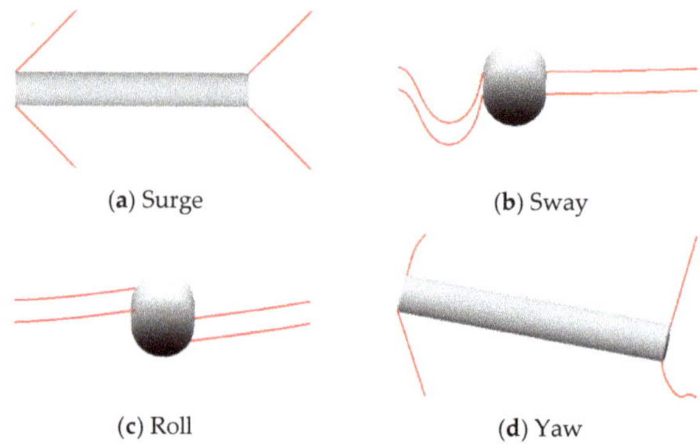

Figure 7. Snapshots of offset test for 4-DOF SFTM motions ($U = 4\ \mathrm{m/s}$).

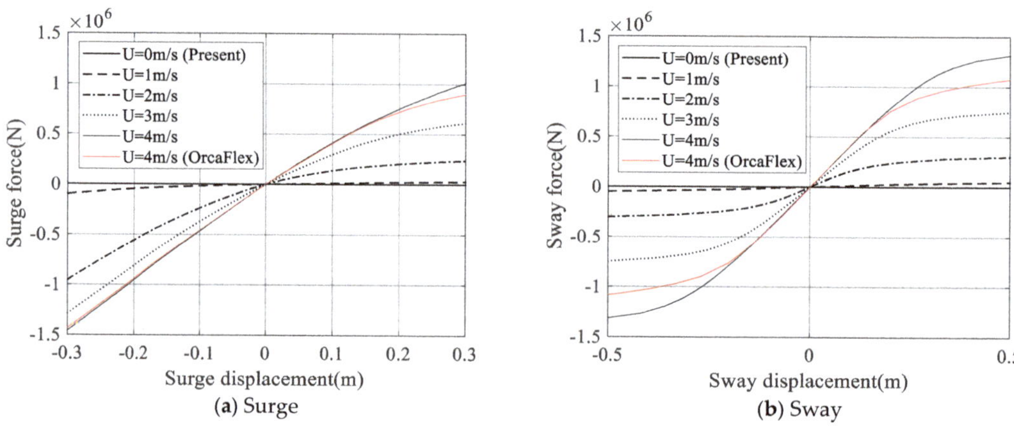

Figure 8. Surge and sway DOF load-displacement curves.

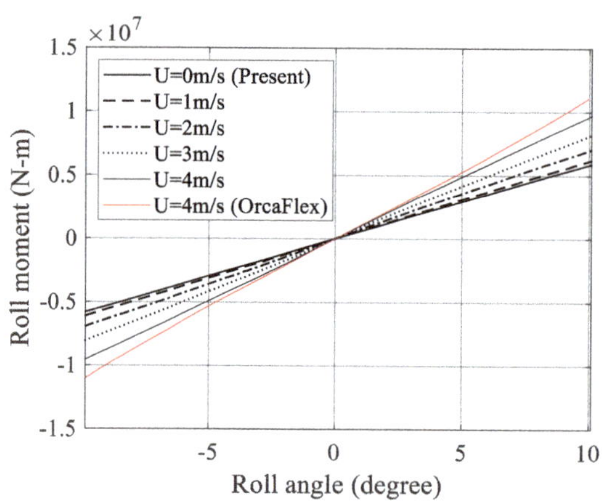

Figure 9. Roll DOF load-displacement curve: The equivalent metacentric height is the sum of the hydrostatic and towing stiffnesses.

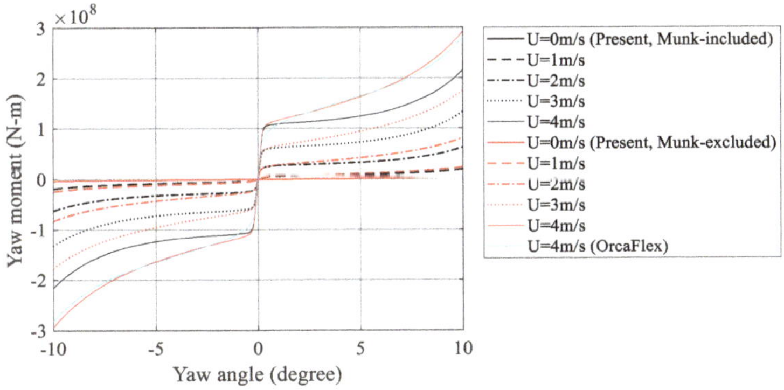

Figure 10. Yaw DOF load-displacement curve: Munk moment effect (black) included and (red) excluded.

Figure 9 shows a load-displacement curve for roll DOF, and the corresponding snapshot is shown in Figure 7c. As discussed, the static stability of the roll motion is slight, i.e., the transverse metacentric height ($\overline{GM}_T$) is nearly zero. In this regard, the equivalent $\overline{GM}_T$, which is calculated by dividing the weight force of the SFTM by the equivalent roll stiffness for varying tow speeds, is summarized in Table 3. The $\overline{GM}_T$ of 0.1 m is initially set as a minimum value to prevent numerical instability. The additional restoring effect from the towlines is minor, reaching at most 0.07 m for a tow speed of 4 m/s. However, the pressure loads acting perpendicular to the SFTM's outer surface are directed toward its geometric center, which is close to the rotation center. As a result, the cross product of the moment arm vector and pressure load vector becomes almost zero. Given that the tangential stresses from viscosity over the SFTM are minimal, the slight increase in restoring force from the towlines seems sufficient to resist the roll moment.

Table 3. Equivalent GM_T (transverse metacentric height) for varying tow speeds.

U (m/s)	0	1	2	3	4
Equiv. GM_T (m)	0.1014	0.1072	0.1257	0.1507	0.1733

The load-displacement curve for the yaw DOF is shown in Figure 10, and the corresponding snapshot is presented in Figure 7d. Since the Munk moment arises from the cross-flow and is proportional to twice the yaw angle, i.e., $\sin 2\theta_3 (= 2\cos\theta_3 \sin\theta_3)$, the larger the yaw motion, the greater the effect of the Munk moment. This trend is well represented by the discrepancies between the black and red solid lines in Figure 10. For validation, the OrcaFlex v11.4 result at a tow speed of $U = 4$ m/s is also illustrated, which shows that the program does not account for the destabilizing moment effect (Munk moment) in the static analysis. Taking a closer look at the x-axis, it can be seen that the relatively high stiffness region is maintained only for a limited range of less than ±0.5 degrees. Beyond this small angle, the stiffness rapidly decreases, which seems to be caused by the instantaneous slack of towlines located on the opposite side of the applied loads, as presented in Figure 7d. In addition, the Munk moment plays a role when the yaw angle is greater than ±0.5 degrees. It turns out that the Munk moment decreases the yaw stiffness, which implies that an instantaneously large yaw DOF motion may affect the restoring force of the system and reduce course stability.

From the results, it can be seen that towing speed also influences the displacement-force curve to some degree, while nonlinear stiffness is observed. Moreover, as shown in Section 4.2, the pre-tension also changes with respect to the towline angle. Therefore, various aspects should be carefully examined to determine an appropriate limit for the stiffness level and resulting static stability.

4.4. Statistical Analysis Under Sea State 4

Statistical analyses are performed based on the pre-calculated system stiffness for varying tow speeds. Sea state 4 is considered a mild sea state of typical wet tow situations. The metocean parameters for sea state 4 are given in Table 4. We again investigate five different tow speeds, and the encounter-frequency wave energy spectra for selected wave headings, β, as shown in Figure 5. Encounter-frequency wave spectra at different wave headings at tow speed $U = 4$ m/s are illustrated in Figure 11. For following and stern seas, the wave energy beyond the critical frequency at $C_g = U\cos\beta$ is artificially removed. The motion response spectra are calculated from Equation (30) and depicted in Figure 12a–f for $\beta = 135°$. The shift of motion energy can be seen as expected associated with changes in wave energy as a function of wave frequency due to the encounter-frequency effect.

Table 4. Metocean parameters for sea state 4.

H_s (m)	T_p (s)	γ	β (°)	U (m/s)
2.44	8.1	2.14	0–360 (15° interval)	0, 1, 2, 3, 4

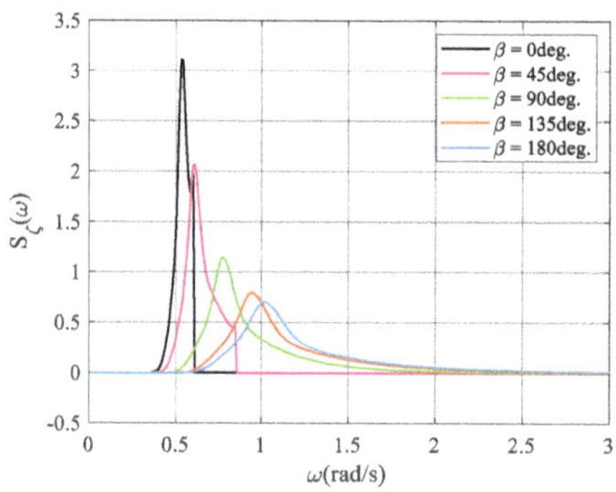

Figure 11. Encounter-frequency wave energy spectra (U = 4 m/s).

Based on the generated motion spectra, the most probable extreme motions for 3 h are estimated and shown in Figure 13a–f. Taking a closer look at Figure 13a, the surge motion increases as the tow speed increases, particularly at the following sea ($\beta = 0°$). However, their amplitudes are inappreciable, and the largest amplitude is at most 0.25 m. On the other hand, Figure 13b shows that the sway motion is largest at beam sea ($\beta = 90°$), which is in a range of 0.30–0.34 m for $U = 0 - 4$ m/s. This variation is due to the change in equivalent sway stiffness. Recalling Figure 8b, even this small amplitude is no longer in the linear duration of the load-displacement curve. Figure 13c also exhibits that the heave response is largest at beam sea. The most probable extreme heave amplitude is 0.59 m, which is likely to impact negatively the roll hydrostatic stability since the equivalent $\overline{GM_T}$ is only 0.1–0.17 m in the present wet tow scenario. Due to the varying waterplane and center of buoyancy, it might experience an instantaneous loss of roll stability. However, as shown in Figure 13d, the roll amplitude in the steady-state analysis seems insignificant. Due to its circular geometry, the wave-exciting moment is rarely generated. This is obvious if one recalls that the cross product of the moment arm vector from its geometric center and the local pressure-load vector acting upon the circular outer surface becomes null. This is the limitation of the present frequency-domain analysis. Nevertheless, instantaneous negative stability due to the coupling effect from towlines and the other rigid DOF modes or nonlinear hydrostatic effect may have a larger response until the SFTM returns to its stable condition. On the other hand, interestingly, the largest pitch extreme motion is 0.71 deg., which occurred under wave heading $\beta = 60°$ and tow speed $U = 4$ m/s. The amplitude variation due to the tow speed is larger around the following sea condition, but it is not for $\beta = 60°$, where the maximum response is observed. Considering the length of the SFTM (L = 160 m), the vertical displacement at the bow (or stern) is at most 0.98 m, which can be significant. Lastly, it can be seen in Figure 13f that the most probable yaw amplitudes are greatest at $\beta = 60°$ for $U = 0 - 3$ m/s, whereas $\beta = 30°$ for U = 4 m/s. The maximum amplitudes are 0.23–0.85 deg. Again, recalling Figure 10, these amplitudes exceed the linear duration of load-displacement curves, similar to the sway case. Beyond the linear region, the stiffness is rapidly decreased, and in particular, the presence of a destabilizing moment (Munk moment) negatively impacts it. Hence, the SFTM response might be much larger. In addition, this is likely to decrease the course stability. It can be seen that the maximum statistical values for all DOFs occur at a tow speed of U = 4 m/s, and they are summarized in Table 5.

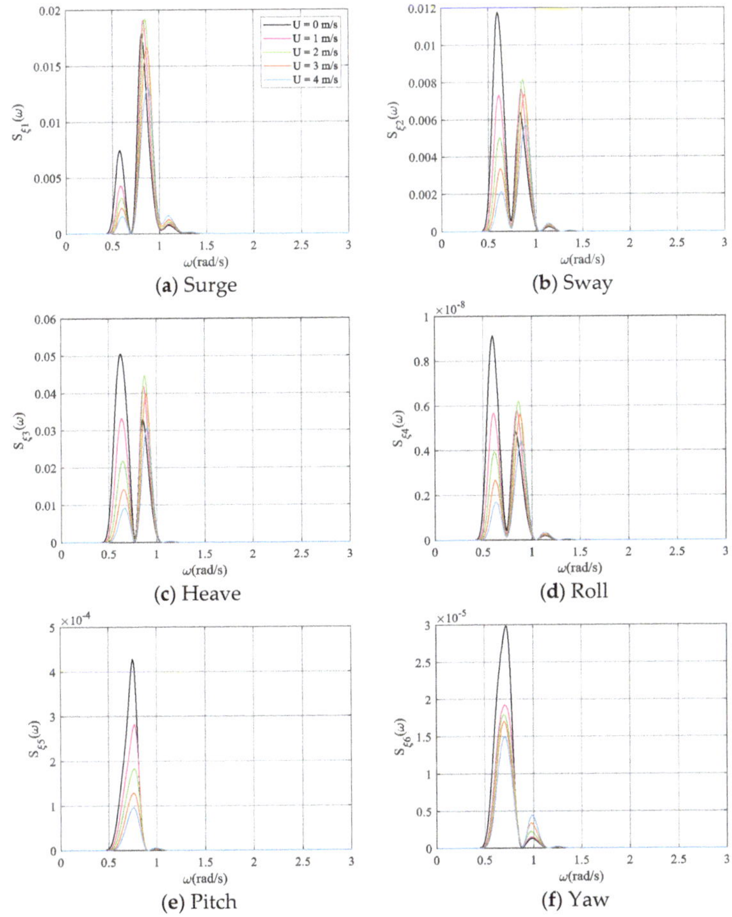

Figure 12. Motion response spectra (β = 135°).

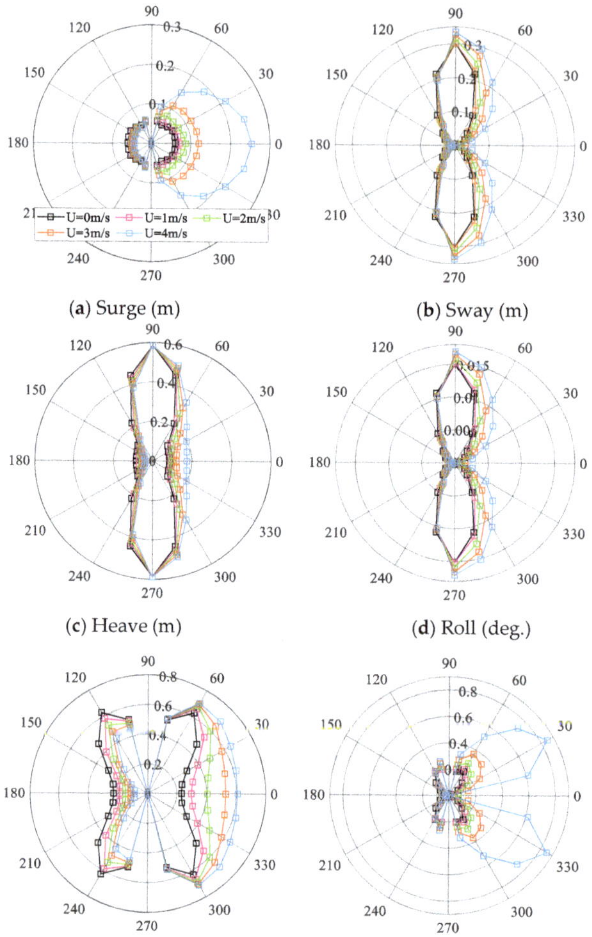

(a) Surge (m) (b) Sway (m) (c) Heave (m) (d) Roll (deg.) (e) Pitch (deg.) (f) Yaw (deg.)

Figure 13. The most probable extreme 6-DOF motions for 3 h (β = 0–360°).

Table 5. Metocean parameters for sea state 4 (U = 4 m/s).

DOF	Max. Value	β (°)
Surge	0.25 m	0
Sway	0.34 m	90
Heave	0.59 m	90
Roll	0.017 deg.	90
Pitch	0.71 deg.	60
Yaw	0.85 deg.	30

The given results are valid for specific wave conditions, seakeeping performance can be much worse if the different SFTM and towline configurations and/or higher sea state are taken into account. The present methodology can give further insight into different configurations/setups.

4.5. Discussion and Remarks

The scope of the present study is restricted to the frequency-domain analysis. As observed from numerical results, the effect of nonlinear system stiffness is likely to exist at relatively high tow speeds, particularly from following to beam sea conditions. In this regard, time-domain analyses are necessary for the present scenarios. Based on the encounter-frequency approximation with shifted energy spectra (user-generated), a coupled floater-mooring time-domain simulator, e.g., OrcaFlex v11.4, can be used to conduct time-domain analysis of coupled wet towing problems. Further, uniform flow or double-body flow approximations [27,32] can be used to improve the accuracy by modifying the body boundary condition in Equation (21) and hydrodynamic coefficients and loads in Equations (23) and (24). In addition, multiple different wet tow configurations and the use of synthetic fiber ropes will be considered in future research. Since numerous wet tow scenarios are available, initial frequency-domain analyses are able to sort out unrealistic or ineffective solutions. Also, based on the assessment of nonlinear system stiffness (e.g., load-displacement curves), critical cases of realistic scenarios can be selected. Subsequently, time-domain analyses can be effectively implemented for a few selected load cases. Future studies will account for these interesting and important topics to address the wet towing problems more accurately.

5. Conclusions

In the present study, a nonlinear static analysis for the coupled towline SFTM system was investigated under varying wet tow conditions. The mathematical details of the coupled towline SFTM system were described. First, static equilibrium states were estimated for different tow speeds, and the static offset tests were subsequently performed for 4-DOFs based on each tow speed's equilibrium state. Corresponding load-displacement curves were provided. The roll stability was discussed based on the equivalent metacentric height. Also, the Munk moment effect on the yaw motion was investigated. Statistical analyses were performed based on the most probable extreme 6-DOF motion amplitudes of 3 h for varying tow speeds, U = 0–4 m/s, and all wave heading angles, β = 0–360°.

From the systematic investigation of static stability and frequency-domain dynamics assessment, the following key conclusions are drawn.

- Nonlinear stiffness, i.e., force-displacement curve, is observed, especially for surge, sway, and yaw DOFs associated with losing stiffness under stack towline.
- The Munk moment is especially important at higher yaw angles where larger yaw angles are seen at the same yaw moment. At lower angles, its effect is minor.
- The most probable extreme heave amplitude is considerable, which may negatively impact the roll stability.
- However, the extreme roll amplitude appears inappreciable, which is the limitation in the present frequency-domain analysis.
- In sea state 4, the extreme sway and yaw amplitudes exceed the linear duration of load-displacement curves, which are rapidly decreased due to the destabilizing moment and instantaneous slack of towlines located on the opposite side of the applied loads, which may reduce the course stability.

Based on the present nonlinear static and frequency-domain analyses, time-domain analysis is critical to better understand the wet tow scenario. In addition, various system configurations and wave conditions need to be considered to better explain various physical phenomena and statistical results associated with wet towing and seakeeping performance. Future studies will account for these interesting and important topics to address the wet towing problems.

Author Contributions: Conceptualization, I.L. and C.J.; methodology, I.L. and S.-J.K.; software, I.L.; validation, I.L.; formal analysis, I.L.; investigation, I.L., C.J. and S.-J.K.; resources, M.K.; data curation, I.L.; writing—original draft preparation, I.L.; writing—review and editing, C.J., S.-J.K. and M.K.; visualization, I.L.; supervision, C.J. and M.K. All authors have read and agreed to the published version of the manuscript.

Funding: This research received no external funding.

Institutional Review Board Statement: Not applicable.

Informed Consent Statement: Not applicable.

Data Availability Statement: The raw data supporting the conclusions of this article will be made available by the authors upon request.

Conflicts of Interest: The authors declare no conflicts of interest.

References

1. Remseth, S.; Leira, B.J.; Okstad, K.M.; Mathisen, K.M.; Haukås, T. Dynamic Response and Fluid/Structure Interaction of Submerged Floating Tunnels. *Comput. Struct.* **1999**, *72*, 659–685. [CrossRef]
2. Kim, G.-J.; Kwak, H.-G.; Jin, C.; Kang, H.; Chung, W. Three-Dimensional Equivalent Static Analysis for Design of Submerged Floating Tunnel. *Mar. Struct.* **2021**, *80*, 103080. [CrossRef]
3. Jin, C.; Chung, W.C.; Kwon, D.-S.; Kim, M. Optimization of Tuned Mass Damper for Seismic Control of Submerged Floating Tunnel. *Eng. Struct.* **2021**, *241*, 112460. [CrossRef]
4. Jeong, K.; Kim, S. Structural Response of Submerged Floating Tunnels with Free-End Boundary Condition Based on an Analytical Approach. *Appl. Ocean. Res.* **2024**, *143*, 103861. [CrossRef]
5. Jeong, K.; Min, S.; Jang, M.; Won, D.; Kim, S. Feasibility Study of Submerged Floating Tunnels with Vertical and Inclined Combined Tethers. *Ocean. Eng.* **2022**, *265*, 112587. [CrossRef]
6. Jin, B.; Kim, J. Considerations on the Behavior of Submerged Floating Tunnel Module During Immersion Stage. In Proceedings of the International Conference on Vibration Problems 2019, Crete, Greece, 1–4 September 2019.
7. Kwon, D.-S.; Jin, C.; Kim, M. Prediction of Dynamic and Structural Responses of Submerged Floating Tunnel Using Artificial Neural Network and Minimum Sensors. *Ocean. Eng.* **2022**, *244*, 110402. [CrossRef]
8. Luo, G.; Zhang, Y.; Ren, Y.; Guo, Z.; Pan, S. Dynamic Response Analysis of Submerged Floating Tunnel Subjected to Underwater Explosion-Vehicle Coupled Action. *Ocean. Eng.* **2021**, *232*, 109103. [CrossRef]
9. Luo, W.; Huang, B.; Tang, Y.; Ding, H.; Li, K.; Cheng, L.; Ren, Q. Numerical Simulation of Dynamic Response of Submerged Floating Tunnel under Regular Wave Conditions. *Shock Vib.* **2022**, *2022*, 4940091. [CrossRef]
10. Min, S.; Jeong, K.; Noh, Y.; Won, D.; Kim, S. Damage Detection for Tethers of Submerged Floating Tunnels Based on Convolutional Neural Networks. *Ocean. Eng.* **2022**, *250*, 111048. [CrossRef]
11. Won, D.; Seo, J.; Kim, S. Dynamic Response of Submerged Floating Tunnels with Dual Sections under Irregular Waves. *Ocean. Eng.* **2021**, *241*, 110025. [CrossRef]
12. Won, D.H.; Park, W.S.; Kim, S. Cyclic Bending Performance of Joint on Precast Composite Hollow Rc for Submerged Floating Tunnels. *Mar. Struct.* **2021**, *79*, 103045. [CrossRef]
13. Lee, I.; Jin, C.; Kim, M. Static Stability of a Submerged Floating Tunnel Module in Wet tow. In Proceedings of the 2022 Structures Congress (Structures22), Seoul, Republic of Korea, 16–19 August 2022.
14. Wang, J.J.; Lu, R.; Lu, N. Truss Spar Strength and Fatigue Analysis for Wet tow. In Proceedings of the ISOPE International Ocean and Polar Engineering Conference 2003, Honolulu, HI, USA, 25–30 May 2003.
15. Sherman, M.; Sablok, A.; Kopsov, I.; Chen, L. Design of a Floating Spar Wind Platform with an Integrated Substructure and Tower. In Proceedings of the Offshore Technology Conference 2019, Houston, TX, USA, 6–9 May 2019.
16. Amaral, G.A.; Pesce, C.P.; Franzini, G.R. Mooring System Stiffness: A Six-Degree-of-Freedom Closed-Form Analytical Formulation. *Mar. Struct.* **2022**, *84*, 103189. [CrossRef]
17. Kim, B.W.; Sung, H.G.; Kim, J.H.; Hong, S.Y. Comparison of Linear Spring and Nonlinear Fem Methods in Dynamic Coupled Analysis of Floating Structure and Mooring System. *J. Fluids Struct.* **2013**, *42*, 205–227. [CrossRef]
18. Orcina. *Orcaflex Manual*; Orcina Ltd.: Ulverston, UK, 2024.
19. Ran, Z. *Coupled Dynamic Analysis of Floating Structures in Waves and Currents*; Texas A&M University: College Station, TX, USA, 2000.
20. Garrett, D.L. Dynamic Analysis of Slender Rods. *J. Energy Resour. Technol.* **1982**, *104*, 302–306. [CrossRef]

21. Yang, M.; Teng, B. Static and Dynamic Analysis of Mooring Lines by Nonlinear Finite Element Method. *China Ocean. Eng.* **2010**, *24*, 417–430.
22. Dvorkin, E.N.; Onte, E.; Oliver, J. On a Non-Linear Formulation for Curved Timoshenko Beam Elements Considering Large Displacement/Rotation Increments. *Int. J. Numer. Methods Eng.* **1988**, *26*, 1597–1613. [CrossRef]
23. Triantafyllou, M.S.; Hover, F.S. *Maneuvering and Control of Marine Vehicles*; Massachusetts Institute of Technology: Cambridge, MA, USA, 2003.
24. Bathe, K.-J. *Finite Element Procedures—Second Edition*; Klaus-Jurgen Bathe: Watertown, MA, USA, 2006.
25. John, F. On the Motion of Floating Bodies II. Simple Harmonic Motions. *Commun. Pure Appl. Math.* **1950**, *3*, 45–101. [CrossRef]
26. Havelock, T.H. Waves Due to a Floating Sphere Making Periodic Heaving Oscillations. *Proc. R. Soc. Lond. Ser. A Math. Phys. Sci.* **1955**, *231*, 1–7.
27. Bakti, F.P.; Jin, C.; Kim, M.-H. Practical Approach of Linear Hydro-Elasticity Effect on Vessel with Forward Speed in the Frequency Domain. *J. Fluids Struct.* **2021**, *101*, 103204. [CrossRef]
28. Hasselmann, K.; Barnett, T.P.; Bouws, E.; Carlson, H.; Cartwright, D.E.; Enke, K.; Ewing, J.; Gienapp, A.; Hasselmann, D.; Kruseman, P. Measurements of Wind-Wave Growth and Swell Decay During the Joint North Sea Wave Project (JONSWAP). *Ergaenzungsheft Dtsch. Hydrogr. Z. Reihe A* **1973**. Available online: https://pure.mpg.de/rest/items/item_3262854_4/component/file_3282032/content (accessed on 11 November 2024).
29. Lee, C.H. *Wamit Theory Manual*; Massachusetts Institute of Technology: Cambridge, MA, USA, 1995.
30. Cao, X.; Zhang, H.; Zhang, N. Research on Dynamic Responses of Multi-Float Systems During Wet towing. *Ocean. Eng.* **2024**, *312*, 119226. [CrossRef]
31. OCIMF. *Static Towing Assembly Guidelines (Stag)*, 1st ed.; OCIMF: London, UK, 2020.
32. Zhang, X.; Bandyk, P.; Beck, R.F. Seakeeping Computations Using Double-Body Basis Flows. *Appl. Ocean. Res.* **2010**, *32*, 471–482. [CrossRef]

Disclaimer/Publisher's Note: The statements, opinions and data contained in all publications are solely those of the individual author(s) and contributor(s) and not of MDPI and/or the editor(s). MDPI and/or the editor(s) disclaim responsibility for any injury to people or property resulting from any ideas, methods, instructions or products referred to in the content.

Journal of
Marine Science and Engineering

MDPI

Article

The Numerical Analysis of Hydrodynamic Response and Structural Stability of an Eccentric Conical Floating Structure

Fei Qu [1], Shengtao Chen [1,2,3,*] and Kang Zhang [1]

1 College of Naval Architecture and Ocean Engineering, Dalian Maritime University, Dalian 116026, China; qufly5@163.com (F.Q.); zhangkang@dlmu.edu.cn (K.Z.)
2 Liaoning Provincial Key Laboratory of Rescue and Salvage Engineering, Dalian Maritime University, Dalian 116026, China
3 International Joint Research Centre for Subsea Engineering Technology and Equipment, Dalian Maritime University, Dalian 116026, China
* Correspondence: dutchenshengtao@sina.com

Abstract: This study examines the hydrodynamic response and structural stability of an eccentric conical floating structure, a return capsule for manned space missions, to ensure safe water landings. Using numerical simulations and experiments, we evaluated how center-of-mass offsets, displacement volume control, and environmental factors, including waves, currents, and wind, affect capsule stability. In still water, lateral center-of-mass offsets strongly affect stability through nonlinear restoring moments, whereas foam-based displacement control reduces motion amplitude and tilt angle. In dynamic sea conditions, wave parameters dominate motion, with surge displacement and pitch angle varying by wavelength and sea state. At higher sea states, nonlinear phenomena, including subharmonic resonance, amplify pitch angle extrema, compromising safety margins. This research offers key insights for evaluating and improving return capsule safety, highlighting the importance of complex multi-physics interactions in marine environments.

Keywords: hydrodynamic characteristics; marine structures; six-degree-of-freedom motion; motion response; numerical simulation

1. Introduction

For manned lunar missions, the return capsule requires both land and sea landing sites to maximize the astronauts' return window to Earth. During ascent, the return capsule must autonomously manage escape within seconds, unable to distinguish between land or sea landing, necessitating an emergency water landing mode adaptable to both scenarios [1–3]. In adverse sea conditions, limited helicopter speed and range delay rescue, requiring astronauts to remain in the capsule for extended periods, necessitating the consideration of its attitude and stability during sea landing to ensure astronaut safety [4].

Research on return capsule water landings began in 1958 with the U.S. Mercury program, initiating safe sea recovery studies [5,6]. In the 1960s, NASA extensively studied the Apollo capsule; Vaughan [7] analyzed rigid body impact acceleration during landing and water entry using two 1/6-scale dynamically similar models. Herting et al. [8] developed the capsule's landing impact system, prioritizing water surface landings. Stubbs [9] investigated impact loads and post-impact floating conditions of the capsule at various attitudes using a scaled model. In 1966, Benson [10] reviewed water landing studies for the Apollo and Mercury capsules, concluding that scaled model tests, by comparing acceleration and pressure in rigid and elastic models, yielded results comparable to full-scale tests, reducing costs and enhancing efficiency. Building on Benson's work, Baker and Westine [11] in 1967 analyzed the Apollo capsule's structural response to water impact, employing complex models for dynamic simulations that included various failure states. In the same year, Thompson [12] extended prior Gemini capsule research, modeling its response to waves at varying yaw angles and horizontal speeds. Building on this, Wang [13] simulated the capsule's water landing using explicit nonlinear dynamic finite element code, incorporating water and air meshes.

By the 1990s, advances in finite element analysis enabled integrated fluid–structure modeling. Anghilleri and Spizzica [14] employed MSC DYTRAN for finite element analysis, focusing on boundary conditions, material properties, and fluid–solid coupling algorithms; their ALE coupling enhanced accuracy and reduced computation time. Although these methods offer rapid calculations, they often neglect fluid viscosity. Consequently, researchers increasingly adopted computational fluid dynamics (CFD) models, which provide high accuracy and comprehensive results. Advances in CFD and computing have broadened its application in structural fluid dynamics; in 2019, Song [15] used CFD wall functions to model barnacle fouling surface roughness, predicting its impact on ship hydrodynamic characteristics. In December 2019, Jasak [16] compared sea trial data from two self-propelled ships with full-scale CFD results, demonstrating that numerical methods can accurately calculate ship-scale flow characteristics, including free surface effects, nonlinearity, turbulence, and propeller–hull–flow interactions. In 2020, Wang [17] simulated KVLCC1 ship stopping maneuvers using the CFD solver naoe-FOAM-SJTU with dynamic overlapping grids, experimentally validating results to identify optimal stopping methods under various conditions. In 2022, Jiao [18] employed a CFD-based unsteady Navier–Stokes and volume-of-fluid solver to estimate ship motion responses in bidirectional transverse waves. Islam [19] compared hull resistance and propulsion power predictions using empirical formulas and CFD schemes, finding strong agreement between the two. Kim [20] applied CFD methods to enhance the understanding of ship maneuverability in irregular waves, aiding captains and pilots in making informed decisions under adverse sea conditions. Andersson et al. [21] used CFD to predict power reductions from pre-delivery PSD installation on ships, establishing reliability indicators and clarifying the impact of computational processes on results. Huang [22] introduced a novel fluid–solid coupling method, bidirectionally integrating CFD and FEA solvers to estimate ship wave loads and hydroelastic responses, offering new approaches for fluid–solid coupling calculations. By late 2024, researchers had moved beyond investigating the hydrodynamics of a single body in isolation. Mi C. [23] employed a VOF–overset–grid technique to simulate two identical barges moored side-by-side and traced the wave–load spectrum from the initial transient stage to steady state. This work provided the first systematic quantification of higher-order harmonic forces generated during transient gap resonance under nonlinear focused-wave groups, offering direct design guidance for extreme-wave scenarios. Building on that advance, Gong S. K. [24] used a two-dimensional viscous CFD wave tank in OpenFOAM to examine a pair of dissimilar boxes. By comparing the fully fixed configuration with one that allowed all six degrees of freedom, Gong demonstrated that steady-state gap resonance is strongly modulated by the coupled motion of non-identical bodies—an aspect neglected in earlier studies. The results supply a more realistic baseline for extreme-load assessments in future side-by-side or mooring arrangements. With the dual-body groundwork established, Song Z. W. [25] reported, in January 2025, the first high-fidelity CFD investigation of three barges in close proximity, where two gaps interact simultaneously. The study revealed the intricate influence of multi-gap coupling on the force spectrum and harmonic generation, providing valuable insight for floating LNG transfer operations and other multi-vessel configurations.

Although research on conical return capsules is limited, advancements in computational fluid dynamics for ship hydrodynamics and offshore structures provide robust methodologies for modeling the complex fluid–structure interactions and wave-induced responses of the Apollo capsule, which have been adapted in this study. This study examines the Apollo capsule, integrating experiments and numerical simulations to analyze its water entry and floating dynamics, focusing on hydrodynamic characteristics in still water and diverse sea conditions. By developing a CFD-based fluid dynamics model, this study constructs a predictive framework for the conical capsule's hydrodynamic characteristics and supports the optimization of its fluid dynamic shape and sea state adaptability for future designs.

2. Theory

2.1. Theoretical Background

Numerical simulations of the conical structure require modeling the flow field to analyze its motion response. Fluid flow adheres to three fundamental conservation laws—mass, momentum, and energy—mathematically expressed through governing equations.

2.1.1. Control Theory

The Reynolds-Averaged Navier–Stokes (RANS) equations [26], closely resembling the original Navier–Stokes equations, are employed to model the flow around the conical structure. Turbulence models supply closure relations for the RANS equations, governing the transport of mean flow momentum. The mean mass and momentum conservation equations are expressed as follows:

$$\frac{\partial \rho}{\partial t} + \nabla \cdot (\rho \overline{v}) = 0 \tag{1}$$

$$\frac{\partial}{\partial t}(\rho \overline{v}) + \nabla \cdot (\rho \overline{v} \otimes \overline{v}) = -\nabla \cdot \overline{p} I + \nabla \cdot (T + T_t) + f_b \tag{2}$$

where ρ denotes density, $\overline{v}$ and $\overline{p}$ represent mean velocity and pressure, respectively, I is the identity tensor, T is the viscous stress tensor, T_t is the Reynolds stress tensor, and f_b is the resultant body force.

The Reynolds-Averaged Navier–Stokes (RANS) equations are discretized via the finite volume method (FVM) with the k-ε turbulence model. The Semi-Implicit Method for Pressure-Linked Equations (SIMPLE) is employed to couple pressure and velocity. The segregated flow model solves the equations independently, discretizing convection terms with a second-order upwind scheme. The volume of fluid (VOF) method accurately models the free surface evolution in two-phase flow. The dynamic fluid–body interaction (DFBI) module and overlapping grid technology account for the conical structure's motion.

2.1.2. Wave Theory

The volume of fluid (VOF) wave model simulates surface gravity waves at the interface of light and heavy fluids [27]. Typically, it is coupled with the six-degree-of-freedom (6DOF) model for ocean engineering applications. The model characterizes fluid velocity, pressure, and wave height using steadily propagating periodic wave trains. These wave trains are defined and solved based on three fundamental quantities: water depth, wavelength, and wave height. The first-order VOF wave velocity is determined as follows:

The wave surface elevation Is given by the following:

$$\eta = A\cos(kx - \omega t) \tag{3}$$

The horizontal velocity is as follows:

$$u = A\omega\cos(kx - \omega t)e^{kz} \quad (4)$$

The vertical velocity is as follows:

$$w = A\omega\sin(kx - \omega t)e^{kz} \quad (5)$$

where A denotes the wave amplitude, ω the wave frequency, k the wave number, and z the vertical distance from the mean water level.

The wave period for the i-th sub-wave is as follows:

$$T = \frac{2\pi}{\omega} \quad (6)$$

Compared to paddle wave making, the VOF wave superposition module generates waves more efficiently and conveniently, with the expression referenced as follows:

$$\eta = \sum_i A_i\cos(k_i x - \omega_i t) \quad (7)$$

where A_i denotes the amplitude, ω_i the frequency, and k_i the wave number of the i-th sub-wave in the superimposed wave.

3. Numerical Model Development and Validation

3.1. Environmental Conditions

This study's environmental conditions are derived from hydrological data in the northern South China Sea, with reference locations marked by red solid circles in Figure 1. The wave parameters in Table 1 represent fifth-order Stokes waves, generated using the high-order volume of fluid (VOF) method, to provide a consistent and controlled wave environment for analyzing the capsule's hydrodynamic characteristics. Wave data from 2012–2013 in this region were thoroughly analyzed, including maximum wave height, mean period, and other parameters. The South China Sea, situated off China's southern coast, is China's largest marginal sea and the world's second-largest sea, making its vast expanse a natural landing site for return capsules.

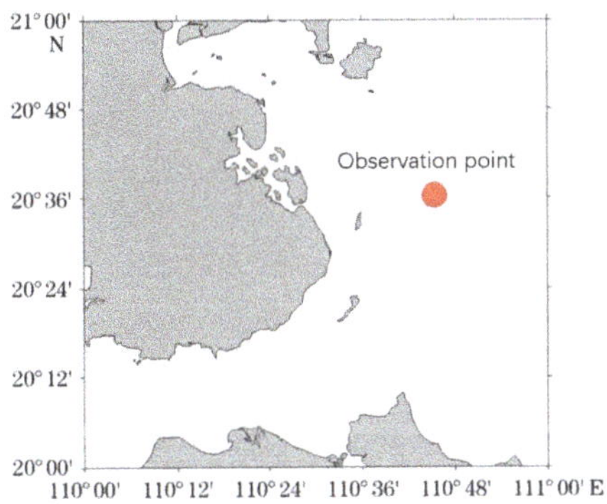

Figure 1. Geographical Locations of Observation Points in the Northern South China Sea.

Table 1. Measured sea state data for fifth-order Stokes waves, derived from hydrological data in the northern South China Sea.

Sea Condition Levels	Level 3	Level 4	Level 5
Wave Height (m)	1	2	3
Wavelength (m)	8	16	24
Wave Period (s)	2.5	3.6	4.7
Wind Speed (m/s)	4	8	11

3.2. Eccentric Conical Structure Model

This study employs the Apollo return capsule model [28], depicted in Figure 2. The floating body, shaped as a conical structure, is hollow with an integrated center-of-mass adjustment mechanism and a skirt containing eight interconnected compartments for gas cylinders, batteries, and other components. The eccentric conical structure, constructed from aluminum alloy (density 2800 kg/m^3), has a cabin thickness of 10 mm. The model's mass parameters are computed using a mass properties plug-in, as presented in Table 2. The center-of-mass coordinates are provided in both the geodetic coordinate system and the structural coordinate system of the eccentric conical structure, with the origin at the center of the platform's top circle, as shown in Figure 1. The x-axis is perpendicular to the model's top plane and directed upward, the y-axis lies parallel to the platform's top plane and points to the cabin's right, and the z-axis lies parallel to the platform's top plane and points to the object's front. These definitions enable calculation of the platform's detailed parameters, including mass, center of mass, and inertia tensor relative to the center of mass.

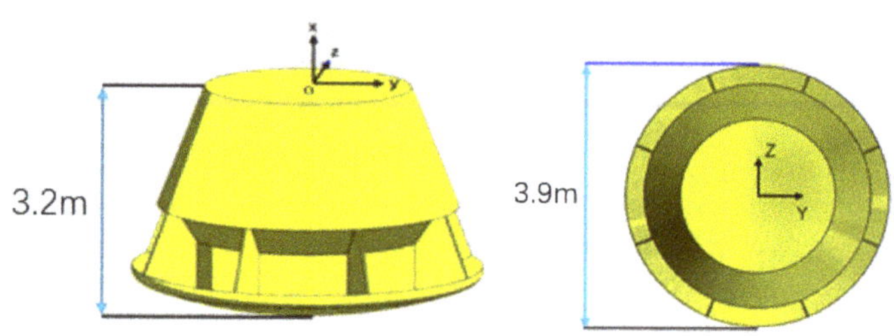

Figure 2. The structural coordinate system for the eccentric conical structure.

Table 2. Mass parameters of the eccentric conical structure.

Parameter Name	Parameter Value
Total Mass (kg)	7800
Center of Mass X_c (mm)	−1750
Center of Mass Y_c (mm)	30
Center of Mass Z_c (mm)	0
Inertia Tensor I_{xx}(kg·m^2)	9675.11
Inertia Tensor I_{yy} (kg·m^2)	11,805.97
Inertia Tensor I_{zz} (kg·m^2)	11,266.03

3.3. Simulation Setup

In this study's numerical simulation, a 1:1 scale model is used, with a computational domain of 50 m × 50 m, a water depth of 10 m, and an air height of 5 m above the water surface. Boundary conditions follow standard numerical simulation methods [29]: velocity inlets at the front and back boundaries, a pressure outlet at the top, symmetry planes at the sides and bottom, and a wall boundary at the conical structure's surface, as depicted in Figure 3.

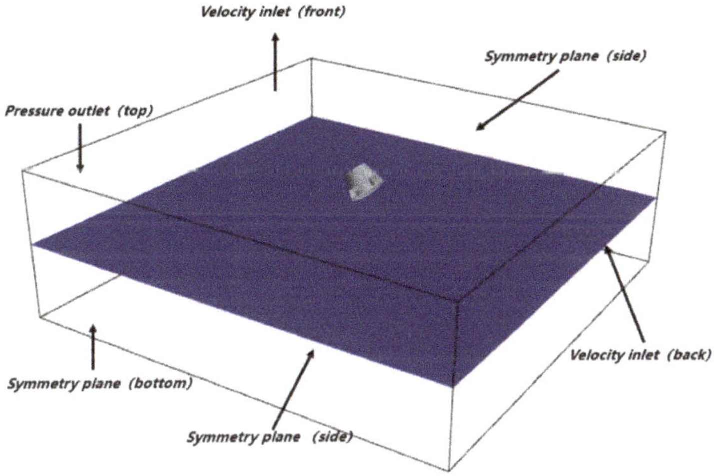

Figure 3. The schematic of the computational domain and boundary conditions.

This study employs structured overlapping grids, refining the mesh near the conical structure to optimize computational resources and enhance simulation accuracy. The mesh size varies across the computational domain: larger far from the platform and smaller near it, with overlapping grids in the central region simulating platform motion, as depicted in Figure 4. Prior to overlapping, the mesh is divided into background and overlapping grids. The computational domain typically comprises one background grid and one or more smaller overlapping grid regions. Overlapping grids categorize cells as active or inactive. Active cells solve the discretized governing equations. Inactive cells do not solve equations, but movement of the overlapping grid region may activate some inactive cells. The relative motion between sub-regions requires only the definition of motion laws, without grid deformation or regeneration. This method excels in handling complex geometries and dynamic object motion. Overlapping grids, due to their ease of generation, are widely used for managing complex mesh layouts.

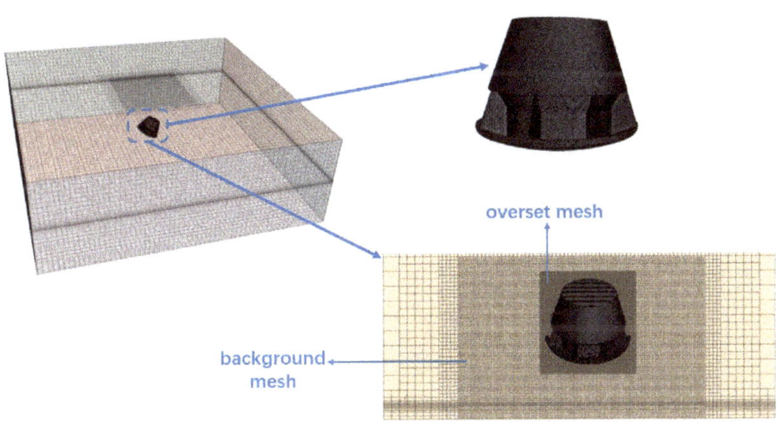

Figure 4. The schematic of the computational mesh.

Grid independence analysis verifies that the selected mesh division and time steps meet hydrodynamic accuracy requirements for wave–structure interactions. Table 3 presents the grid independence study for coarse, medium, and fine mesh types under

Sea State 3 conditions, detailing maximum roll, pitch, and yaw responses of the conical structure. Comparisons reveal significant differences between coarse and fine meshes, whereas medium and fine meshes exhibit a maximum difference of 4.5% for pitch, with other key parameters varying by less than 5%, satisfying acceptable accuracy thresholds. Compared to fine meshes, medium meshes reduce computation time by approximately 25%, and coarse meshes by 46%. Considering computational efficiency, medium-density meshes were selected for numerical simulations.

Table 3. Grid independence analysis.

Group	Cells (Millions)	$Pitch_{max}$	$Pitch_{diff}$	Yaw_{max}	Yaw_{diff}	$Roll_{max}$	$Roll_{diff}$
Coarse	1.5	4.70	5.86%	1.31	4.8%	28.13	1.7%
Medium	2.6	4.50	1.35%	1.27	1.6%	27.72	0.22%
Fine	3.3	4.44	0%	1.25	0%	27.66	0%

From the perspective of grid computing and meshing principles, this grid resolution demonstrates robust generality across the simulated wave conditions (wave heights 1–5 m). The mesh was designed to resolve the spatial and temporal scales of nonlinear wave-structure interactions, governed by consistent wave steepness across all sea states. The grid size and time step adhere to the following:

$$\Delta x \leq \frac{\lambda}{N}, \Delta t \leq \frac{T_w}{M} \tag{8}$$

where Δx and Δt are the spatial and temporal grid sizes, λ is the wavelength, T_w is the wave period, and N and M are constants ensuring adequate resolution. This resolution captures dominant hydrodynamic phenomena, such as boundary layer dynamics and vortex shedding, which scale primarily with wave steepness rather than wave height. The stable convergence of motion responses, including roll, pitch, and yaw, across all sea states, with errors below 5%, confirms the grid's applicability to higher sea states without requiring additional studies. The volume-of-fluid (VOF) method and k-ε turbulence model ensure numerical stability by resolving free-surface dynamics and turbulent flows, maintaining accuracy across varying wave heights. This generality extends to multiple motion modes, as the grid effectively captures the coupled dynamics of roll, pitch, and yaw induced by eccentric structural configurations, ensuring reliable simulations across the studied conditions.

3.4. Experimental Pool Testing and Validation

3.4.1. Subsubsection Experimental Design

The eccentric conical structure was tested in a 50 m × 30 m × 4 m rectangular pool under still water conditions. As depicted in Figure 5, a bridge crane system lifted the platform to a preset height and released it, with the vertical water entry speed controlled by adjusting the lifting height. The initial vertical speed was calculated using the free fall formula $v = \sqrt{2gh}$, where the lifting height h was measured as the vertical distance from the platform's bottom to the water surface using a ruler.

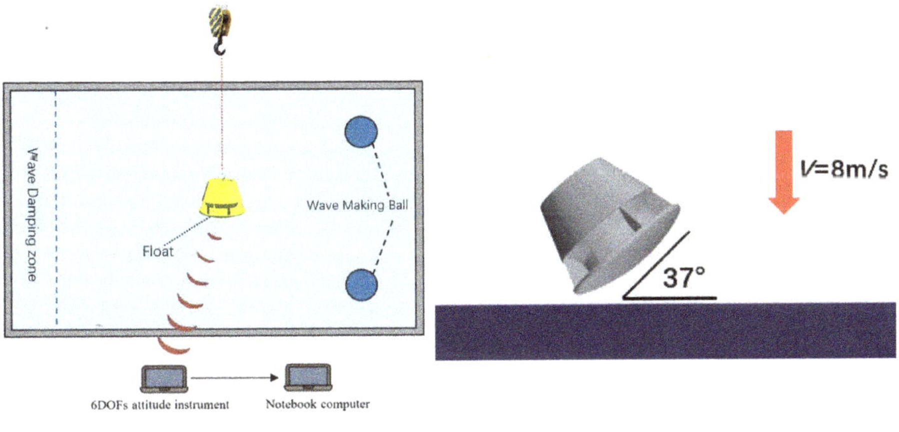

Figure 5. The schematic of the pool experiment: (**a**) experimental flow chart; (**b**) the schematic diagram of the angle of entry into the water.

During the water entry experiment, a three-axis attitude sensor monitored the platform's pitch, roll, and yaw angles in real time. Time-domain low-pass filtering (cutoff frequency: 20 Hz) was applied to denoise the data, effectively removing high-frequency interference. The experiment verified floating stability under initial conditions: a 37° tilt angle, 8 m/s vertical speed, and zero horizontal speed.

3.4.2. Results and Analysis

This experiment validates the hydrodynamic numerical model's accuracy for the eccentric conical structure. Figure 6 shows the dynamic response of the platform's three-degree-of-freedom attitude angles (pitch, roll, yaw) during the experiment. Experimental data indicate that, during early water entry ($t < 1.5$ s), the platform experiences significant angular displacement around the z-axis ($\theta_{max} = 129°$), stabilizing to an equilibrium state ($\theta_s = 55.2°$) via self-adjustment, with transient roll oscillations around the x-axis ($\Delta\Phi < 5°$).

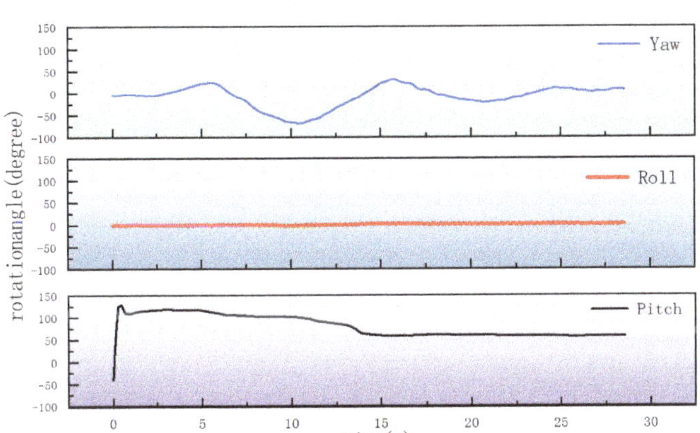

Figure 6. Attitude angle variations in the eccentric conical structure during the pool experiment.

Figure 7 compares steady-state pitch angles, revealing a 3.2% relative deviation between experimental and numerical results, confirming the model's accuracy in predicting equilibrium states. Figure 8 analyzes the dynamic response, showing the experimental system reaches the maximum tilt angle 4.4 s earlier than the numerical model (t_{exp} = 0.5 s vs. t_{cfd} = 4.9 s), with a 23% shorter stabilization time. These differences arise from two numerical model assumptions: (1) a fluid viscosity coefficient set to the theoretical value $\mu = 1.0 \times 10^{-3}$ Pa·s, lower than the experimental water's measured viscosity, and (2) a rigid body model excluding structural elastic damping, leading to an 18% reduction in fluid damping due to viscosity parameter deviations. The 18% reduction in fluid damping, due to viscosity parameter deviations, primarily causes the delayed dynamic response in the numerical solution.

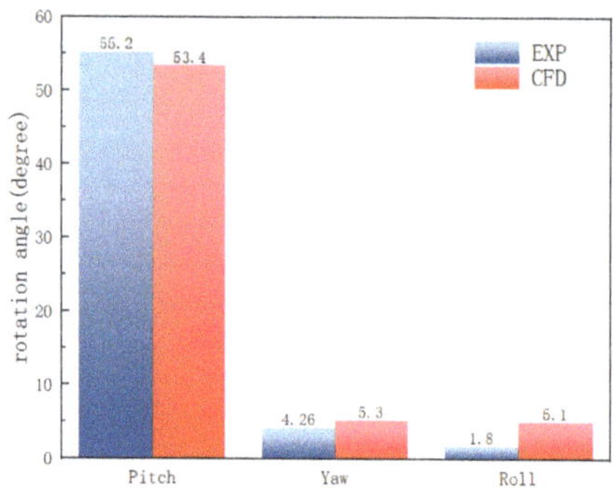

Figure 7. The steady-state comparison of pitch, roll, and yaw for the eccentric conical structure.

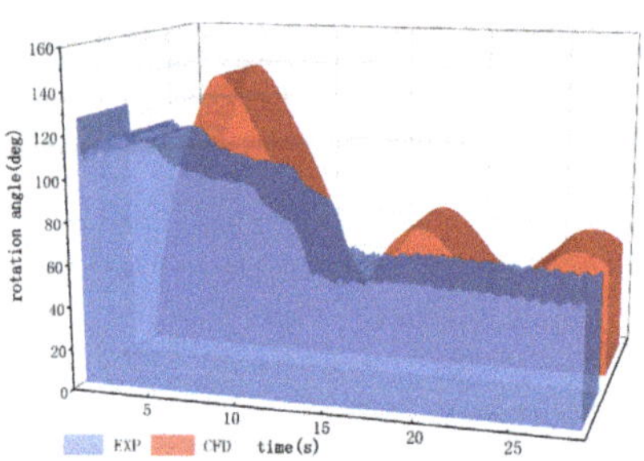

Figure 8. Dynamic pitch angle variation comparison.

Figure 7 compares steady-state pitch angles, revealing a 3.2% relative deviation between experimental and numerical results, confirming the model's accuracy in predicting equilibrium states. Figure 8 analyzes the dynamic response, showing the experimental system reaches the maximum tilt angle 4.4 s earlier than the numerical model (t_{exp} = 0.5 s vs. t_{cfd} = 4.9 s), with a 23% shorter stabilization time. These differences arise from two numerical model assumptions: (1) a fluid viscosity coefficient set to the theoretical value $\mu = 1.0 \times 10^{-3}$ Pa·s, lower than the experimental water's measured viscosity (approximately $\mu = 1.22 \times 10^{-3}$ Pa·s, accounting for temperature and impurities), and (2) a rigid-body model excluding structural elastic damping. The 18% reduction in fluid damping, primarily due to viscosity parameter deviations, causes the delayed dynamic response in the numerical solution. The fluid damping force is modeled as follows:

$$F_d = C_d \mu v A \tag{9}$$

where C_d is the damping coefficient (determined via CFD calibration), μ is the fluid viscosity, v is the relative velocity, and A is the wetted surface area. The viscosity deviation results in a proportional reduction in damping force:

This closely aligns with the observed 18% damping reduction.

$$\frac{\Delta F_d}{F_d} \approx \frac{\Delta \mu}{\mu} = \frac{1.22 \times 10^{-3} - 1.0 \times 10^{-3}}{1.22 \times 10^{-3}} \approx 16.67\% \quad (10)$$

Despite transient phase differences, relative errors of key parameters remain below 10% across the operating range, validating the numerical model's accuracy in (1) predicting steady-state equilibrium angles (error band $\pm$ 5°); (2) reproducing nonlinear decay oscillation modes (attenuation index β error < 8%); and (3) characterizing multi-degree-of-freedom motion coupling (covariance matching degree > 91%). These validation results ensure reliability for subsequent numerical studies.

4. Steady-State Analysis

In return capsule design, the alignment of cabin mass characteristics with displacement volume is a critical parameter governing hydrodynamic stability. In practice, uncertainties from assembly tolerances ($\pm$1.5% mass moment of inertia) and trim errors ($\Delta V \leq 0.3$ m^3) cause systematic deviations between actual and theoretical mass distributions. This section develops a dynamic characteristic model under multi-parameter coupling to investigate the influence of mass and displacement volume on the platform's stability domain.

4.1. Impact of Mass Characteristics on Platform Floating Stability

Using the previously developed parameterized numerical model, six design sample spaces with mass deviations ($\Delta y \in [30, 180]$ mm) and ($\Delta x \in [-1750, -1670]$ mm) are generated via Latin Hypercube Sampling (LHS) [30]. A six-degree-of-freedom coupled dynamics solver quantifies the effects of each parameter on (1) steady-state equilibrium angle and (2) dynamic response amplitude.

Numerical simulations, accounting for the platform's asymmetric mass distribution, reveal a significant +y-axis tilt under hydrostatic equilibrium (tilt angle θ = 24.3°, see Figure 9). To meet manned spacecraft attitude tolerance requirements, maintaining a stable upward cabin orientation is critical, with the lateral center-of-mass offset (y-axis coordinate) as the primary control parameter influencing the attitude angle.

Figure 9. Stability state diagram of eccentric cone.

Using the viscous fluid–rigid body coupled dynamics model from Section 3.2, this study conducts a parameterized scan of the y-axis center-of-mass offset, varying from 30 mm to 180 mm along the +y-axis (Δy = +150 mm, step size 25 mm). Transient numerical simulations yield the pitch angle time history, steady-state equilibrium angle, and dynamic response amplitude (see Figure 10). Within the operating range (Δy = 30–180 mm), the platform maintains a single stable equilibrium without capsizing instability, as shown by the numerical results.

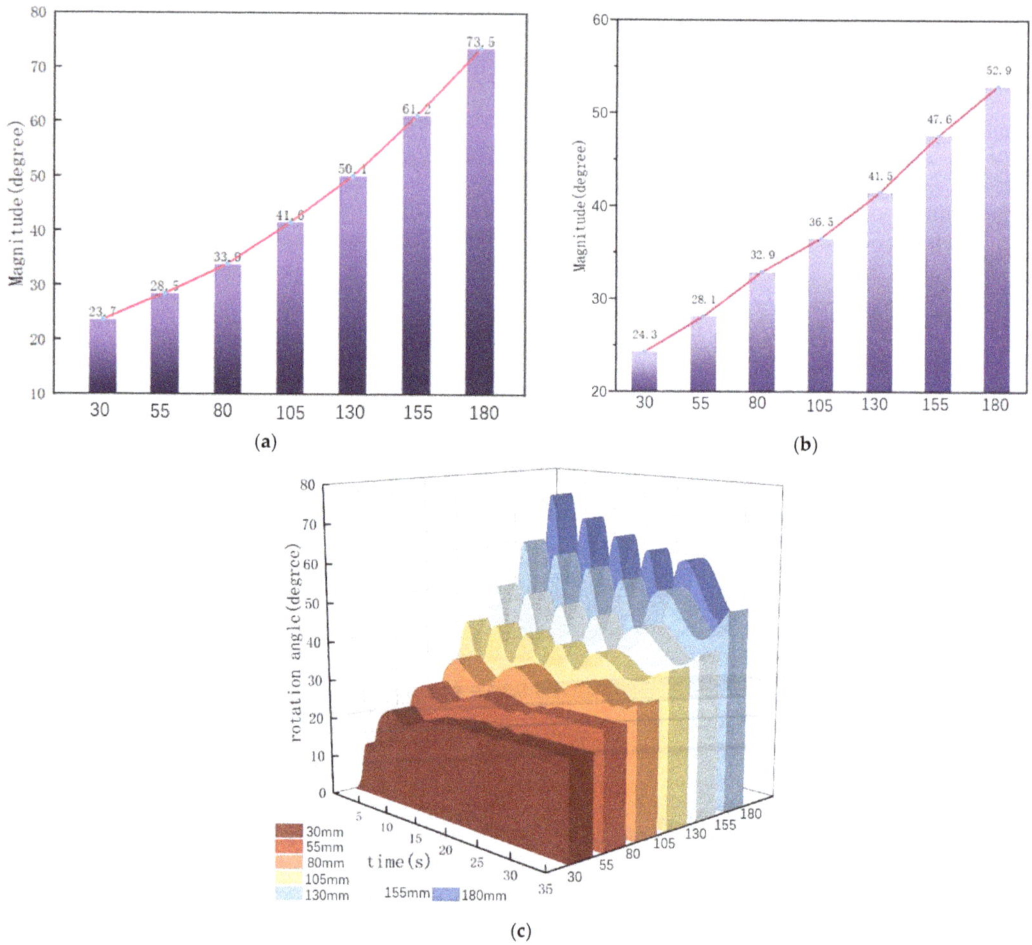

Figure 10. The effects of y-axis center-of-mass offset on the eccentric conical structure: (**a**) maximum pitch angle; (**b**) stable attitude; (**c**) pitch angle response.

The longitudinal center-of-mass offset (x-axis) also significantly affects the platform's hydrodynamic stability, alongside the lateral offset (y-axis). This study employs the same six-degree-of-freedom coupled numerical model, shifting the center of mass 80 mm along the positive x-axis (from −1750 mm to −1670 mm). Parameterized numerical simulations yield the pitch angle dynamic response curve, equilibrium attitude angle, and extreme parameters (see Figure 11). Numerical results indicate that, as the x-axis center-of-mass shifts forward linearly, the pitch angle response amplitude in stable equilibrium increases monotonically by 33.5%, with no capsizing instability observed. Comparative analysis reveals that the y-axis mass eccentricity influence coefficient (0.155 deg/mm) on the steady-state equilibrium angle exceeds that of the x-axis (0.095 deg/mm), highlighting the dominant role of lateral center-of-mass distribution in hydrodynamic stability design. Notably, this analysis assumes an ideal fluid medium (Sea State 0). The presence of multi-stable equilibrium points and parametric roll risks in real wave conditions will be evaluated through subsequent restoring moment analysis.

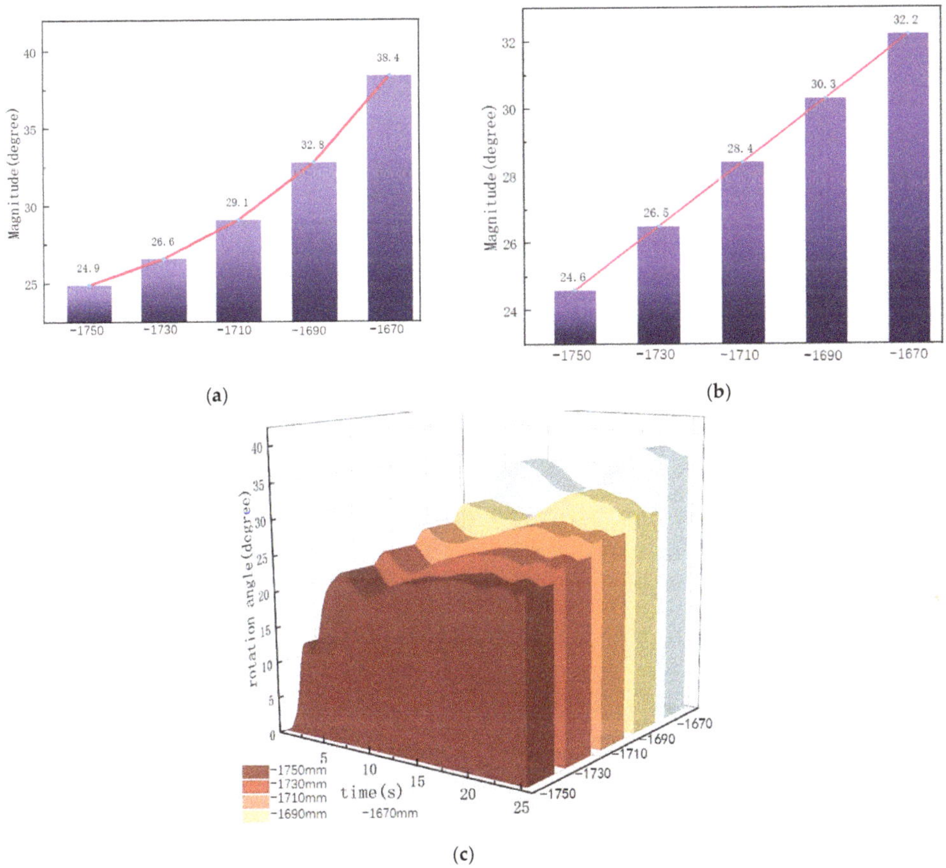

Figure 11. The effects of x-axis center-of-mass offset on the eccentric conical structure: (**a**) maximum pitch angle; (**b**) stable attitude; (**c**) pitch angle response.

4.2. Restoring Moment Analysis of the Eccentric Conical Structure

To assess if the platform has a single stable equilibrium under the designed center-of-mass configuration, its restoring moment is calculated [31]. This method identifies all possible stable equilibria while excluding water entry effects on the platform's stable attitude [32]. Restoring moment analysis for offshore structures is derived from ship hydrodynamics and fluid dynamics. A ship's restoring moment depends on its moment of inertia and the relative positions of its center of gravity and center of buoyancy. For small heel angles, a ship is stable when the metacenter is vertically above the center of gravity. For large heel angles, stability is assessed by the restoring moment, arising from the coupling of gravity and buoyancy, expressed as follows:

$$M_R = W_{ship} \times \overline{GZ} \tag{11}$$

where W_{ship} is the ship's weight, M is the righting moment, and $\overline{GZ}$ is the horizontal distance between the center of buoyancy and center of gravity. Figure 12 illustrates that the restoring moment's magnitude primarily depends on $\overline{GZ}$. For small heel angles, the restoring arm is calculated using the metacentric height:

$$\overline{GZ} = \overline{GM}\sin\phi \tag{12}$$

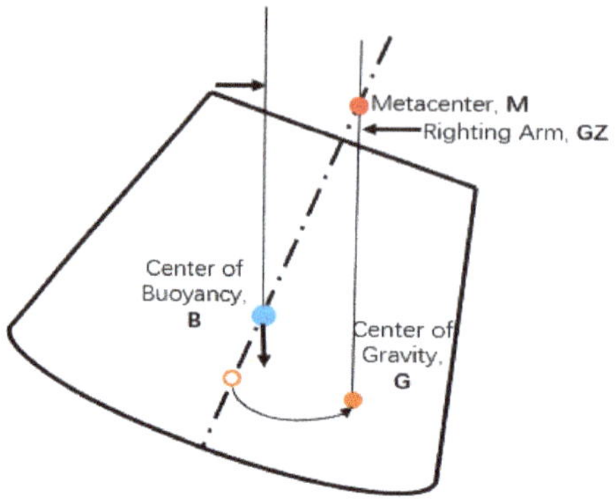

Figure 12. Restoring arm.

Using the stability calculation module, multi-angle static stability simulations are conducted on the platform, systematically evaluating the restoring moment response at various tilt angles (see Figure 13). Numerical results indicate that, under the designed center-of-mass configuration, the platform exhibits a single stable equilibrium, satisfying the restoring moment condition $M_R = 0$. Parameterized scans reveal that, at a 26° tilt angle, the restoring moment is zero ($M_R(26°) = 0$ N·m), confirming the platform's static equilibrium, with the center of buoyancy aligning with the center of mass. This solution aligns closely with theoretical predictions in Section 4.2 (relative error < 2%), validating the hydrostatic equilibrium calculation method.

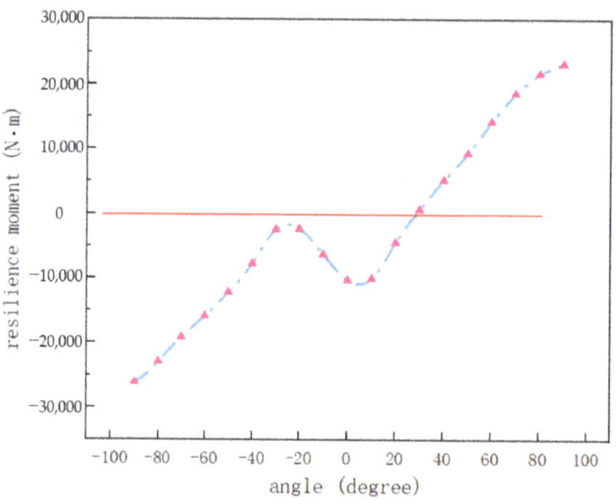

Figure 13. Restoring moment of the eccentric conical structure.

4.3. Impact of Displacement Volume on Platform Floating Stability

The return capsule's skirt structure comprises eight circumferentially distributed compartments with asymmetric configurations. The +y-axis compartment houses the deceleration parachute system, while the remaining seven compartments contain battery packs, high-pressure gas cylinders, and other life support equipment. During water entry, parachute ejection from the +y-axis compartment causes significant variations in water ingress among compartments. This non-uniform fluid intrusion alters the cabin's displacement volume distribution, significantly impacting its hydrostatic stability. To investigate the effect of skirt displacement volume distribution on the cabin's hydrostatic equilibrium, three conditions were designed (see Figure 14): (1) baseline: no foam filling; (2) uniform: 0.5 m^3 of closed-cell foam evenly distributed across eight compartments; and (3) directional: 0.5 m^3 of closed-cell foam in the $+y$-axis compartment only. Controlling the foam's spatial distribution enables a systematic study of displacement volume effects on cabin attitude.

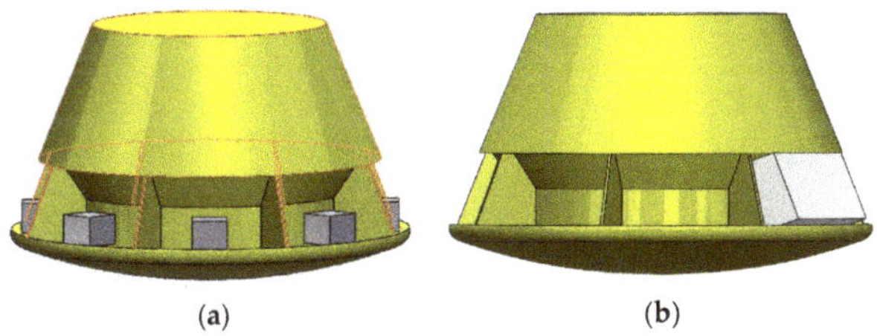

Figure 14. Cabin states for (**a**) uniform filling and (**b**) directional filling.

Simulations (Figure 15) show that increasing displacement volume reduces the platform's maximum pitch angle and stable tilt angle. Compared to the no-foam baseline, uniform foam filling in eight compartments reduces the maximum tilt angle by 12.1% and the stable tilt angle by 18.3%. Directional filling in the $+y$-axis compartment reduces the maximum tilt angle by 31.9% and the stable tilt angle by 56%. This effect likely stems from cabin tilt, which prevents foam in the $-y$-axis empty compartment from fully detaching from the water surface, reducing the effective displacement volume compared to directional filling. This suggests that a filling device could be designed to increase the $+y$-axis compartment's displacement volume post-parachute ejection, aiding capsule righting.

Simulations (Figure 15) reveal significant differences in the capsule's hydrodynamic characteristics across the three conditions: (1) uniform filling reduces the maximum pitch angle by 12.1% and the steady-state angle by 18.3%, and (2) directional filling in the $+y$-axis compartment achieves a 31.9% reduction in the maximum pitch angle and a 56% reduction in the steady-state angle. This effect results from buoyancy redistribution due to the platform's asymmetric mass distribution. The reduced waterline surface in the $-y$-axis empty compartment, caused by the tilt angle, lowers the effective displacement volume below the theoretical value. To quantify the damping enhancement of the directional foam-filling strategy, we analyzed the energy dissipation rate (Γ), defined as follows:

$$\Gamma = \frac{2\beta E}{\Delta t} \tag{13}$$

where β represents the decay index of pitch angle oscillations, E is the initial oscillation energy, and Δt is the oscillation period. Using numerical simulations from Figure 15, Table 4 presents the initial pitch angles, oscillation energies, and energy dissipation rates for baseline, uniform, and directional foam-filling conditions. The baseline condition exhibits a higher dissipation rate due to larger initial oscillations, while uniform and directional filling show progressively lower rates, reflecting enhanced damping through buoyancy redistribution. The reduced dissipation rate in directional filling indicates faster oscillation decay, aligning with its substantial reduction in pitch angles.

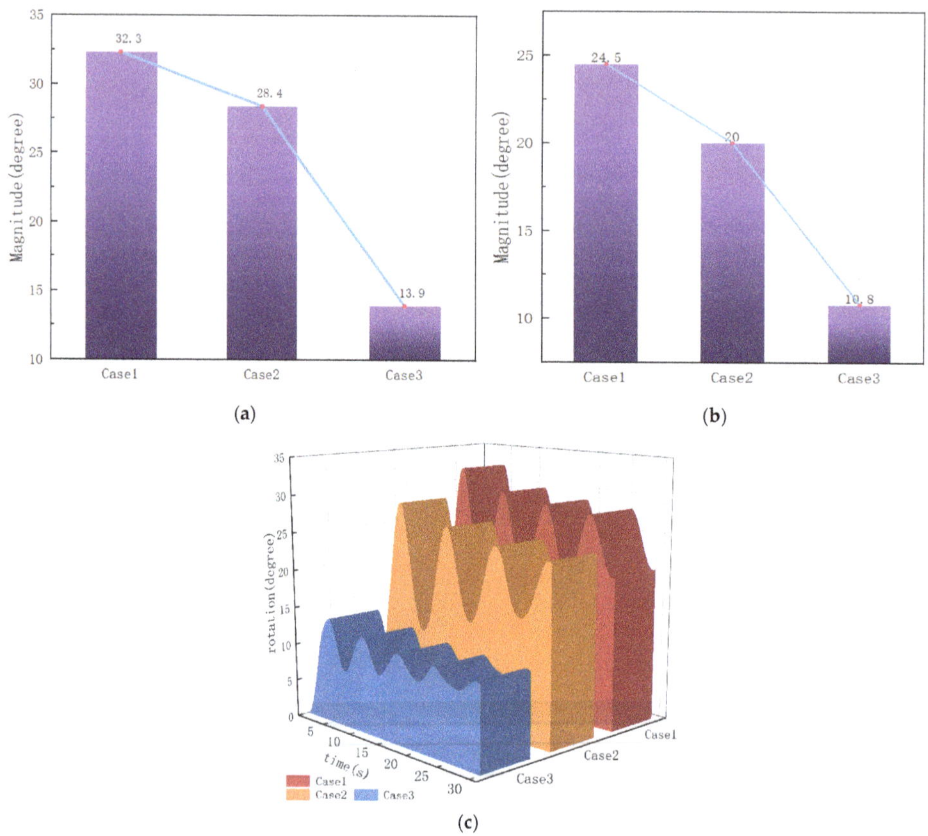

Figure 15. The effects of displacement volume on the eccentric conical structure: (**a**) maximum pitch angle; (**b**) stable attitude; (**c**) pitch angle response.

Table 4. Foam-filling damping metrics.

Condition	Initial Pitch Angle (°)	Oscillation (J)	Energy Dissipation Rate (J/s)
Baseline	24.5	1030.3	41.2
Uniform Filling	20.0	686.7	31.6
Directional Filling	10.8	200.2	10.4

Analysis reveals that directional filling enhances hydrostatic moment balance by generating a +y-axis buoyancy correction moment, increasing the restoring moment coefficient by 64%. This finding suggests engineering optimization: post-parachute separation, a +y-axis controllable ballast system (capacity ≥ 0.5 m^3) can maintain the steady-state angle within safety limits via active buoyancy compensation. Numerical validation confirms that this approach shortens the capsule's righting time and enhances astronaut safety margins during water entry.

5. Multi-Physics Coupling Analysis

The marine environment is dynamic. After landing, the return capsule is influenced by waves, wind, and ocean currents, exhibiting complex coupling mechanisms. This section analyzes sea state conditions from Table 1 using three load combinations, i.e., waves only, waves with ocean currents, and waves with ocean currents and wind, to investigate the platform's dynamic response mechanisms. The analysis of wind and current effects in this section focuses on their interactions with waves from the -y-axis direction, as this maximizes their impact on the eccentric conical floating structure's motion responses due to its lateral center-of-mass offset along the +y-axis. Section 4.1 demonstrates that a +y-axis offset induces a significant pitch angle, making winds and currents from the opposite direction most critical for amplifying surge displacement and pitch angle extrema, as validated by CFD simulations.

Waves induce surge (y-axis translation), heave (x-axis translation), and pitch (z-axis rotation) in the floating platform. Once stabilized, the platform's motion responses are analyzed (see Figure 16). The analysis reveals the following: (1) The platform's motion responses are primarily governed by wave period. With waves alone, the motion period is twice the wave period. This occurs because, at this wavelength, the platform cannot return to a stable state before the next wave's impact. (2) Ocean currents reduce the motion period from 5.0 s to 4.9 s, altering pitch angular velocity and surge extrema by 2–8%, while increasing surge displacement by 27%. (3) Wind primarily increases pitch angle extrema by up to 32%, due to the platform's 3 m height and large windward area. (4) The platform's motion responses vary across cycles, primarily due to the fifth-order Stokes waves used in this study.

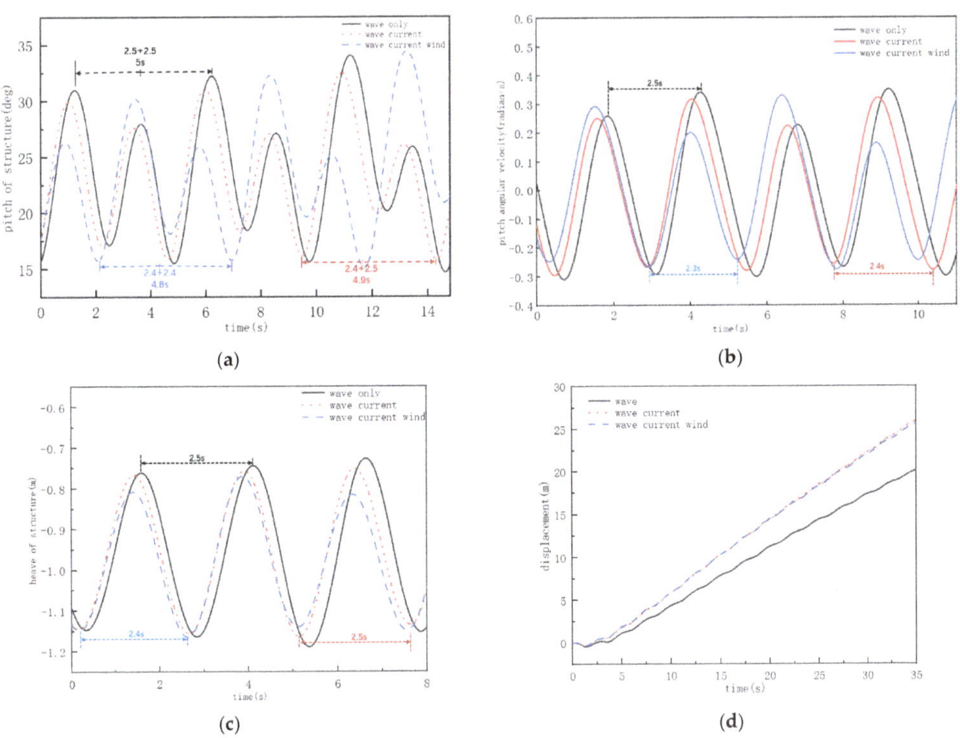

Figure 16. The motion responses of the eccentric conical structure: (**a**) pitch variation; (**b**) pitch velocity variation; (**c**) heave variation; (**d**) surge displacement.

The analysis of wave-coupled dynamics (Figure 16) identifies three key platform motion responses, i.e., surge (y-axis translation), heave (x-axis translation), and pitch (z-axis rotation): (1) fifth-order Stokes waves induce 1/2 subharmonic resonance in the platform, extending the motion period to 5.0 s, driven by frequency multiplication coupling between wave excitation and structural natural frequency. (2) Ocean currents reduce the motion period by 2% (T = 4.9 s) and increase surge displacement by 27%, due to the flow-induced added mass effect altering the system's kinetic energy transmission. (3) Wind load amplifies pitch angle extrema by 32%, resulting from aerodynamic pressure center offset caused by the 3 m structural height. (4) Fifth-order Stokes waves excitation induces unsteady responses, with broadband characteristics causing quasi-periodic motion trajectories.

This study elucidates the nonlinear hydrodynamic response of the platform under multi-load coupling, confirming that wave–current–wind interactions significantly alter its energy transmission and hydrodynamic behavior. These findings support the optimization and control of the return capsule system and can be extended to analyze structural parameters under complex sea conditions across various geographical locations.

6. Wave Parameter Effects

Section 6 extends the findings of Section 5, which established that wind and current effects on the motion responses of the eccentric conical floating structure are negligible compared to wave parameters. Consequently, this study prioritizes wave height and wavelength as the primary drivers of hydrodynamic characteristics and dynamic stability. Using wave parameters from Table 1, this section systematically analyzes their relationship with the platform's nonlinear hydrodynamic responses, elucidating their dominant role in shaping motion patterns under diverse wave conditions.

6.1. Wavelength Effects

Using the high-order volume of fluid (VOF) method, fifth-order Stokes waves are generated (surface reconstruction accuracy $\eta \geq 0.97$) with a baseline wave environment of 1 m wave height and 0° incident angle (head-on). Parameterized analysis investigates the effects of wavelengths (λ = 8–32 m, 8 m increments) on the platform's hydrodynamic characteristics, focusing on surge-heave coupled motion and pitch angle velocity amplitudes.

Figure 17 shows that wavelength (λ = 8–32 m) significantly influences the platform's hydrodynamic responses. The key findings include the following: (1) Motion period evolution: As λ increases from 8 m to 32 m, the pitch angle period T_p shifts from 2 T_w (where T_w is the wave period) to T_w. This results from reduced phase differences in fluid excitation, enabling faster dynamic balance between the platform's restoring moment and wave excitation moment. (2) Nonlinear surge displacement growth: Surge amplitude increases by 70% as λ rises from 8 m to 16 m, then slows to a 7% increase beyond 16 m. This is driven by a fluid–solid coupling mode transition at the critical wavelength ratio. At λ = 8 m, surge amplitude exhibits strong nonlinear growth in the subcritical region. Beyond 16 m, it enters the supercritical region, with growth slowing to 7%. The wave crest action time under these conditions is expressed as follows:

$$t_p = \frac{\lambda}{2c} = \frac{\lambda}{2\sqrt{\frac{g\lambda}{2\pi}}} = \sqrt{\frac{\pi\lambda}{2g}} \tag{14}$$

where t_p is the wave crest action time, λ is the wavelength, and c is the water flow speed. When t_p is less than the structural natural period, kinetic energy input is incomplete, yielding an energy transmission efficiency η_e < 45%. Otherwise, a quasi-steady state is reached with $\eta_e \approx$ 82%. (3) Pitch angle amplitude dependence on wavelength: Pitch angle extrema increase linearly with λ, while angular velocity extrema remain stable at 0.82 ± 0.05 rad/s. This suggests wavelength primarily affects displacement response amplitude, with minimal impact on inertial forces. This subsection elucidates the multi-scale effects of wavelength on the platform's hydrodynamic characteristics via phase and energy transmission adjustments. Wave height, another critical parameter, influences fluid–structure nonlinear coupling and will be examined next.

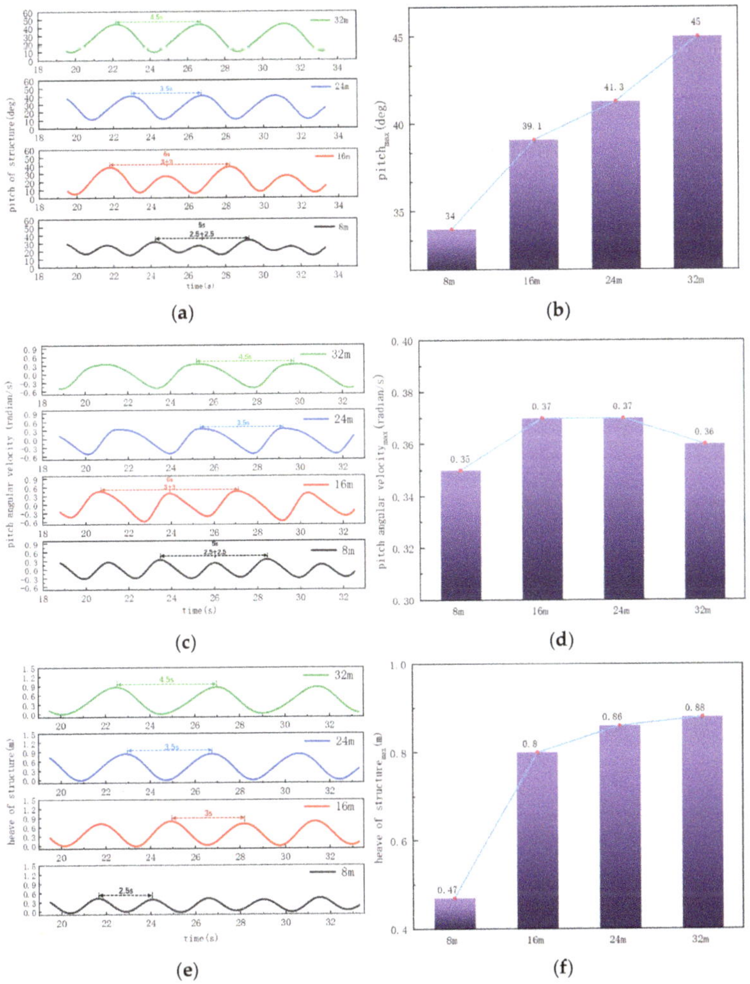

Figure 17. Platform motion responses at varying wavelengths: (**a**) pitch variation; (**b**) pitch angle extremum; (**c**) pitch velocity variation; (**d**) pitch angular velocity extremum; (**e**) heave variation; (**f**) heave extremum.

6.2. Wave Height Effects

Using the high-order VOF method, fifth-order Stokes waves are generated (surface reconstruction accuracy $\eta \geq 0.97$) with a baseline deep-water wave environment of λ = 40 m and 0° incident angle (head-on) [33]. Wave heights (H = 1–5 m), covering WMO Sea States 3–6, are analyzed to elucidate the nonlinear effects of height gradients on the platform's hydrodynamic characteristics [34].

Numerical analysis (Figure 18) reveals that wave height significantly influences the platform's hydrodynamic characteristics through nonlinear regulation [35]. In Sea States 3–5 (H = 1–3 m, δ = 0.125), motion parameters (e.g., pitch angle, surge frequency) align closely with the wave fundamental frequency (f_s/f_w = 1.00 ± 0.05), with stable pitch angle extrema (48.6° ± 4.5°, CV = 8.5%), indicating predictable linear responses. At H = 5 m (Sea State 6, δ = 0.125), the system enters a strongly nonlinear regime, exhibiting subharmonic resonance and broadband random vibrations, with pitch angle extrema rising to 150°. In this state, the platform is critically stable, with the longitudinal distance between the center of buoyancy and center of mass reduced to 0.12 m. The restoring moment ($M_R = \rho g \nabla GZ$) falls to 1.3 times the wave capsizing moment's safety margin. To clarify the origin of the subharmonic resonance observed at Sea State 6, we analyzed whether it is driven by structural or fluid dynamics. The capsule's pitching natural frequency (f_n) is determined as follows:

$$f_n = \frac{1}{2\pi}\sqrt{\frac{k_\theta}{I_z}} \tag{15}$$

where k_θ is the pitching restoring stiffness, and I_z is the moment of inertia about the z-axis. Using structural parameters from Table 2 and restoring moment data from Section 4.2, the capsule's natural frequency is approximately 0.143 Hz, corresponding to a period of 7.0 s. In contrast, the wave excitation frequency for Sea State 6 is approximately 0.200 Hz (period 5.0 s), with its subharmonic component at 0.100 Hz (period 10.0 s), aligning with the observed resonance period. The proximity of the capsule's natural frequency (0.143 Hz) to the subharmonic frequency (0.100 Hz) suggests a coupled response, but the resonance is primarily fluid-driven, resulting from nonlinear wave–structure interactions at high wave steepness (δ = 0.125). This is supported by CFD simulations, which capture the nonlinear fluid force spectrum exhibiting the subharmonic component.

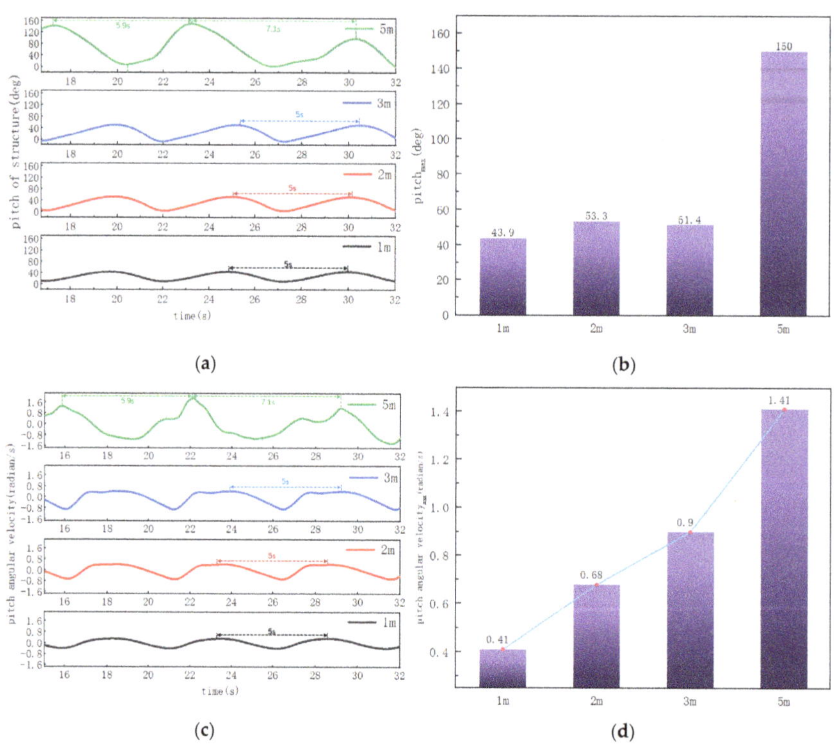

Figure 18. *Cont.*

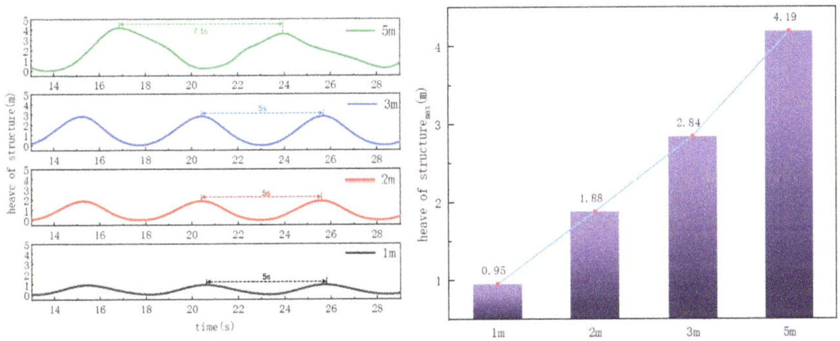

(e) (f)

Figure 18. Platform motion responses at varying waveheights: (**a**) pitch variation; (**b**) pitch angle extremum; (**c**) pitch velocity variation; (**d**) pitch angular velocity extremum; (**e**) heave variation; (**f**) heave extremum.

The subharmonic resonance significantly impacts capsule design, as amplified pitch angles (up to 150°) reduce stability margins and increase astronaut physiological risks. To mitigate this, active damping systems, such as controllable ballast mechanisms (Section 4.3), can dynamically adjust the capsule's natural frequency to avoid resonance. Additionally, optimizing the eccentric structure (Section 4.1) to minimize center-of-mass offsets and enhancing nonlinear damping through directional foam-filling (Section 4.3) can further suppress resonance effects. These strategies enhance capsule safety under extreme wave conditions, ensuring robust performance during water landings.

For H = 1–3 m, heave displacement extrema grow linearly ($heave_{max}$ = $0.95H \pm 6\%$). At H = 5 m, they decrease to $0.84H$, a reduction of 11.6%. This phenomenon primarily results from reduced energy transmission efficiency driven by wave steepness. When $\delta > 0.1$, the wave crest action time (t_p) is shorter than the structural dynamic response time (t_d), limiting effective wave energy input to 82%.

7. Conclusions and Discussion

This study investigates manned return capsule dynamics using an eccentric conical structure, employing potential flow theory and integrating experimental and numerical methods to analyze wind, wave, and current interactions. It also examines the effects of wave height and wavelength on the platform's dynamics. The key conclusions are as follows:

(1) In still water, the lateral (y-axis) center-of-mass offset sustains a single stable equilibrium via a nonlinear restoring moment driven by geometric asymmetry. Its stability sensitivity exceeds that of the longitudinal (x-axis) offset. The x-axis offset increases pitch response amplitude but does not exceed the instability threshold. Biaxial center-of-mass disturbances do not cause capsizing in still water, but wave excitation in real sea conditions may induce multi-stable equilibria.

(2) For displacement volume control, uniform foam filling across all compartments reduces motion amplitude and steady-state tilt angle compared to the no-foam baseline. Directional filling in the $+y$-axis compartment provides superior stability correction, reducing the maximum tilt angle more effectively than uniform filling. In engineering applications, post-parachute $+y$-axis ballast control can rapidly right the capsule by adjusting the center of buoyancy and center of mass.

(3) Waves induce significant surge, heave, and pitch coupling in the platform, with response periods strictly synchronized with the wave period. Ocean currents and wind loads slightly alter motion amplitude and phase, but their effects are less significant than wave periodicity. Wave parameters are the primary determinants of the platform's dynamic stability, while current and wind disturbances have limited effects on local motion modes.
(4) As wavelength increases, the platform's restoring moment and wave action moment balance more rapidly, reducing the motion period from twice the wave period to wave period synchronization. Surge displacement amplitude exhibits nonlinear growth, rising sharply up to a wavelength threshold, then leveling off. Pitch angle amplitude increases linearly with wavelength, while angular velocity remains stable, suggesting that wavelength primarily affects displacement response potential energy, not inertial forces.
(5) In moderate sea conditions (Sea States 3–5), the platform's motion frequency aligns with the wave fundamental frequency, maintaining stable pitch angle amplitudes. At a wave height of 5 m (Sea State 6), the system enters a strongly nonlinear regime, with subharmonic resonance and broadband vibrations causing a sharp increase in pitch angle extrema and a reduced safety margin. Notably, increasing wave steepness reduces energy input efficiency, attenuating heave displacement responses at extreme wave heights.

This study addresses water landing stability requirements for manned return capsules, elucidating the effects of wave parameters and structural characteristics on capsule hydrodynamic responses and highlighting the critical role of parameter thresholds in dynamic stability. The platform model verifies self-stability characteristics under still water conditions; however, it insufficiently characterizes the complexity of multi-physics field coupling effects in actual wave conditions, potentially underestimating the risk of multi-stable instability, such as parametric roll, under extreme conditions. Future research will explore capsule dynamics in complex wave environments, emphasizing adaptive center-of-mass regulation, wave spectrum sensitivity optimization, rapid righting technologies, and the influence of structural elasticity to enhance model applicability under extreme conditions, such as through elastic models analyzing damping and motion responses. Additionally, studies will investigate a broader range of wind and current directions to elucidate their coupling with waves and address practical challenges of directional foam-filling, including deployment timing and reliability under dynamic wave conditions, to advance its engineering feasibility. These advancements aim to improve water landing safety and reduce astronaut physiological risks in high wave conditions.

Author Contributions: Conceptualization, S.C. and F.Q.; methodology, F.Q.; software, F.Q.; validation, S.C. and F.Q.; formal analysis, F.Q. and K.Z.; investigation, F.Q. and K.Z.; resources, S.C.; data curation, F.Q.; writing—original draft preparation, F.Q. and K.Z.; writing—review and editing, S.C. and F.Q.; visualization, S.C. and F.Q.; supervision, S.C.; project administration, S.C.; funding acquisition, F.Q. and K.Z. All authors have read and agreed to the published version of the manuscript.

Funding: This research was funded by the Fundamental Research Funds for the Central Universities (grant number: 3132023513).

Data Availability Statement: Due to confidentiality restrictions, the data used in this study are not publicly available. However, certain relevant data may be made available from the corresponding author upon reasonable request.

Acknowledgments: The author contribution and funding sections have covered all the received support.

Conflicts of Interest: The authors declare that they have no known competing financial interests or personal relationships that could have appeared to influence the work reported in this paper.

References

1. Wang, Y.; Shi, X. Current status and progress of research on water impact problems. *Explos. Shock Waves* **2008**, *28*, 276–282.
2. Xuan, J.; Miao, Y.; Cheng, J. Experimental study and theoretical calculation of water impact characteristics of return capsules. *Hydrodyn. Res. Prog. (Ser. A)* **2000**, *3*, 276–286.
3. Mei, C.; Zhang, J.; Pan, G. Multi-objective optimization design of airbag cushioning for manned spacecraft return capsules. *Manned Spacefl.* **2015**, *21*, 444–449.
4. Takasawa, H.; Fujii, T.; Takahashi, Y.; Moriyoshi, T.; Takayanagi, H.; Nagata, Y.; Yamada, K. Static attitude stability of deep space sample return capsule with thin aeroshell. *CEAS Space J.* **2025**, *17*, 55–69. [CrossRef]
5. Lin, W.; Wang, N.; Gao, Z. Optimization of joint search and rescue by rescue ships and helicopters in remote sea areas. *J. Transp. Eng.* **2021**, *21*, 187–199.
6. Peng, Z.; Yang, H.; Wei, J. Current status and prospects of China's maritime search and rescue equipment development. *World Shipp.* **2023**, *46*, 26–29.
7. Vaughan, V.L. *Landing Characteristics and Flotation Properties of a Reentry Capsule. Unpublished report, 1961. Vaughan, V. L. Landing Characteristics and Flotation Properties of a Reentry Capsule*; NASA Langley Research Center: Hampton, VA, USA, 1961.
8. Herting, D.; Pohlen, J.; Pollock, R. Analysis and design of the Apollo landing impact system. In Proceedings of the AIAA 3rd Manned Space Flight Meeting, Houston, TX, USA, 4–6 November 1964. [CrossRef]
9. Stubbs, S.M. *Landing Characteristics of the Apollo Spacecraft with Deployed-Heat-Shield Impact Attenuation Systems*; NASA Langley Research Center: Hampton, VA, USA, 1966.
10. Benson, H.E. Water impact of the Apollo spacecraft. *J. Spacecr.* **1966**, *3*, 1282–1284. [CrossRef]
11. Baker, W.E.; Westine, P.S. Model tests for determination of the structural response of the Apollo command module to water impact. *Arch. Int. Pharmacodyn. Ther.* **2013**, *23*, 165–174. [CrossRef]
12. Thompson, W.C. *Dynamic Model Investigation of the Rough-Water Landing Characteristics of a Spacecraft*; NASA Langley Research Center: Hampton, VA, USA, 1967.
13. Wang, J.T.; Lyle, K.H. Simulating Space Capsule Water Landing with Explicit Finite Element Method. In Proceedings of the AIAA Conference, San Antonio, TX, USA, 7–10 January 2007. [CrossRef]
14. Anghileri, M.; Spizzica, A. Experimental validation of finite element models for water impacts. *J. Spacecraft Rockets* **1995**, *32*, 621–628.
15. Song, S.; Demirel, Y.K.; Atlar, M. An investigation into the effect of biofouling on the ship hydrodynamic characteristics using CFD. *Ocean Eng.* **2019**, *175*, 122–137. [CrossRef]
16. Jasak, H.; Vukevi, V.; Gatin, I.; Lalović, I. CFD validation and grid sensitivity studies of full scale ship self propulsion. *Int. J. Nav. Archit. Ocean Eng.* **2019**, *1*, 45–58. [CrossRef]
17. Wang, J.; Wan, D. CFD study of ship stopping maneuver by overset grid technique. *Ocean Eng.* **2020**, *197*, 106895. [CrossRef]
18. Jiao, J.; Huang, S.; Soares, C.G. Numerical investigation of ship motions in cross waves using CFD. *Ocean Eng.* **2021**, *223*, 108711. [CrossRef]
19. Islam, H.; Ventura, M.; Guedes Soares, C.; Tadros, M.; Abdelwahab, H.S. Comparison between Empirical and CFD Based Methods for Ship Resistance and Power Prediction. *J. Mar. Sci. Eng.* **2022**, *10*, 258. [CrossRef]
20. Kim, D.; Tezdogan, T. CFD-based hydrodynamic analyses of ship course keeping control and turning performance in irregular waves. *Ocean Eng.* **2022**, *248*, 110808. [CrossRef]
21. Andersson, J.; Shiri, A.A.; Bensow, R.E.; Jin, Y.; Wu, C.; Qiu, G.; Deng, G.; Queutey, P.; Yan, X.-K.; Horn, P.; et al. Ship-scale CFD benchmark study of a pre-swirl duct on KVLCC2. *Appl. Ocean Res.* **2022**, *123*, 103134. [CrossRef]
22. Huang, S.; Jiao, J.; Soares, C.G. Uncertainty analyses on the CFD–FEA co-simulations of ship wave loads and whipping responses. *Mar. Struct.* **2022**, *82*, 103129. [CrossRef]
23. Mi, C.; Gao, J.; Song, Z.; Liu, Y. Hydrodynamic Wave Forces on Two Side-by-Side Barges Subjected to Nonlinear Focused Wave Groups. *Ocean Eng.* **2025**, *317*, 120056. [CrossRef]
24. Gong, S.; Gao, J.; Yan, M.; Song, Z.; Shi, H. Effects of Floater Motion on Wave Loads during Steady-State Gap Resonance Occurring between Two Non-Identical Boxes. *Ocean Eng.* **2025**, *323*, 120649. [CrossRef]
25. Song, Z.; Jiao, Z.; Gao, J.; Mi, C.; Liu, Y. Numerical Investigation on the Hydrodynamic Wave Forces on the Three Barges in Proximity. *Ocean Eng.* **2025**, *316*, 119941. [CrossRef]
26. Lari, K.S. Comparison and verification of turbulence Reynolds-averaged Navier–irregular waves closures to model spatially varied flows. *Sci. Rep.* **2020**, *10*, 76128. [CrossRef]
27. Zhan, J.; Li, Y. A hybrid turbulence model for wave breaking simulation. *Chin. J. Theor. Appl. Mech.* **2019**, *51*, 1712–1719. [CrossRef]

28. Babaji, B. Computational Aerodynamic Analysis of Flow Around Apollo Reentry Capsule with Anisotropic Mesh Adaptation. Ph.D. Thesis, Cranfield University, Bedford, UK, 2021.
29. Wheeler, M.P.; Matveev, K.I.; Xing, T. Numerical study of hydrodynamics of heavily loaded hard-chine hulls in calm water. *J. Mar. Sci. Eng.* **2021**, *9*, 184. [CrossRef]
30. Yuan, R.; Guo, B.; Liu, M.Q. Flexible sliced Latin hypercube designs with slices of different sizes. *Stat. Pap.* **2021**, *62*, 137–157. [CrossRef]
31. Ma, Y.; Chen, C.; Fan, T.; Lu, H. Research on the Dynamic Behaviors of a Spar Floating Offshore Wind Turbine with an Innovative Type of Mooring System. *Front. Energy Res.* **2022**, *10*, 853448. [CrossRef]
32. Arredondo-Galeana, A.; Brennan, F. Floating Offshore Vertical Axis Wind Turbines: Opportunities, Challenges and Way Forward. *Energies* **2021**, *14*, 8000. [CrossRef]
33. Fang, H.; Liu, L.F.; Tang, L.; Lin, P. The theory of fifth-order Stokes waves in a linear shear current. *Proc. R. Soc. A* **2023**, *479*, 19. [CrossRef]
34. Ahn, H.; Shin, H. Experimental and numerical analysis of a 10 MW floating offshore wind turbine in regular waves. *Energies* **2020**, *13*, 2608. [CrossRef]
35. Jana, O.; Taylor, P.H.; Wolgamot, H.A.; Madsen, F.J.; Pegalajar-Jurado, A.M.; Bredmose, H. Wave- and drag-driven subharmonic responses of a floating wind turbine. *J. Fluid Mech.* **2021**, *929*, A50.

Disclaimer/Publisher's Note: The statements, opinions and data contained in all publications are solely those of the individual author(s) and contributor(s) and not of MDPI and/or the editor(s). MDPI and/or the editor(s) disclaim responsibility for any injury to people or property resulting from any ideas, methods, instructions or products referred to in the content.

Journal of
Marine Science and Engineering

MDPI

Article

Hyperbolic Paraboloid Free-Surface Breakwaters: Hydrodynamic Study and Structural Evaluation

Sam Smith *, Gaoyuan Wu, Krisna A Pawitan and Maria Garlock

Department of Civil and Environmental Engineering, Princeton University, Princeton, NJ 08544, USA; gaoyuanw@princeton.edu (G.W.); krisna.pawitan@princeton.edu (K.A.P.); mgarlock@princeton.edu (M.G.)
* Correspondence: hse@princeton.edu

Abstract: This study investigates the potential of hyperbolic paraboloid (hypar) shapes for enhancing wave attenuation and structural efficiency in Free-Surface Breakwaters (FSBW). A decoupled approach combining Smoothed Particle Hydrodynamics (SPH) and Finite Element Method (FEM) is employed to analyze hypar-faced FSBW performance across varying hypar warping values and wave characteristics. SPH simulations, validated through experiments, determine wave attenuation performance and extract pressure values for subsequent FEM analysis. Results indicate that hypar-faced FSBW produces increased wave attenuation compared to traditional flat-faced designs, particularly for shorter wave periods and smaller drafts. Furthermore, hypar surfaces exhibit up to three times lower principal stresses under wave loading compared to the flat counterpart, potentially allowing for thinner surfaces. The study also shows that peak-load static stress values provide a reasonable approximation for preliminary design, with less than 6% average difference compared to dynamic analysis results. In summary, this research presents hypar-faced FSBW as a promising alternative in coastal defense strategies, offering effective wave attenuation and structural efficiency in the context of rising sea levels and increasing storm intensities.

Keywords: hydrodynamics; marine structures; wave load; fluid–structure interaction; hyperbolic paraboloid; free-surface breakwaters; wave attenuation; structural efficiency; SPH; FEM; coastal defense; climate change adaptation

1. Introduction

The Intergovernmental Panel on Climate Change (IPCC) has reported a strong indication of the rise in sea levels, attributed to the warming of sea temperatures resulting from global warming [1]. This rise in sea levels, coupled with a projected increase in the intensity of tropical storms [2], underscores the critical need for enhanced coastal resilience, especially in densely populated coastal regions [3]. Traditional coastal defense systems, specifically gravity-type bottom-standing breakwaters, while effective in wave attenuation, may present several drawbacks, such as requiring significant material use, disrupting marine habitats, and altering sediment transport. These drawbacks can potentially lead to erosion and other ecological impacts, and can be costly to build and maintain in deeper waters [4–6]. For these reasons, Free-Surface Breakwaters (FSBW) offer a promising alternative to traditional gravity-type breakwaters [4].

FSBW, also known as open breakwaters, are barriers positioned near the water surface, where the wave energy flux is dominant [4]. FBSW can be mounted on a series of piles or jacket structures, or they may be maintained in a floating state using mooring cables [4].

FBSW are designed to attenuate the height of incoming waves predominantly through reflection and energy dissipation, proving most effective in environments exposed to wave periods of up to 5 s [7]. Among the various types of FSBW, the box-type breakwater constructed from reinforced concrete (RC) is the most common due to its simple design [4,8]. There has been increasing interest in parametric optimization for Free-Surface Breakwaters (FSBW) designs, particularly in exploring various shapes and configurations such as U- and T-shapes or quadrant front faces [5,7,9–13], as well as alternative approaches like fluid-filled membranes [14], primarily to improve wave attenuation performance, with fewer studies also discussing structural performance [8,15,16]. This presents an opportunity to explore new FSBW forms that are potentially efficient in both wave attenuation and structural performance. One such form is the hyperbolic paraboloid (hypar) surface.

Hypars are doubly curved shell structures featuring "arches" that carry compressive forces from the corners to the vertex, balanced by "cables" in tension in the opposite direction (Figure 1), thereby contributing to structural efficiency [17,18]. Additionally, the straight-line generators of the hypars in two directions, as shown in Figure 1, aid in construction by facilitating the use of linear elements in both formwork and reinforcing steel [19].

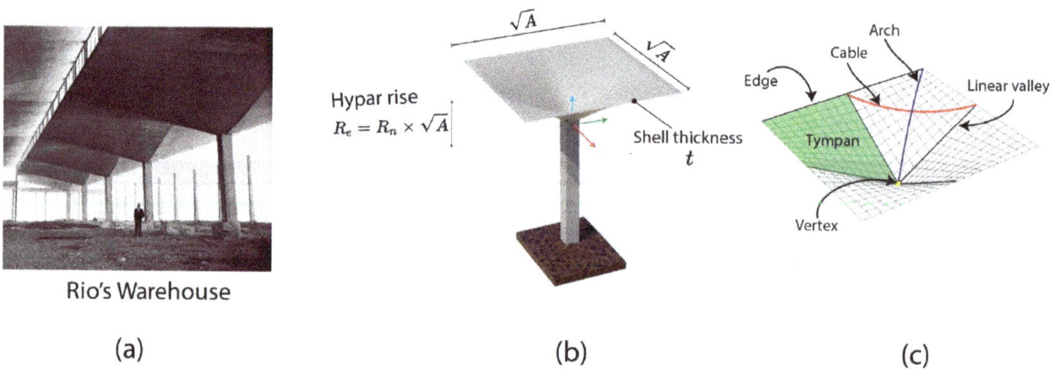

Figure 1. Four-edged hypar umbrella. (**a**) A real-world example of this type of umbrella in Mexico, image courtesy of Princeton University Library's Department of Special Collections. (**b**) Rendering defining the geometric parameters. (**c**) Hypar shell showing straight-line generators.

The most notable designer of hypar-formed roofs in the 20th century is Felix Candela [20], who designed hundreds of hypar-roofed structures, mostly in Mexico [20]. In the last two decades, research interest in hypar surfaces within architecture and engineering extended to the near coast. For example, Wang et al. [21–23] have proposed using hypar umbrellas as deployable flood barriers that function as a canopy during normal weather, without restricting beach access, and deploy into impermeable barriers during severe weather conditions. In one study, Wang et al. [23] investigated the structural behavior of these umbrellas under hydrostatic inundation, considering factors such as hypar warping (R_e/A), inundation height, and lateral column stiffness. Smoothed particle hydrodynamics (SPH) were employed to calculate fluid forces, which were then applied to a structural model analyzed using finite element method (FEM) to evaluate stresses and deformations in the shell. The results demonstrated that employing hypar geometry significantly reduces bending and shear stress compared to flat plates, underscoring the structural benefits of hypar shapes. Furthermore, the research investigated the effect of longitudinal asymmetry of hypars on structural behavior, indicating that increased asymmetry increases the demands on the shell and its supporting structures [23]. In a subsequent study, Wang et al. [23]

validated a decoupled SPH-FEM model through a series of dam-break experiments on 140 × 140 mm 3D-printed hypar umbrella specimens and by comparing the numerical results with Goda's formula for wave pressure on a flat plate. The study confirmed the effectiveness of the SPH-FEM model in capturing the complex hydrodynamic behavior of the hypar umbrellas. Validated models were then used to assess the feasibility of Kinetic Umbrellas as deployable flood barriers during a realistic hurricane scenario based on Hurricane Sandy (2012) [23].

Furthermore, Wu et al. [24] revealed that the hydrodynamic wave pressure on the hypar follows a bilinear-like shape along the height and increases gradually from the edge to the longitudinal spine. The same study found that the maximum displacement and tensile forces of the hypar shell were significantly underestimated by Goda's formula, underscoring the importance of dynamic analyses in accurately assessing the structural response of hypar umbrellas. For other critical demands, such as maximum shell moment and shear force, the difference between static and dynamic analyses was less than 20% [24].

Additionally, ElDarwich et al. [25] investigated the wave attenuation performance of hypar surfaces as faces of box-type FSBW, as illustrated in Figure 2, under regular waves with high-steepness. The study employed a 2D numerical wave flume using SPH to simulate the interaction of regular waves with a FBSW. The results demonstrated that hypar-faced FSBW, due to their curvature, can reduce wave transmission more effectively than their flat-faced counterpart [25]. This finding suggests that hypar geometry can be a viable alternative for improving the wave attenuation performance of FSBW [25].

Figure 2. Artistic rendering of a hypar FSBW, showcasing a protected region (such as a harbor or marina) on the left and the seaside on the right.

Previous work on hypar-faced FSBW has been limited to 2D numerical simulations, focusing solely on wave attenuation performance without addressing structural response. The current study builds on the promising results of the preliminary study. The novel objective of this paper is thus to present a comprehensive 3D analysis of fixed hypar-faced FSBW, evaluating both their hydrodynamic performance and structural response across a wide range of wave periods and drafts, utilizing both SPH numerical simulations and experimental validation.

2. Methodology

This study employs a decoupled SPH-FEM methodology, previously applied to hypar structures in coastal engineering [21,23,24], as illustrated in Figure 3. In most coastal engineering applications, it is reasonable to model structures as stationary objects in numerical simulations of wave-structure interactions [26]. The process begins with parametrically generating the geometry in Rhino Grasshopper [27], followed by SPH analysis using DualSPHysics [28], a popular open-source software for SPH simulations. Based on the SPH analysis, wave attenuation performance is determined through post-processing. Pressure values on the hypar FSBW surface, probed using SPH, are then applied to a FEM model in SAP2000 v20 [29] with only the seaside surface of the FSBW considered for FE analysis, since this surface is subjected to the highest wave loading. This section elaborates on each step of this methodology.

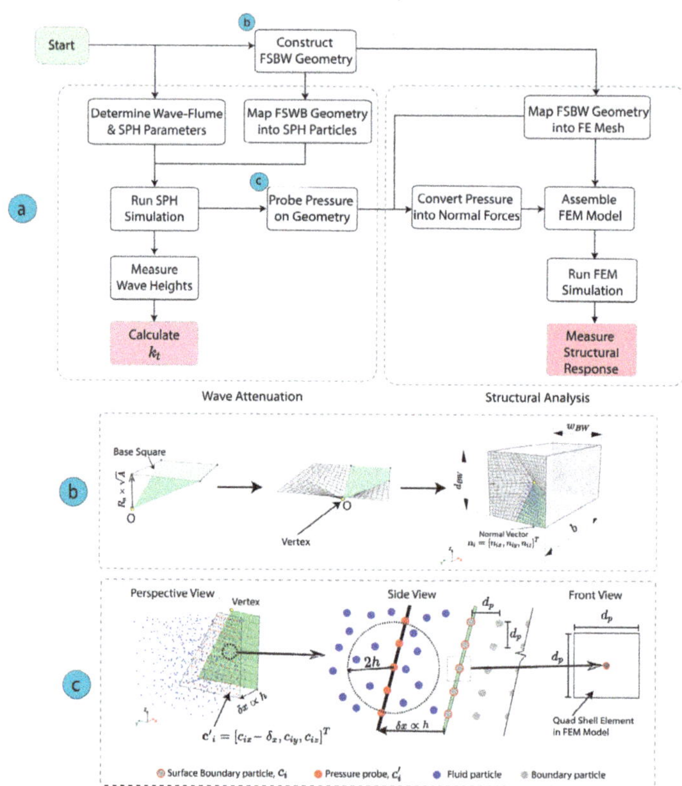

Figure 3. Decoupled SPH-FEM methodology for hypar FSBW analysis. (**a**) Process flowchart from FSBW geometry generation to wave attenuation and structural analysis. (**b**) Hypar FSBW geometry generation in Rhino Grasshopper. (**c**) SPH pressure probing technique.

2.1. Smoothed Particle Hydrodynamics Principle

SPH is a meshless method that discretizes a continuum into a set of particles, where the fluid dynamics equations are then solved [30]. Unlike mesh-based methods, SPH is particularly advantageous for handling free-surface flows, large deformations, and complex boundary geometries [21,24,28,31], such as the hypar. In SPH, any physical quantity F—such as pressure or velocity—at a given point is determined through the interpolation of contributions from all particles within the kernel function's compact support [28]. This interpolation can be expressed:

$$F(\boldsymbol{r}_a) \approx \sum_b F(\boldsymbol{r}_b) \frac{m_b}{\rho_b} W(\boldsymbol{r}_a - \boldsymbol{r}_b, h), \quad (1)$$

where the subscripts a and b denote the target particle and neighboring particles, respectively, $\boldsymbol{r}$ represents the position vector of a particle where F is computed. For each particle, m stands for its mass, and ρ represents its density. The term $W(\boldsymbol{r}_a - \boldsymbol{r}_b, h) = W(\boldsymbol{r}_{ab}, h)$ is the weighting function, also known as the kernel, which determines the influence of neighboring particles on the target particle. The parameter h is the smoothing length, which defines the area of influence for the kernel. The smoothing length h is related to the initial particle spacing d_p by $h = 1.2\sqrt{3}\, d_p$ [31,32]. In this study, the Quintic Wendland kernel [33] is employed, and it has an area of influence that vanishes beyond a radius of $2h$ around the target particle (see Figure 3c).

2.2. Governing Equations and Boundary Treatment

The fluid dynamics system is governed by the discrete form of the Navier–Stokes equations, presented in Lagrangian form [26]:

$$\frac{d\boldsymbol{v}_a}{dt} = -\sum_b m_b \left(\frac{p_b + p_a}{\rho_b \rho_a} \right) \nabla_a W_{ab} + \boldsymbol{g} + \boldsymbol{\Gamma}_a, \quad (2)$$

$$\frac{d\rho_a}{dt} = \sum_b m_b \boldsymbol{v}_{ab} \cdot \nabla_a W_{ab} + D. \quad (3)$$

Equation (2) is the momentum equation describing the change in particle velocity over time, while Equation (3) is the continuity equation representing mass conservation. In these equations, t is the time of the simulation, $\boldsymbol{v}$ is the velocity, p is the pressure,ρ is the density, W_{ab} is the kernel function, g is the gravitational acceleration, Γ_a is the artificial viscosity term proposed by Monaghan [30], and D is a density diffusion term added to mitigate density field fluctuations, with the formulation proposed by Fourtakas et al. [34] being used in the current study.

To solve these equations numerically, the study employs a symplectic position Verlet scheme with a predictor–corrector stage [35] as implemented in DualSPHysics. This scheme incorporates a variable time step approach, as proposed by Monaghan et al. [36], to satisfy the Courant–Friedrichs–Lewy (CFL) stability condition. DualSPHysics utilizes a weakly compressible SPH requiring an equation of state to calculate fluid pressure from particle density [37]:

$$p = \frac{c_s^2 \rho_0}{\gamma} \left[\left(\frac{\rho}{\rho_0} \right)^{\gamma} - 1 \right], \quad (4)$$

where $c_s = \sqrt{\partial p / \partial \rho}$ is the numerical speed of sound, γ is the polytropic constant, and the $\rho_0 = 1000\ \mathrm{kg/m^3}$ is the reference density.

In the present study, the dynamic boundary condition (DBC) implemented in the DualSPHysics code was adopted for handling fluid–boundary interactions [28]. DBC has been applied in multiple studies in coastal engineering [31,38,39], including those involving hypar structures [21,23]. In DBC, boundaries are represented by a discrete set of boundary particles governed by the same equations as fluid particles, but they are not moved by the forces exerted on them. Instead, boundary particles either remain stationary (fixed boundary) or move according to predefined external motions (moving boundaries like gates or flaps) [31,40]. When a fluid particle comes within the range of influence of a boundary particle (closer than the distance covered by the kernel support), the density at the boundary particle increases, causing a rise in pressure. This increase in pressure generates a repulsive force on the fluid particle, driven by the pressure component in the momentum equation [40].

2.3. Wave Generation

Waves are generated using Madsen's second order wavemaker theory [41] implemented in DualSPHysics, which models the wave piston as a moving boundary. This approach was chosen to prevent the generation of spurious secondary waves and to ensure that the waves maintain a consistent shape as they propagate [28,42]. The wave piston stroke S_0 can be determined using Equation (5). Once the stroke is determined, the wave piston displacement $e(t)$ at any time t is given by Equation (6).

$$m_1 = \frac{H}{S_0} = \frac{2\sinh^2(kd)}{\sinh(kd)\cosh(kd) + kd}, \tag{5}$$

$$e(t) = \frac{S_0}{2}\sin(\omega t + \delta) + \left[\left(\frac{H^2}{32d}\right) \cdot \left(\frac{3\cosh(kd)}{\sinh^3(kd)} - \frac{2}{m_1}\right)\right]\sin(2\omega t + 2\delta). \tag{6}$$

In these equations, H is the wave height, d is water depth, $k = 2\pi/\lambda$ with T equal to the wave period and λ is the wave number, $\omega = 2\pi/T$ is the wave angular frequency, and δ is the initial phase.

2.4. Geometric Modeling and SPH-FEM Mapping

To generate the hypar FSBW geometry, a parametric approach was implemented using Rhino Grasshopper. Grasshopper—a visual programming environment in Rhino—was chosen as it enables the parametric generation and manipulation of complex geometric shapes [27]. The process begins by defining hypar surface geometry through the hypar geometric parameters: plan area (A), normalized rise (R_n), number of edges (N), and shell thickness (t) as shown in Figure 1. Though the number of edges can be any integer equal or greater than three, this study focuses exclusively on four-sided hypars. As illustrated in Figure 3b, to generate a single tympan, a base square with a plan of $A/4$ is constructed. One corner of this square, designated as the vertex "O", is then displaced vertically downward by a magnitude of $R_e = R_n \times \sqrt{A}$, where R_e is the hypar rise. For example, $R_n = 0$ is just a flat surface. The resulting tympan is then arrayed four times around a central point to form the complete hypar surface, with a total plan area of A. The hypar surface is then assigned to the faces on the landside and seaside of a box-type FSBW with the specified dimensions of $length\ (b) \times depth\ (d_{BW}) \times width\ (w_{BW})$, as shown in Figure 3b. Constructing the geometry in Grasshopper aids in parametrically generating the shapes, extracting normal vectors n_i, shown in Figure 3b, and applying loads onto the model [18]. The generated hypar FBSW is then imported into DualSPHysics for SPH analysis. For the subsequent FEM analysis, the pressure mapped from SPH is then applied to the seaside facing hypar surface, which is subjected to higher wave loading compared to the leeside hypar surface.

In the SPH model, a pressure probe c_i' is placed at a distance δx (see Figure 3c) in front (seaside) of the surface boundary particle c_i. The pressures at these probes are applied as normal loads on the FEM model on the corresponding shell element, where one SPH cell (with c_i at the center) is then mapped into one FEM Quad shell element with uniform thickness [21,24].

2.5. Wave Attenuation

Wave attenuation performance is quantified by wave transmission coefficient, k_t, defined as the ratio of the transmitted wave height H_t (waves at the leeside of the FSBW) to the incident wave height $H_{i,t}$, as shown in Equation (7):

$$k_t = \frac{H_t}{H_{i,t}}. \quad (7)$$

A lower k_t value indicates better breakwater performance, with values smaller than 0.5 generally regarded as "good" [38]. To accurately measure the incident wave height, $H_{i,t}$ is measured without the FSBW in place ("control" case) on the leeside (i.e., behind) of where it would be [43]. To be consistent, H_t is measured in the same location but with the FSBW in place. This measurement approach is to account for the excessive numerical wave decay in SPH [28,44]. Note that all wave heights were determined from wave surface elevation using the zero-up-crossing method, i.e., a single wave starts and ends when the water elevation crosses the still-water level from below [45].

The wave attenuation coefficient is a key measure of breakwater performance. To further evaluate the performance of a FSBW design, two additional parameters can be examined: the wave reflection coefficient, k_r, and the wave dissipation coefficient, k_d [38,46]. To calculate these parameters, the reflected wave elevation, η_r, at time t and at a specific location x can be expressed as [46]:

$$\eta_r(t,x) = \eta_{Tot}(t,x) - \eta_i(t,x), \quad (8)$$

where $\eta_{Tot}(t,x)$ represents the wave elevation recorded in simulation with the FSBW in place, while $\eta_{i,r}(t,x)$ corresponds to the "control" case, the same simulation but without the FSBW, as described above. The wave reflection coefficient, k_r, can then be computed as:

$$k_r = \frac{H_r}{H_{i,r}}, \quad (9)$$

where H_r is the reflected wave height, calculated from $\eta_r(t,x)$, and $H_{i,r}$ is the incident wave height, calculated from $n_{i,r}(t,x)$. Note that $H_{i,r}$ and $H_{i,t}$ are measured at different locations. For calculating H_r and $H_{i,r}$, measurements are taken in front of the FSBW (on the wave-facing side).

Additionally, the wave dissipation coefficient, k_d, can then be calculated based on wave energy conservation [38]:

$$k_d = \sqrt{1 - k_r^2 - k_t^2}. \quad (10)$$

2.6. Static and Dynamic Structural Analysis

A time history analysis of the total force acting on the FSBW front facing surface is performed. From this time series, the peak time step t_{peak} is identified, defined as the instant when the maximum total hydrodynamic force occurs, a similar method to that applied for hypar structures in previous studies [21]. The wave pressure values at each pressure probe at t_{peak} are then extracted and applied as a static load to the FSBW surface in subsequent structural analyses.

For the dynamic analysis, in contrast, the entire time history of pressure values is applied to the FEM model. Direct integration is employed using the Newmark method (Average Acceleration Method) for time stepping [24,47]. Proportional damping is not considered since [24] demonstrates that damping plays an insignificant role. Further, excluding damping will result in a slightly more conservative evaluation of the dynamic effects [24]. SAP2000 v20 [29] was used for both static and dynamic analysis.

3. SPH Validation for Hypar FSBW

3.1. Experiment Setup

To validate the SPH scheme for hypar FSBW wave-structure interaction (WSI), an experiment was conducted in the Coastal and Hydraulic Engineering Research Laboratory (CHERL) Lab facility at Stony Brook University (SBU), utilizing a wave flume measuring 25 m in length, 1.5 m in width, and 1.5 m in height, equipped with a wet-back piston wave generator.

Scaling based on the Froude number, commonly used in WSI studies [48], is employed in this experiment. The Froude number is the ratio of the inertial forces to the external force field (gravity in this case), so maintaining the same Froude number ensures that the relative importance of inertial forces to the particle's weight is preserved [49]. The full scale FSBW is 5 m wide (w_{BW}) and deep (d_{BW}), and 10 m long (b), see Figure 3b. Based on the limits of the flume, the model was 0.15 m wide and deep, and 0.3 m long, as shown in Figure 4. Thus, a Froude length scaling factor of $\frac{0.15}{5} = 0.03$ was used. Hypar FSBW with $R_n = 0.25$ is selected for the validation. This scaled model was fabricated using Polylactic Acid (PLA) material on an Ultimaker S5 3D printer. Table 1 summarizes the wave characteristics and FSWB draft used for experimental validation. The experiment investigated two wave periods: 0.52 s and 1.04 s, corresponding to full-scale periods of 3 and 6 s, respectively. These periods represent the minimum and maximum ranges in the current parametric study described in the next section. The wave height was scaled to 0.054 m from 1.8 m, with a water depth of 0.7 m to maintain intermediate-deep water conditions.

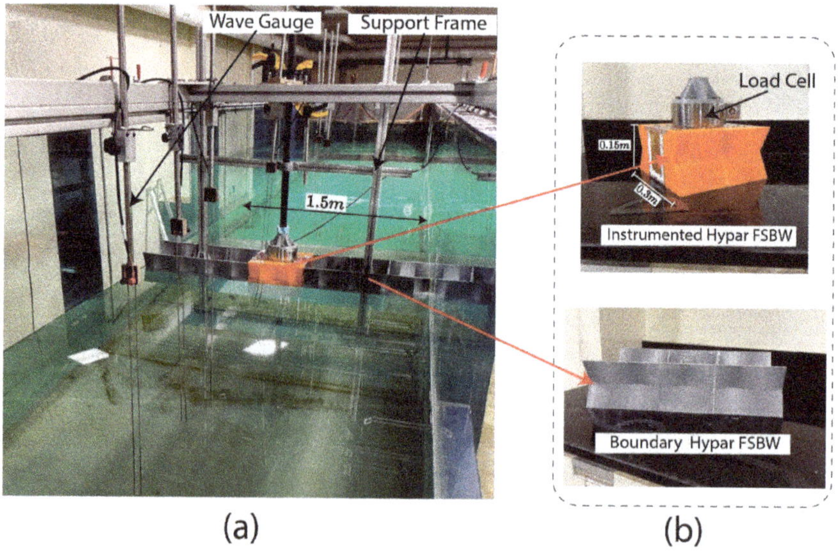

Figure 4. Experimental setup: (**a**) Wave flume showing support frame for hypar FSBW, capacitance wave gauges, and load cell. (**b**) Close-up views of the instrumented hypar FSBW with load cell (top) and boundary hypar FSBW (bottom).

Table 1. Wave characteristics and FSWB draft used for experimental validation.

Parameter (Units)	Full-Scale Value	Model Scale Value
Wave Period, T (s)	3, 6	0.52, 1.04
Wave Height, H (m)	1.8	0.054
Breakwater draft, d_r (m)	2.5	0.075

Figures 4 and 5 show the experimental setup. Multiple capacitance wave gauges (WG1 to WG8) along the wave flume were used to measure wave heights at a sampling frequency of 128 Hz. Force measurements were recorded at a sampling frequency of 1000 Hz. A low-pass Butterworth filter [50] was applied to remove the effect of support frame's natural vibration frequency and high-frequency noise. A hammer test determined the natural frequency of the support frame to be approximately 8 Hz. The low-pass filter effect was found negligible in the current study, likely due to minimal frame vibration amplitude given the submerged FSBW with non-breaking waves.

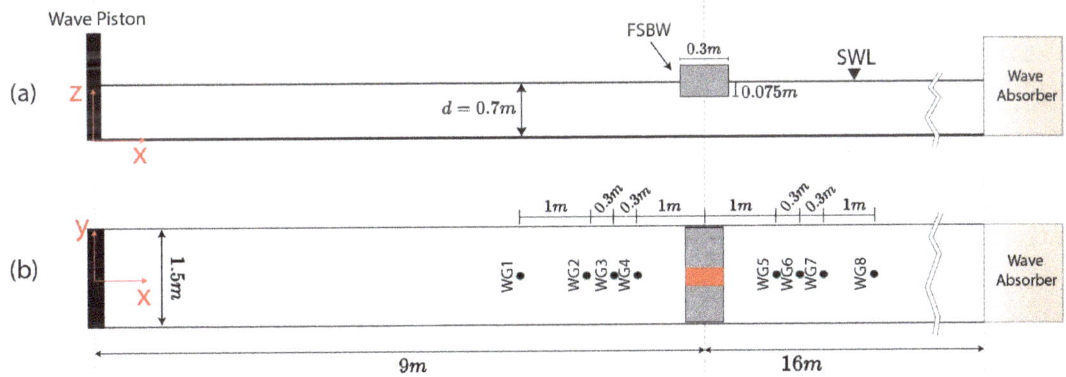

Figure 5. Experimental setup: (**a**) an elevation view, (**b**) top view.

3.2. SPH Numerical Model

A numerical wave flume, similar to that shown in Figures 4 and 5, was created in DualSPHysics. In the numerical model, however, the flume width was limited to a single hypar FSBW due to computational and memory constraints. Waves are generated by a piston using second order wave theory, as described in Section 2.3. Validation results and numerical simulations settings are presented in the next section.

3.3. Model Validation Study

For initial interparticle distance (d_p), values ranging from 3 mm to 10.8 mm were considered in the sensitivity study for experimental validation. Since the scaled H remains constant at 54 mm, this range implies a ratio of wave height to initial interparticle distance (H/d_p) of 5 to 18. These values fall within the range suggested in the literature, which typically recommends an H/d_p of 10 [31,38]. The range of d_p is constrained by the computational capabilities [51], i.e., RAM, Graphics Processing Unit (GPU), and storage, with smaller d_p values increasing computational demand [52]. Hence, a balance between accuracy and computational efficiency is required. Note that for the current study, $H/d_p = 18$ was the finest possible due to computational limits in DualSPHysics.

The d_p values were studied by examining wave profiles as shown in Figure 6, which compares experimental and numerical free-surface elevations, η $[m]$, at Wave Gauge 1

(WG1, before the FSBW) and Wave Gauge 8 (WG8, after the FSBW) for wave periods $T = 0.52$ and 1.04 s. The numerical results with the different d_p values generally show good agreement with experimental data, with $d_p = 3$ mm showing best agreement. To quantify the accuracy, the Root Mean Square Error (RMSE) is calculated between experimental and SPH results [51]:

$$RMSE = \sqrt{\frac{\sum_{i=1}^{n} (S_i - E_i)^2}{n}}, \tag{11}$$

where S_i is the SPH value, E_i is the experimental value, and n is the number of time steps.

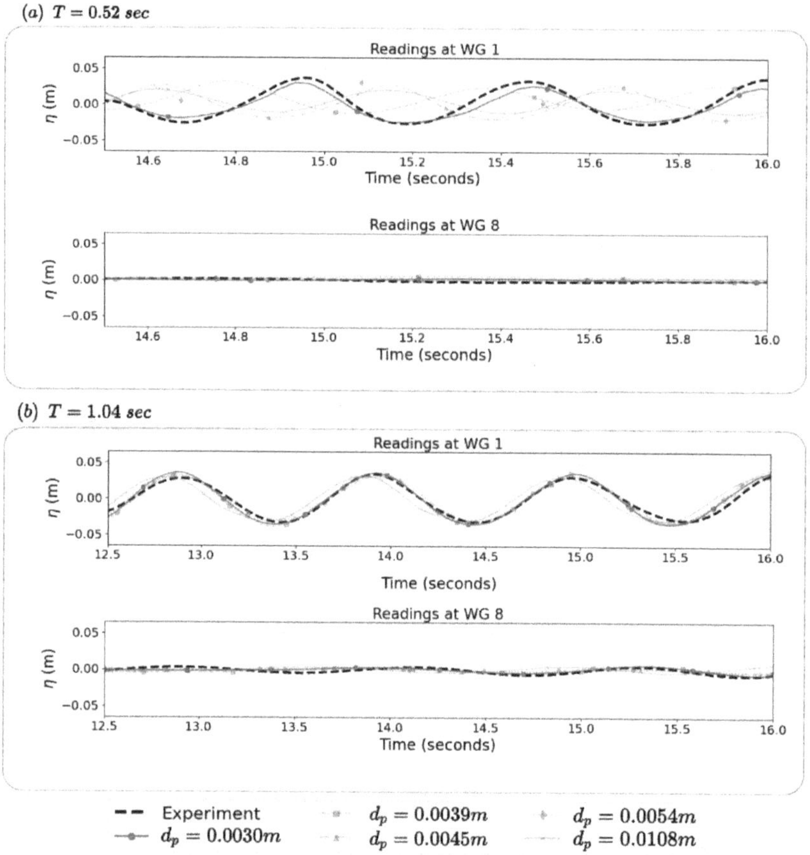

Figure 6. Comparison of experimental and numerical wave gauge readings for different d_p at two wave periods: (**a**) T = 0.52 s and (**b**) T = 1.04 s. The top plots in each panel show readings at Wave Gauge 1 (WG 1), located before the FSBW, while the bottom plots show readings at Wave Gauge 8 (WG 8), positioned after the FSBW.

For $d_p = 3$ mm and WG1, which represents waves approaching the FSBW, RMSE values were 0.003 m and 0.006 m for $T = 1.04$ s and $T = 0.52$ s, respectively. These results indicate relatively small deviation and hence, good matching between experimental and SPH incident wave results. Similar results were observed for other wave gauges.

Additionally, different d_p values were studied through force measurements for the same wave periods on the FSBW, with results shown in Figure 7. The SPH simulation with

$d_p = 3$ mm shows the best agreement with experimental data, matching within 6.5% and 1.1% for $T = 0.52$ s and $T = 1.04$ s, respectively.

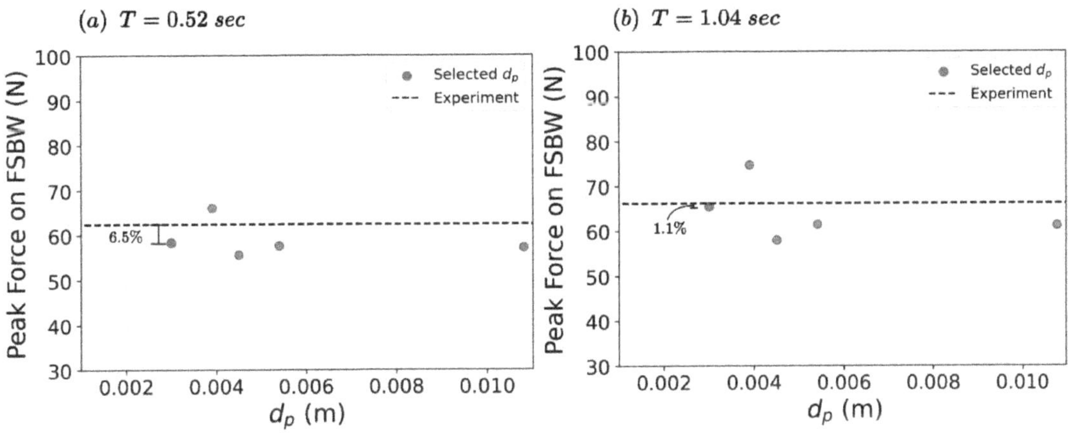

Figure 7. Peak force on the FSBW model ($R_n = 0.25$) for different d_p for two wave periods: (**a**) T = 0.52 s and (**b**) T = 1.04 s.

Thus, d_p studies indicate that $d_p = 3$ mm is a reasonable selection for this study. This d_p value corresponds to $\frac{H}{d_p} = 18$ (recall $H = 54$ mm), which is finer than the typically recommended ratio of 10 [31,38]. Figure 8 provides additional validation of this selection of d_p. Figure 8a shows a visual comparison between experimental observations and SPH visualizations are shown at two-time steps: wave trough and crest at FSBW front for $T = 1.04$ s for the selected $d_p = 3$ mm $\left(\text{i.e., } \frac{H}{d_p} = 18\right)$ value. Additionally, Figure 8b presents a time-history comparison of total force between the SPH simulation and experimental results. The comparison reveals good agreement in both the wave profile and the forces during the FSBW interaction.

A sensitivity study for the probing distance δx (see Figure 3) varying from 2 h to 4 h, indicates that 3 h is adequate. Probing pressure too close to the structure boundary resulted in density and pressure fluctuations due to unphysical gaps inherent to SPH, as discussed in previous studies [38,52,53].

Table 2 summarizes the numerical settings used in the parametric study simulations. The simulations required approximately 600 s of real time to simulate 1 s of SPH time. These simulations were executed on high-performance hardware at Princeton Research Computing, utilizing an NVIDIA A100 GPU and an Intel Xeon Platinum 8380 CPU.

Table 2. SPH numerical simulation and device settings.

Attribute	Values
Kernal type	Quintic Wendland kernel
Boundary Method	Dynamic Boundary Condition
Interparticle distance (d_p)	$H/18$
Artificial viscosity coefficient	0.01 (near irrotational flows)
Time per second	~600
GPU	NVIDIA A100
CPU	Intel(R) Xeon(R) Platinum 8380

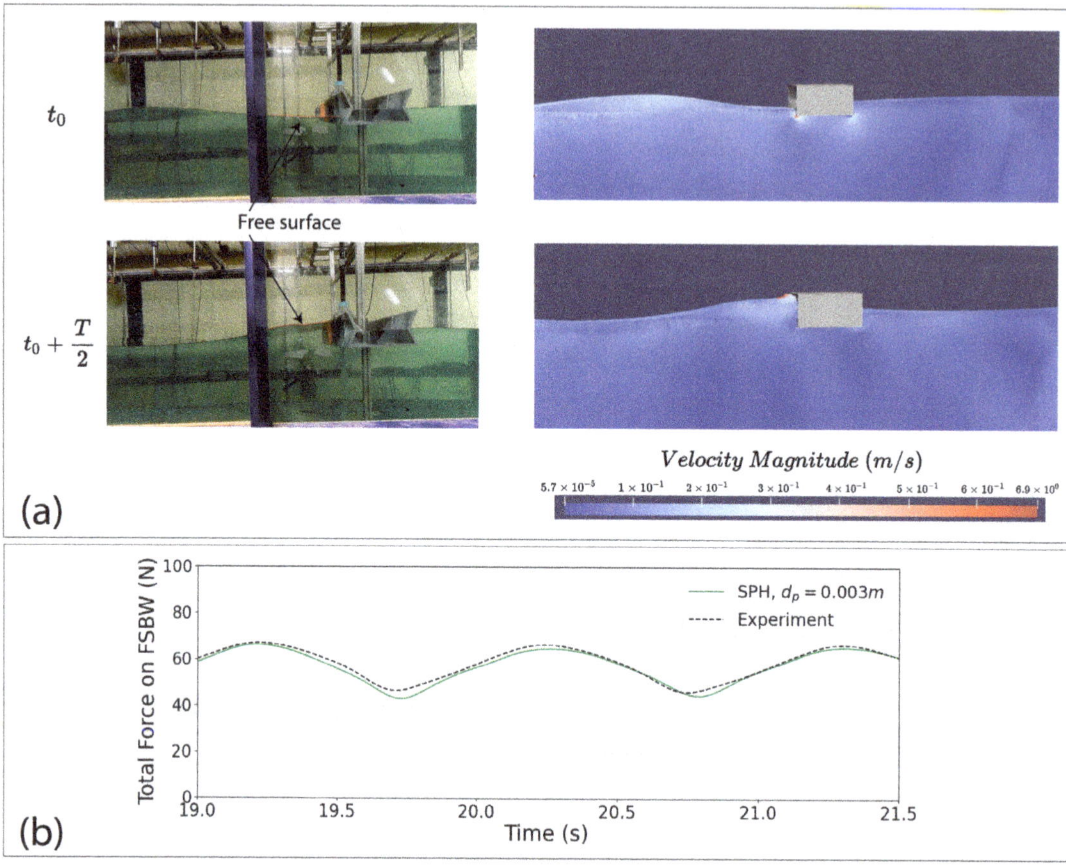

Figure 8. Comparison of experimental and SPH results for selected $d_p = 0.003$ m: (**a**) Wave elevation visualization for SPH (right) and experiment (left) during wave interaction ($T = 1.04$ s) with a hypar FSBW ($R_n = 0.25$) at two-time steps: t_0, wave trough at FSBW front, and $t_0 + T/2$, half a wave period later where wave crest is at FSBW front. (**b**) Time-history comparison of total force from SPH and experiment results.

4. Parametric Study

4.1. Parameters of the Study

Table 3 presents the geometric parameters selected for this study. The normalized rise R_n describes the warping of the hypar. In the parametric study, three values of R_n were selected: 0 (a flat surface), 0.25, and 0.50. The dimensions of the FSBW, shown in Figure 3b, were selected to be 10 m in length, b, and 5 m in depth, d_{BW}, and width, w_{BW}. These parameters were selected to fall within the typical range for FSBW designs [8,15,54], and remained constant in all simulations.

For the FE structural analysis, the hypar surface is discretized into 900 quadrilateral Mindlin/Reissner-type elements with a uniform thickness of 0.25 m, with the selected number of elements being finer than those used in [24,55], which is based on a sensitivity study. Additionally, the hypar surface is assumed to be fixed around its parameter. Given its common application in FSBW [8,15,54], concrete material properties were utilized in this study. High-strength concrete with a compressive strength f_c' of 60 MPa and an elastic modulus E_c of 44,000 MPa was selected. The structural analysis assumes an elastic behavior, which is valid for concrete when stresses remain significantly below the failure threshold [56]—a condition that will be demonstrated in the analysis section.

Table 3. Parameters of the study.

	Parameter	Values
Geometric Properties (See Figure 3)	Normalized rise, R_n	0, 0.25, 0.5
	Breakwater length, b (m)	10
	Breakwater width, w_{BW} (m)	5
	Breakwater depth, d_{BW} (m)	5
	Shell thickness, t (m)	0.25
Wave and Bathymetry (See Figure 9)	Wave Period, T (s)	3, 4, 5, 6
	Wave Height, H (m)	1.8
	Breakwater draft, d_r (m)	2.5, 3.5

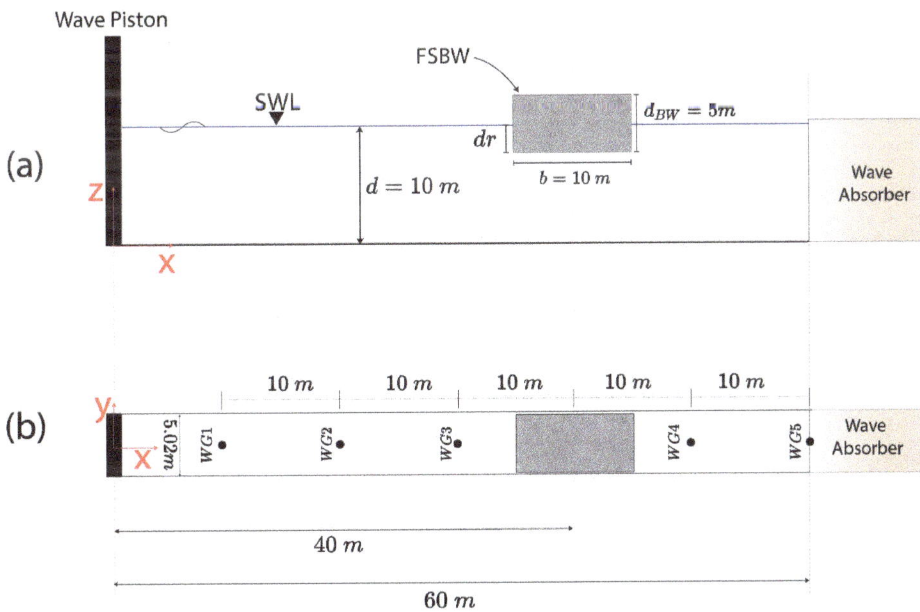

Figure 9. Numerical wave flume setup (**a**) side view and (**b**) top view.

The wave and bathymetry parameters are shown in Table 3. Incident wave periods, T, ranging from 3 to 6 s were considered to represent a typical operating range for FSBW [54]. A water depth of 10 m is selected to ensure transitional (or intermediate) to deep wave conditions (see Figure 9). The draft, d_r, was set to two values: 2.5 m (corresponding to the FSBW being half immersed) and 3.5 m (resulting in 70% immersion of the FSBW). All applied waves in this study are second order regular, non-breaking waves with a constant wave height, H, of 1.8 m; this wave height is selected to prevent breaking for all waves periods considered and satisfy second order wave maker theory [41,57], as described in Section 2.3.

The numerical setup is illustrated in Figure 9. The configuration consists of a wave piston positioned on the far left, with an Active Wave Absorption System (AWAS) implemented to mitigate wave reflection. The AWAS aims to represent real open sea conditions by eliminating unphysical wave re-reflection [31]. The FSBW is situated 60 m in front of the piston. On the right side of the domain, a 1:10 sloped beach serves as a wave absorber. Wave height along the flume is measured using five wave gauges (WG1 to WG5) installed at different locations, as illustrated in Figure 9b.

4.2. Parametric Study Results

This section presents the results of the parametric study. First, the wave attenuation performance is examined, comparing hypar and flat FSBW designs. Then, for the structural response, the difference in principal stress values between the different designs is analyzed.

4.2.1. Wave Attenuation Performance

The wave transmission coefficient, k_t, was calculated using Equation (7) with the method presented in Section 2.5 and average of wave elevation values taken at WG4 and WG5, on the leeside of the breakwater (see Figure 9).

Figure 10 presents the wave transmission coefficient, k_t, results for the hypar FSBW designs across various wave periods and two draft values. For all conditions, k_t increases with the wave period from 3 to 6 s, indicating reduced wave attenuation effectiveness for longer waves. This trend aligns with FSBW performance observed in the literature [15,54,58].

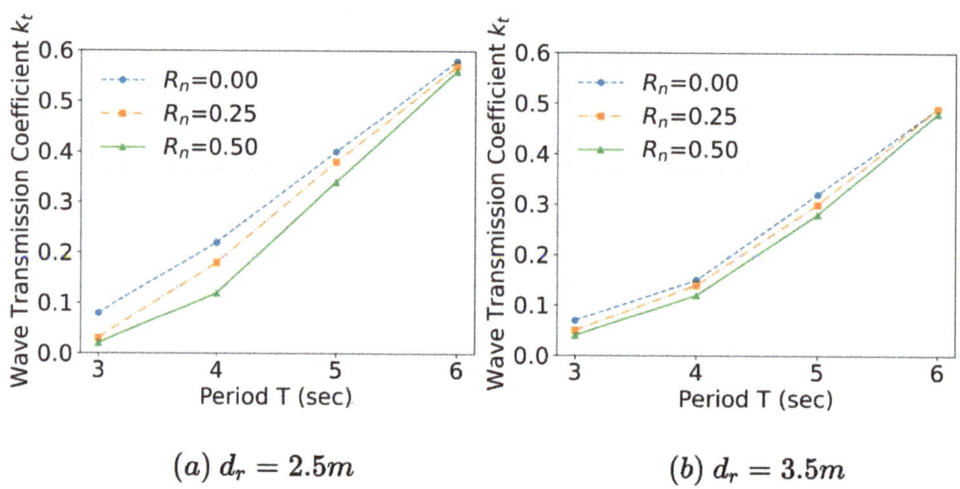

Figure 10. Wave transmission coefficient, k_t as R_n, values. Results are shown for two draft values: (**a**) $d_r = 2.5$ m and (**b**) $d_r = 3.5$ m.

The draft significantly influences wave attenuation performance. The $d_r = 3.5$ m configurations show lower k_t values compared to the 2.5 m draft across all designs (i.e., R_n values) and wave periods. This observation aligns with previous studies showing that increased immersion depth, or draft, results in better wave attenuation [4,54,58,59]. For the larger draft, $d_r = 3.5$ m, the performance gap between flat and hypar designs narrows, especially for longer wave periods, as shown in Figure 10.

Hypar FSBW designs ($R_n = 0.25$ and 0.50) generally show lower k_t values compared to the flat FSBW ($R_n = 0.00$), particularly for shorter wave periods. This improvement is most noticeable for wave periods of 3 and 4 s. Increasing the normalized rise R_n from 0 to 0.50 increases wave attenuation performance, with $R_n = 0.50$ exhibiting the lowest k_t values. This increased performance is up to 5 times for $T = 3$ s and $d_r = 2.5$ m, and 1.8 times for $T = 3$ s and $d_r = 3.5$ m.

There are three possible explanations for how increased warping, R_n, improves wave attenuation performance (i.e., decreases transmission): (1) increased warping may enhance reflection (k_r, Equation (9)), (2) warping may increase wave energy dissipation (k_d, Equation (10)), or (3) both. For k_r and k_d, average wave elevation measurements are taken at WG1 and WG2, located in front of the breakwater. Refer to Section 2.5 for further details. Note that WG3 wave reflection readings were generally found to be unreliable for $T \geq 4$ s because the wave gauge is too close, relative to wavelength, to the FSBW.

To determine the primary wave attenuation mechanism, reflection and dissipation are evaluated and compared for different warping values. Three scenarios are considered: (1) $T = 4$ s, $d_r = 2.5$ m, which demonstrates the best relative performance of the hypar compared to a flat surface as shown in Figure 10; (2) $T = 4$ s, $d_r = 3.5$ m (which is like the first scenario but a larger draft), and (3) $T = 5$ s, $d_r = 2.5$ m (which is like the first scenario but a larger period). The results for these cases are summarized in Table 4.

Table 4. Transmission, k_t reflection, k_r and dissipation k_d coefficients for different warping values (R_n) for representative wave scenarios.

Wave Scenario	Parameter	R_n = 0.00	R_n = 0.25	R_n = 0.50
$T = 4$ s, $d_r = 2.5$ m	k_t	0.22	0.18	0.12
	k_r	0.76	0.60	0.40
	k_d	0.61	0.78	0.91
$T = 4$ s, $d_r = 3.5$ m	k_t	0.15	0.14	0.12
	k_r	0.75	0.61	0.43
	k_d	0.64	0.78	0.89
$T = 5$ s, $d_r = 2.5$ m	k_t	0.40	0.38	0.34
	k_r	0.10	0.10	0.08
	k_d	0.91	0.92	0.94

For scenario 1 ($T = 4$ s and $d_r = 2.5$ m), hypar surface ($R_n = 0.50$) shows significantly better attenuation performance compared to flat surface ($R_n = 0.00$), where smaller k_t indicates better performance. The reflection coefficient for the flat surface ($R_n = 0.00$) is around 0.76 which decreases to 0.40 for hypar surface ($R_n = 0.50$). Therefore, reflection cannot explain why the hypars exhibit better wave attenuation. The wave dissipation coefficients k_d for the above-mentioned wave scenario, indicate that dissipation increases with warping. In particular, dissipation coefficient k_d is 0.61 for $R_n = 0.00$ and reaches up to 0.91 for $R_n = 0.50$.

The same observation—that increased warping enhances dissipation—applies to scenario 2, which has the same wave period ($T = 4$ s) but a larger draft ($d_r = 3.5$ m) compared to scenario 1. In scenario 2, although transmission k_t slightly decreases from 0.15 to 0.12 as R_n increases from 0.00 to 0.50, dissipation k_d increases substantially from 0.64 to 0.89 over the same range of R_n. For scenario 3, which differs from case 1 by an increased wave period ($T = 5$ s) and, consequently, a longer wavelength while maintaining the same draft ($d_r = 2.5$ m), wave attenuation is primarily driven by dissipation regardless of the warping value; k_d increases only slightly, from 0.91 to 0.94, as R_n increases from 0.00 to 0.50.

Table 4 thus indicates that the improved attenuation performance of the hypar FSBW can be attributed to wave dissipation induced by the hypar curvature. Further, these results indicate that the relative wave dissipation efficiency of the hypar compared to a flat surface depends more on wavelength (obtained from wave period) and less on the draft, within the range of the study parameters.

4.2.2. Structural Response

The structural response of the FSBW designs under peak-static and dynamic wave loading was analyzed by examining the maximum principal stress values. For instance, Figure 11 shows maximum principal stress distribution, σ_1, for varying normalized rise, R_n, values, under a wave period of $T = 6$ s and a draft of $d_r = 3.5$ m. This wave scenrio is critical, as it resulted in the largest stresses in the FSBW compared to the other period and draft scenarios. For the flat FSBW ($R_n = 0.00$), a single region of high compression stress is observed slightly below the center of the top surface, with maximum tensile and compressive stresses of around 3.50 MPa and −1.5 MPa, respectively. As R_n increases to 0.25 and 0.50, the compression stress shifts to two regions of lower magnitude stresses closer to the bottom corners. For $R_n = 0.50$, the maximum tensile stress reduces to 1.2 MPa and the maximum compressive stress to −0.20 MPa, indicating a significant stress reduction compared to the flat design—specifically 3 times lower for tension and 7.5 times for compression on the top surface. The trends shown in Figure 11 are the same for other wave scenarios. Indeed, stress values are dependent on thickness, and these results are for a shell thickness of 25 cm. Next, the effect of thickness on stress values, which informed the choice of thickness value, is presented.

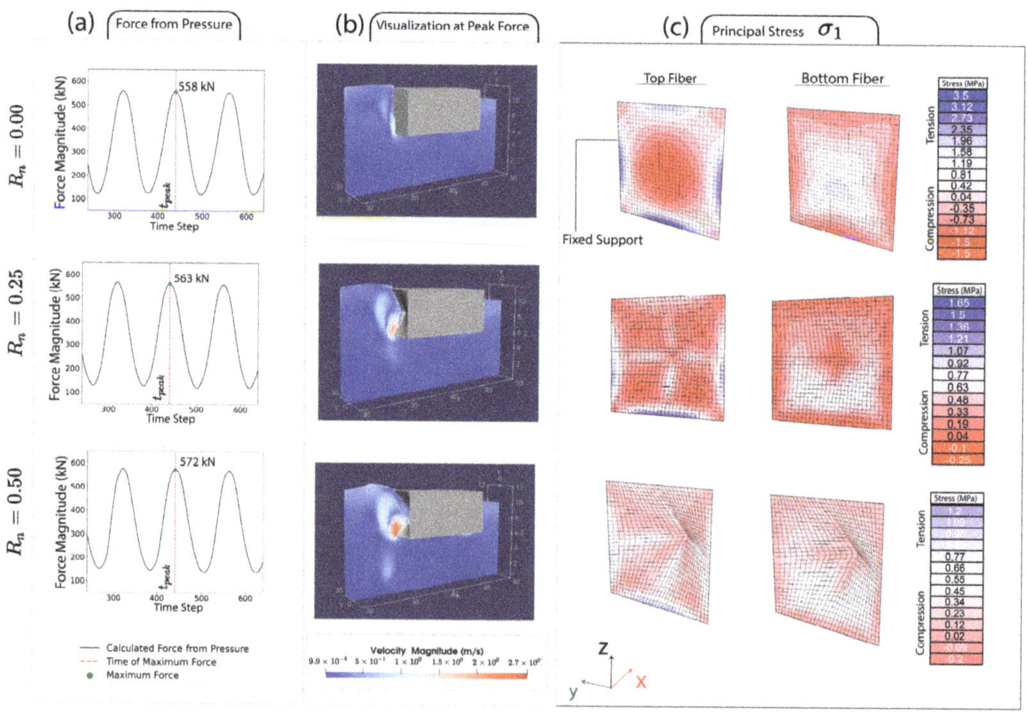

Figure 11. Comparison of structural response for FSBW designs with varying normalized rise, R_n, for a wave period $T = 6$ s and draft, $d_r = 3.5$ m, and shell thickness = 25 cm: (**a**) time-history of force acting on FSBW surface, (**b**) SPH visualization at peak force, and (**c**) principal stress σ_1 distribution on the top and bottom fibers of the FSBW surface.

Elastic analyses were conducted to examine the effect of varying FSBW thickness on the maximum principal tensile stress for different R_n under critical wave scenario, i.e., $T = 6$ s and $d_r = 3.5$ m, with the results shown in Figure 12. The flat FSBW ($R_n = 0.00$) consistently exhibits the highest stress levels across all thicknesses, while the hypar designs

($R_n = 0.00$ and 0.50) show significantly lower stress levels especially at smaller thickness, t, value; for example, at a thickness $t = 5$ cm, the tensile stress for $R_n = 0.50$ is around 13 times lower than for $R_n = 0.00$. Additionally, while increasing R_n from 0.25 to 0.5 further decreases stress, this effect is less significant. For instance, for $t = 5$ cm, decreasing R_n from 0.00 to 0.25 results in 8 times lower tensile stress, but increasing R_n further from 0.25 to 0.5 only reduces tensile stress by a factor of 1.6. The same trend of higher R_n resulting in lower principal stresses is observed for other materials (specifically aluminum and steel) under elastic assumption as well as for the other draft value $d_r = 2.5$ m.

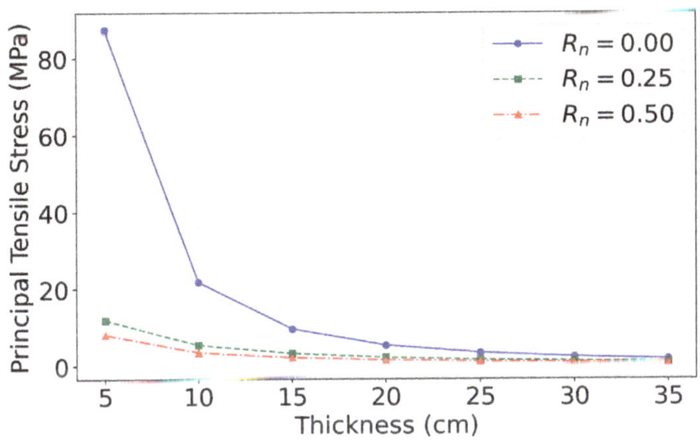

Figure 12. Maximum principal tensile stress, σ, as a function of FSBW thickness for different normalized rise, R_n, values for $T = 6$ s, and $d_r = 3.5$ m.

Figure 13 shows the dynamic effects of wave loading conditions by comparing the maximum principal stress values for different R_n to peak-static values for wave periods of $T = 3$ s and 6 s and drafts $d_r = 2.5$ m and 3.5 m. It is seen that although the dynamic analysis results in higher stress values compared to static analysis, the difference on average is less than 6%. This suggests that a static analysis provides a reasonable approximation for hypar FSBW preliminary design under the regular wave conditions studied herein, a conclusion previously presented by Wu et al. [24]. In addition, the stress values presented under all conditions for FSBW with a thickness of 25 cm are well below the assumed compressive strength f_c' of 60 MPa (tensile strength assumed to equal approximately one-tenth of the compressive strength), validating the applicability of elastic analysis [56].

In all cases, increasing R_n results in a substantial reduction in maximum principal stress magnitudes, in both tension and compression. For instance, for $T = 3$ s and $d_r = 3.5$ m, the maximum tensile stress decreases from 2.8 MPa for $R_n = 0.00$ to 1.4 MPa for $R_n = 0.25$, a 2 times reduction. For $T = 6$ s and $d_r = 3.5$ m, the reduction is from about 3.6 MPa to 1.7 MPa, a similar 2 times reduction. However, as R_n increases from 0.25 to 0.50, the maximum tensile stress decreases from 1.4 MPa to 1.1 MPa, representing only a 1.3 times reduction for $T = 3$ s and $d_r = 3.5$ m. Similarly, for $T = 6$ s and $d_r = 3.5$ m, the reduction is from 1.7 MPa to 1.2 MPa, a 1.4 times decrease. This indicates that the stress reduction effect of increasing warping diminishes as R_n becomes larger, as was previously observed in thickness study in Figure 12. Additionally, the same trend is evident for the smaller draft value, $d_r = 2.5$ m. Principal compressive stresses exhibit a similar trend.

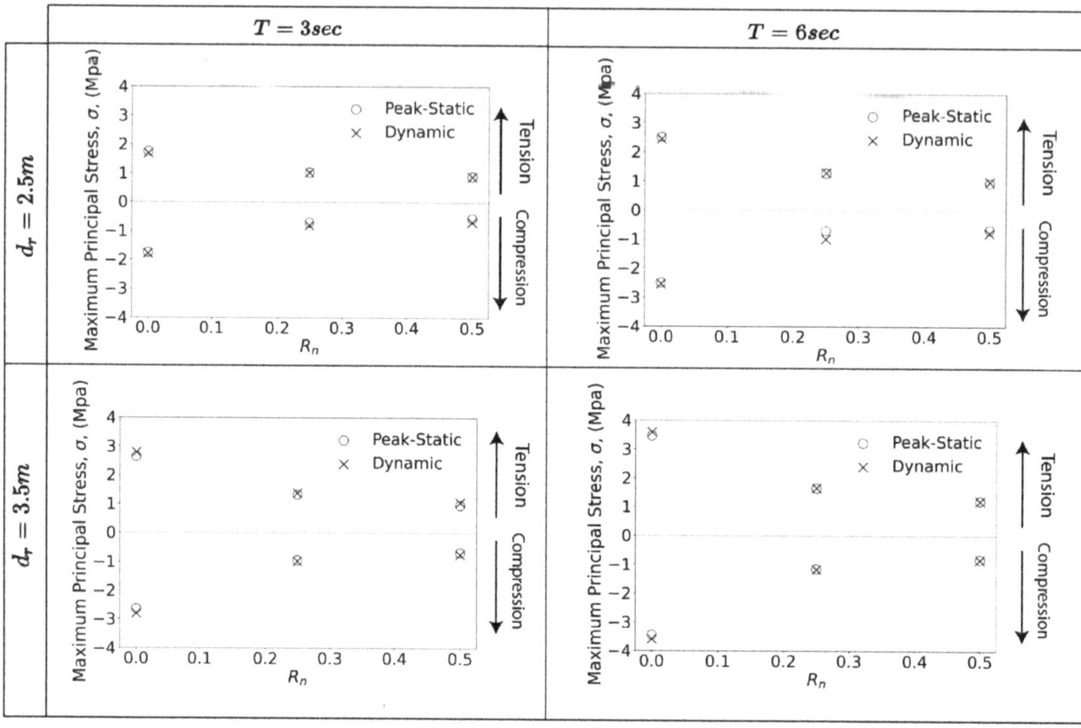

Figure 13. Comparison of peak-static and dynamic loading conditions via maximum principal stress, σ, for different normalized rise, R_n ,values (shell thickness $t = 25$ cm).

5. Summary and Conclusions

This study analyzed the performance of Free-Surface Breakwaters (FSBW) with hypar-faces of varying warping values across various wave characteristics. Experimental studies validated numerical simulations used in a parametric study that investigated the potential of hypar geometry in FSBW for enhancing wave attenuation and structural efficiency via a decoupled SPH-FEM analysis. Wave attenuation performance was examined through the wave transmission coefficient, while structural response was analyzed through comparing principal stress values. The conclusions drawn from this study are:

- Hypar-faced FSBW designs improve wave attenuation performance compared to flat-faced designs, primarily through wave dissipation. The improved performance is more pronounced for shorter wave periods and lower draft values.
- Hypar geometry significantly reduces maximum principal stress values compared to flat designs. This effect is most pronounced for FSBWs with thinner shell walls.
- Peak-static analysis provides a reasonable approximation for the structural stresses of hypar FSBW under regular wave conditions, with less than 6% difference from dynamic analysis results.

While this study provides valuable insights into the potential of hypar-faced FSBW, it is noted that the study was limited to regular waves with a constant height, and one material, concrete, which did not account for long-term effects or environmental degradation. Regardless, the findings still remain significant. Since the wave period (or wavelength) is identified as the primary factor influencing wave attenuation [8,38,59], the results cover the performance of hypar FSBWs across a wide range of wave conditions, potentially providing insight about behavior under irregular waves. Additionally, the elastic assumption used in the analysis ensures that the conclusions are applicable to other materials, as stress values remain essentially unaffected under elastic behavior.

Despite its limitations, the study demonstrates the practical benefits of hypar FSBWs, including improved wave attenuation and reduced structural stresses, and it encourages further examination of hypar-faced FSBWs as efficient and effective breakwaters that reduce both attenuation and structural stresses.

Author Contributions: S.S.: conceptualization, methodology, software, validation, formal analysis, writing—original draft, review and editing. G.W.: conceptualization, methodology, software, validation, formal analysis, review and editing. K.A.P.: conceptualization, validation, review and editing. M.G.: conceptualization, validation, writing—review and editing, supervision, funding acquisition. All authors have read and agreed to the published version of the manuscript.

Funding: This work was funded by Project X at Princeton University and the National Science Foundation (NSF) under grant CMMI-2227489. All opinions expressed in this paper are the authors' and do not necessarily reflect the policies and views of the sponsors.

Institutional Review Board Statement: Not applicable.

Informed Consent Statement: Not applicable.

Data Availability Statement: The data used for this paper is publicly available at the following repository: https://github.com/hse95/Hypar-FSBW.

Acknowledgments: The authors thank Ali Farhadzadeh, Tao Xiang (Stony Brook University), and Zhaoyang Song (Princeton University) for their valuable insights and support.

Conflicts of Interest: The authors declare no conflicts of interest.

References

1. Vardy, M.; Oppenheimer, M.; Dubash, N.K.; O'Reilly, J.; Jamieson, D. The intergovernmental panel on climate change: Challenges and opportunities. *Annu. Rev. Environ. Resour.* **2017**, *42*, 55–75. [CrossRef]
2. Pérez-Alarcón, A.; Fernández-Alvarez, J.C.; Coll-Hidalgo, P. Global Increase of the Intensity of Tropical Cyclones Under Global Warming Based on Their Maximum Potential Intensity and CMIP6 Models. *Environ. Process.* **2023**, *10*, 36. [CrossRef]
3. Rambabu, N.; Srineash, V. A Review on Methodologies to Upgrade the Coastal Structures to Enhance the Coastal Resilience. In Proceedings of the OCEANS 2022-Chennai, Chennai, India, 21–24 February 2022; pp. 1–9.
4. Teh, H. Hydraulic performance of free surface breakwaters: A review. *Sains Malays.* **2013**, *42*, 1301–1310.
5. Lorenzoni, C.; Soldini, L.; Brocchini, M.; Mancinelli, A.; Postacchini, M.; Seta, E.; Corvaro, S. Working of defense coastal structures dissipating by macroroughness. *J. Waterw. Port Coast. Ocean. Eng.* **2010**, *136*, 79–90. [CrossRef]
6. He, F. Experimental Investigation of Pile-Supported/Floating Breakwaters Integrated with Oscillating-Water-Column Converters. Ph.D. Thesis, Nanyang Technological University, Singapore, 2013.
7. Isaacson, M.; Whiteside, N.; Gardiner, R.; Hay, D. Modelling of a circular-section floating breakwater. *Can. J. Civ. Eng.* **1995**, *22*, 714–722. [CrossRef]
8. Elchahal, G.; Lafon, P.; Younes, R. Design optimization of floating breakwaters with an interdisciplinary fluid–solid structural problem. *Can. J. Civ. Eng.* **2009**, *36*, 1732–1743. [CrossRef]
9. Brossard, J.; Jarno-Druaux, A.; Marin, F.; Tabet-Aoul, E. Fixed absorbing semi-immersed breakwater. *Coast. Eng.* **2003**, *49*, 25–41. [CrossRef]
10. Günaydın, K.; Kabdaşlı, M. Performance of solid and perforated U-type breakwaters under regular and irregular waves. *Ocean. Eng.* **2004**, *31*, 1377–1405. [CrossRef]
11. Wang, Y.; Wang, G.; Li, G. Experimental study on the performance of the multiple-layer breakwater. *Ocean. Eng.* **2006**, *33*, 1829–1839. [CrossRef]
12. Sundar, V. Hydrodynamic pressures and forces on quadrant front face pile supported breakwater. *Ocean. Eng.* **2002**, *29*, 193–214. [CrossRef]
13. Koutandos, E.; Prinos, P. Hydrodynamic characteristics of semi-immersed breakwater with an attached porous plate. *Ocean. Eng.* **2011**, *38*, 34–48. [CrossRef]

14. Lv, C.; Zhao, X.; Zheng, K.; Zong, Y.; Zheng, S.; Jiang, H.; Yao, S.; Luan, H. Hydrodynamic performance of a floating fluid-filled membrane breakwater: Experimental and numerical study. *J. Fluid Mech.* **2024**, *1001*, A21. [CrossRef]
15. Elchahal, G.; Younes, R.; Lafon, P. Optimization of coastal structures: Application on detached breakwaters in ports. *Ocean. Eng.* **2013**, *63*, 35–43. [CrossRef]
16. Cebada-Relea, A.J.; López, M.; Claus, R.; Aenlle, M. Short-term analysis of extreme wave-induced forces on the connections of a floating breakwater. *Ocean. Eng.* **2023**, *280*, 114579. [CrossRef]
17. Wang, S.; Garlock, M.; Glisic, B. Geometric and area parameterization of N-edged hyperbolic paraboloidal umbrellas. *Eng. Struct.* **2022**, *250*, 113499. [CrossRef]
18. ElDarwich, H.; Mansouri, I.; Garlock, M.; Wang, S. Predicting Maximum Deflection of N-Edged Thin-Shelled Hyperbolic-Paraboloid Umbrella Using Machine Learning Techniques. *Thin-Walled Struct.* **2024**, *205*, 112412. [CrossRef]
19. Gergely, P.; Banavalkar, P.V.; Parker, J.E. *The Analysis and Behavior of Thin-Steel Hyperbolic Paraboloid Shells*; Report No. 338; Wei-Wen Yu Center for Cold-Formed Steel Structures, Civil, Architectural and Environmental Engineering Department, Missouri University of Science and Technology: Rolla, MO, USA, 1971.
20. Garlock, M.E.M.; Billington, D.P. *Félix Candela: Engineer, Builder, Structural Artist*; Princeton University Art Museum: Princeton, NJ, USA; Yale University Press: New Haven, CT, USA, 2008.
21. Wang, S.; Garlock, M.; Deike, L.; Glisic, B. Feasibility of Kinetic Umbrellas as deployable flood barriers during landfalling hurricanes. *J. Struct. Eng.* **2022**, *148*, 04022047. [CrossRef]
22. Wang, S.; Garlock, M.; Glisic, B. Kinematics of deployable hyperbolic paraboloid umbrellas. *Eng. Struct.* **2021**, *244*, 112750. [CrossRef]
23. Wang, S.; Notario, V.; Garlock, M.; Glisic, B. Parameterization of hydrostatic behavior of deployable hypar umbrellas as flood barriers. *Thin-Walled Struct.* **2021**, *163*, 107650. [CrossRef]
24. Wu, G.; Garlock, M.; Wang, S. A decoupled SPH-FEM analysis of hydrodynamic wave pressure on hyperbolic-paraboloid thin-shell coastal armor and corresponding structural response. *Eng. Struct.* **2022**, *268*, 114738. [CrossRef]
25. ElDarwich, H.S.; Pawitan, K.A.; Garlock, M.E. Conceptual Investigation on the Effectiveness of Hyperbolic Paraboloid Surfaces for Floating Breakwaters. In Proceedings of the IASS 2022 Symposium Affiliated with APCS 2022 Conference, Beijing, China, 19–23 September 2022, ISSN 2518–6582.
26. Antoci, C.; Gallati, M.; Sibilla, S. Numerical simulation of fluid–structure interaction by SPH. *Comput. Struct.* **2007**, *85*, 879–890. [CrossRef]
27. McNeel, R. Rhinoceros 3D, Version 7 2020. Robert McNeel & Associates, Seattle, WA, USA, 2015. Available online: https://www.rhino3d.com.
28. Crespo, A.; Domínguez, J.; Rogers, B.; Gómez-Gesteira, M.; Longshaw, S.; Canelas, R.; Vacondio, R.; Barreiro, A.; García-Feal, O. DualSPHysics: Open-source parallel CFD solver based on Smoothed Particle Hydrodynamics (SPH). *Comput. Phys. Commun.* **2015**, *187*, 204–216. [CrossRef]
29. CSI. *SAP2000 CSI Analysis Reference Manual 2014*; Computers and Structures, Inc.: Walnut Creek, CA, USA, 2014. Available online: https://www.csiamerica.com.
30. Monaghan, J.J. Smoothed particle hydrodynamics. *Rep. Prog. Phys.* **2005**, *68*, 1703. [CrossRef]
31. Altomare, C.; Domínguez, J.; Crespo, A.; González-Cao, J.; Suzuki, T.; Gómez-Gesteira, M.; Troch, P. Long-crested wave generation and absorption for SPH-based DualSPHysics model. *Coast. Eng.* **2017**, *127*, 37–54. [CrossRef]
32. Martínez-Estévez, I.; Tagliafierro, B.; El Rahi, J.; Domínguez, J.; Crespo, A.; Troch, P.; Gómez-Gesteira, M. Coupling an SPH-based solver with an FEA structural solver to simulate free surface flows interacting with flexible structures. *Comput. Methods Appl. Mech. Eng.* **2023**, *410*, 115989. [CrossRef]
33. Wendland, H. Piecewise polynomial, positive definite and compactly supported radial functions of minimal degree. *Adv. Comput. Math.* **1995**, *4*, 389–396. [CrossRef]
34. Fourtakas, G.; Vacondio, R.; Alonso, J.D.; Rogers, B.D. Improved Density Diffusion Term for Long Duration Wave Propagation. In Proceedings of the 2020 SPHERIC Harbin International Workshop, Harbin, China, 13 January–16 September 2020.
35. Grubmüller, H.; Heller, H.; Windemuth, A.; Schulten, K. Generalized Verlet algorithm for efficient molecular dynamics simulations with long-range interactions. *Mol. Simul.* **1991**, *6*, 121–142. [CrossRef]
36. Monaghan, J.J.; Cas, R.A.; Kos, A.; Hallworth, M. Gravity currents descending a ramp in a stratified tank. *J. Fluid Mech.* **1999**, *379*, 39–69. [CrossRef]
37. Batchelor, G.K. *An Introduction to Fluid Dynamics*, 1st ed.; Cambridge University Press: Cambridge, UK, 2000. [CrossRef]
38. Liu, Z.; Wang, Y. Numerical studies of submerged moored box-type floating breakwaters with different shapes of cross-sections using SPH. *Coast. Eng.* **2020**, *158*, 103687. [CrossRef]
39. Wu, G.; Garlock, M. Investigating the effects of box girder bridge geometry on solitary wave force using SPH modeling. *Coast. Eng.* **2024**, *187*, 104430. [CrossRef]

40. Cabrera Crespo, A.J.; Gómez Gesteira, R.; Dalrymple, R.A. Boundary conditions generated by dynamic particles in SPH methods. *Comput. Mater. Contin.* **2007**, *5*, 173–184.
41. Madsen, O.S. On the generation of long waves. *J. Geophys. Res.* **1971**, *76*, 8672–8683. [CrossRef]
42. Estevez, I.M. Coupling Between the DualSPHysics Solver and Multiphysics Libraries: Implementation, Validation and Real Engineering Applications. Ph.D. Thesis, Universidade of Vigo, Pontevedra, Spain, 2024.
43. Pawitan, K.A. Wave Loadings and Scaling Effects on, and Within, an Oscillating Water Column (OWC) Caisson Breakwater. Ph.D. Thesis, University of Edinburgh, Edinburgh, UK, 2019.
44. Zago, V.; Schulze, L.J.; Bilotta, G.; Almashan, N.; Dalrymple, R. Overcoming excessive numerical dissipation in SPH modeling of water waves. *Coast. Eng.* **2021**, *170*, 104018. [CrossRef]
45. Mathisen, J.; Bitner-Gregersen, E. Joint distributions for significant wave height and wave zero-up-crossing period. *Appl. Ocean. Res.* **1990**, *12*, 93–103. [CrossRef]
46. Goda, Y.; Suzuki, Y. Estimation of incident and reflected waves in random wave experiments. *Coast. Eng.* **1976**, *1976*, 828–845. [CrossRef]
47. Newmark, N.M. A method of computation for structural dynamics. *J. Eng. Mech. Div.* **1959**, *85*, 67–94. [CrossRef]
48. Hughes, S.A. *Physical Models and Laboratory Techniques in Coastal Engineering*, 7th ed.; World Scientific: Singapore, 1993. [CrossRef]
49. Munson, B.R.; Okiishi, T.H.; Huebsch, W.W.; Rothmayer, A.P. *Fundamentals of Fluid Mechanics*, 7th ed.; John Wiley & Sons, Inc.: Hoboken, NJ, USA, 2013.
50. Butterworth, S. On the theory of filter amplifiers. *Wirel. Eng.* **1930**, *7*, 536–541.
51. Reis, C.; Barbosa, A.R.; Figueiredo, J.; Clain, S.; Lopes, M.; Baptista, M.A. Smoothed particle hydrodynamics modeling of elevated structures impacted by tsunami-like waves. *Eng. Struct.* **2022**, *270*, 114851. [CrossRef]
52. Domínguez, J.; Fourtakas, G.; Cercós-Pita, J.; Vacondio, R.; Rogers, B.D.; Crespo, A. Evaluation of reliability and efficiency of different boundary conditions in a SPH code. In Proceedings of the 10th International SPHERIC Workshop, Parma, Italy, 16–18 June 2015.
53. Ferrand, M.; Laurence, D.R.; Rogers, B.D.; Violeau, D.; Kassiotis, C. Unified semi-analytical wall boundary conditions for inviscid, laminar or turbulent flows in the meshless SPH method. *Int. J. Numer. Methods Fluids* **2013**, *71*, 446–472. [CrossRef]
54. Dai, J.; Wang, C.M.; Utsunomiya, T.; Duan, W. Review of recent research and developments on floating breakwaters. *Ocean. Eng.* **2018**, *158*, 132–151. [CrossRef]
55. Wang, S.; Garlock, M.; Glisic, B. Hydrostatic Response of Deployable Hyperbolic-Paraboloid Umbrellas as Coastal Armor. *J. Struct. Eng.* **2020**, *146*, 04020096. [CrossRef]
56. Robutti, G.; Ronzoni, E.; Ottosen, N. Failure Strength and Elastic Limit for Concrete: A Comparative Study. In Proceedings of the 5th International Conference on Structural Mechanics in Reactor Technology, Berlin, Germany, 12–16 August 1979.
57. Le Méhauté, B.; Koh, R.C.Y. On the Breaking of Waves Arriving at an Angle to the Shore. *J. Hydraul. Res.* **1967**, *5*, 67–88. [CrossRef]
58. Martin, W.; Adee, B.H. Theoretical Analysis of Floating Breakwater Performance. In Proceedings of the Fifth Indian National Conference on Harbour and Ocean Engineering, Goa, India, 4–6 February 1974.
59. Macagno, E. Fluid mechanics: Experimental study of the effects of the passage of a wave beneath an obstacle. In Proceedings of the Academic Des Sciences, Paris, France, 28 February 1953.

Disclaimer/Publisher's Note: The statements, opinions and data contained in all publications are solely those of the individual author(s) and contributor(s) and not of MDPI and/or the editor(s). MDPI and/or the editor(s) disclaim responsibility for any injury to people or property resulting from any ideas, methods, instructions or products referred to in the content.

Journal of
Marine Science
and Engineering

MDPI

Article

Wave Attenuation Performance of a Floating Breakwater Integrated with Flexible Wave-Dissipating Structures

Xianlin Jia [1], Su Guo [2], Kangjie Wang [1], Sai Fu [1], Xintong Yu [3] and Wei Peng [3,*]

1 Power China Hua Dong Engineering Co., Ltd., Hangzhou 311122, China; jia_xl@hdec.com (X.J.); wang_kj2@hdec.com (K.W.); fu_s@hdec.com (S.F.)
2 School of Renewable Energy, Hohai University, Nanjing 211000, China; guosu81@126.com
3 College of Harbour, Coastal and Offshore Engineering, Hohai University, Nanjing 211000, China; yuxintong@hhu.edu.cn
* Correspondence: pengweihhu@hhu.edu.cn; Tel.: +86-025-8378-6648

Abstract

This study develops a two-dimensional numerical model to investigate the hydrodynamic performance of a floating breakwater coupled with flexible wave-dissipating structures (FWDS). The model integrates the immersed boundary method with a finite element structural solver, enabling accurate simulation of fluid–structure interactions under wave excitation. Validation against benchmark cases, including cantilever beam deflection and flexible vegetation under waves, confirms the model's reliability. Parametric analyses were conducted to examine the influence of the elastic modulus and height of the FWDS on wave attenuation efficiency. Results show that structural flexibility plays a crucial role in modifying wave reflection, transmission, and dissipation characteristics. A lower elastic modulus enhances energy dissipation through large deformation and vortex generation, while higher stiffness promotes reflection with reduced dissipation. Increasing the height of the FWDS improves overall wave attenuation but exhibits diminishing returns for long-period waves. The findings highlight that optimized flexibility and geometry can effectively enhance the energy-dissipating capacity of floating breakwaters. This study provides a theoretical basis for the design and optimization of hybrid floating breakwaters integrating flexible elements for coastal and offshore wave energy mitigation.

Keywords: flexible wave-dissipating structure; floating breakwater; fluid–structure interaction; elastic modulus; wave attenuation

1. Introduction

In recent decades, the development of offshore engineering has advanced rapidly, with offshore wind farms, aquaculture facilities, and floating photovoltaic (FPV) platforms gradually moving toward deeper waters. These offshore structures are inevitably exposed to complex and harsh ocean environments throughout their service life. Floating breakwaters have been widely deployed to mitigate wave impacts and provide protection for these offshore systems. However, the long-term operation of floating breakwaters often leads to material fatigue, degradation, and potential detachment of structural components. Detached debris can adversely affect marine ecosystems, posing risks to marine organisms and water quality. As environmental concerns have gained increasing attention worldwide, there has been a strong demand for eco-friendly and sustainable coastal protection solutions that can simultaneously provide wave attenuation and promote ecological restoration [1].

To address these challenges, new concepts of environmentally friendly and biologically compatible coastal structures have been proposed. In particular, flexible structures, inspired by natural vegetation such as kelp and seagrass, have attracted growing interest. These structures can provide habitats and food sources for marine organisms, contributing to biodiversity conservation and ecosystem restoration. Moreover, they can help mitigate environmental issues such as eutrophication and ocean acidification by enhancing local water quality. Beyond ecological benefits, flexible structures also play an important role in altering local hydrodynamics by influencing current patterns and wave propagation characteristics [2]. This dual functionality makes them attractive as auxiliary components for floating breakwaters, potentially offering both effective wave energy dissipation and ecological enhancement.

The capacity of flexible structures to dissipate wave energy and reduce current velocity has been extensively investigated over the past few decades. Early field studies revealed the significant potential of submerged vegetation in wave attenuation. For instance, Knutson et al. [3] observed that wave height decayed exponentially within vegetated areas, with most of the wave energy being dissipated at the leading edge of the vegetation field. Jadhav and Chen [4,5] and Jadhav et al. [6] analyzed field data collected during tropical storms and demonstrated that salt marsh vegetation caused substantial dissipation of random wave energy. They further examined the statistical distribution of zero-crossing wave heights to validate their observations.

On the theoretical side, Dalrymple et al. [7] were among the first to adapt the Morison equation to describe wave–plant interactions under the assumption of rigid cylinders and regular waves, laying the foundation for subsequent modeling studies. Building on this, Mendez and Losada [8] extended Dalrymple's work to predict irregular wave attenuation in vegetated regions. Hu et al. [9] developed a theoretical model based on energy conservation to estimate wave attenuation under combined wave-current conditions, which they later refined to incorporate breaking waves and opposing currents [10]. Magdalena et al. [11] introduced frictional and diffusive terms into the momentum equation to represent vegetation effects and derived analytical solutions for both long and short waves, validating the model against physical experiments. More recently, Liu et al. [12] proposed a new theoretical framework accounting for changes in group velocity and energy dissipation due to currents, further improving predictions of wave attenuation in emerging vegetation zones.

Collectively, these studies established that flexible vegetation-like structures can effectively dissipate wave energy through drag forces, turbulence generation, and vortex shedding. They also highlighted that wave attenuation is strongly dependent on vegetation properties such as density, stiffness, and submergence depth, as well as incident wave characteristics. This body of work provides a strong theoretical and experimental foundation for integrating flexible elements into engineered coastal protection systems. While the flexible elements are inspired by marine vegetation and may offer ecological co-benefits, the current study is limited to hydrodynamic and structural performance. Detailed evaluation of ecological effects (e.g., sediment transport, habitat development, colonization) requires dedicated field and laboratory studies and is beyond the scope of this paper.

To leverage the hydrodynamic benefits of flexible structures in engineering practice, researchers have explored integrating flexible elements into breakwater designs. Initial efforts focused on comparing the performance of rigid and flexible breakwaters. Stamos et al. [13] conducted physical experiments to examine the reflection and transmission characteristics of semi-cylindrical and rectangular breakwater models, both rigid and water-filled flexible types. Similarly, Diamantoulaki et al. [14] used numerical simulations

https://doi.org/10.3390/jmse14010097

to investigate pile-constrained floating breakwaters, comparing the performance of flexible and rigid configurations under monochromatic wave forcing.

Subsequent studies explored more sophisticated structural forms. Koley and Sahoo [15] applied a coupled boundary element–finite difference method based on three-dimensional linear wave theory to analyze the interaction between oblique incident waves and vertical, permeable, flexible membranes. They also derived energy balance equations to verify model accuracy. Guo et al. [16] experimentally studied vertical flexible multi-layer porous membrane breakwaters under regular waves, examining reflection, transmission, and energy dissipation behaviors. Expanding on this work, Guo et al. [17] later investigated a hybrid breakwater system combining horizontal and vertical flexible membranes, systematically analyzing the effects of membrane porosity, submergence depth, length, and spacing on hydrodynamic performance.

Innovative designs inspired by marine vegetation have also emerged. Sun et al. [18] employed a coupled finite volume–finite element numerical model to study the performance of vertical plate-type flexible breakwaters under solitary waves. In another study, Sun et al. [19] proposed a kelp-box floating breakwater, demonstrating that increased kelp length, density, and number of rows significantly enhanced wave attenuation through turbulent energy dissipation and vortex shedding. Wu et al. [20] experimentally investigated the hydrodynamic performance of side-by-side box membrane breakwaters, showing that both narrow and wide gaps between boxes, combined with flexible membranes and increased draft depth, could substantially improve wave dissipation, especially under long-wave conditions.

These advances highlight the potential of flexible elements in improving breakwater performance. However, most research has focused on directly replacing rigid materials with flexible ones, emphasizing structural optimization and hydrodynamic performance, rather than investigating the synergistic interaction between flexible elements and traditional floating breakwaters.

While substantial progress has been made in understanding flexible structures and flexible breakwaters individually, studies on integrated systems that combine floating breakwaters with flexible wave-dissipating elements remain scarce [16,21–24]. Most existing work treats flexibility as a material property substitution rather than a functional addition to enhance wave energy dissipation. The mechanisms through which flexible structures and floating breakwaters interact to achieve superior performance are still poorly understood.

Addressing this knowledge gap is essential, as such combined systems could simultaneously reduce wave energy, minimize environmental impacts, and improve the resilience of offshore infrastructure. In particular, understanding the coupled hydrodynamics of these systems could inform design strategies for next-generation eco-friendly offshore protection structures.

In addition to conventional coastal protection applications, the proposed flexible wave-dissipating structure also shows great potential for integration with emerging floating photovoltaic (FPV) platforms, which are increasingly deployed in nearshore and offshore environments. Floating PV systems are often exposed to persistent wave action, which can lead to excessive motion of the floating foundation, reduced energy conversion efficiency, and potential damage to the mooring and electrical systems. Installing flexible wave-dissipating elements beneath the floating foundation provides a practical solution to mitigate these issues. The flexible structure can effectively attenuate incoming wave energy before it reaches the main platform, thereby reducing structural loads and improving the operational stability and safety of the FPV system. Figure 1 illustrates a conceptual design where flexible wave-dissipating structures are mounted below a floating PV platform, demonstrating how this integration can enhance both wave energy dissipation and renew-

https://doi.org/10.3390/jmse14010097

able energy production reliability. The flexible wave-dissipating elements studied here may be integrated beneath floating photovoltaic platforms to reduce platform motions and structural loads. The present study treats the floating body generically; explicit modeling of FPV modules, cables and electrical components is left for future applied studies.

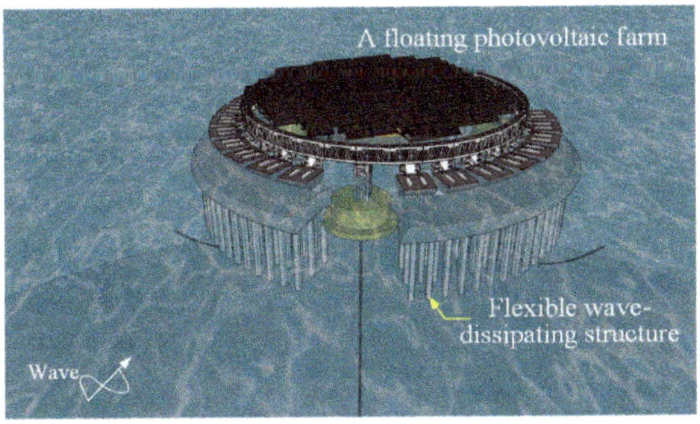

Figure 1. Conceptual illustration of a floating photovoltaic farm coupled with flexible wave-dissipating structures.

Motivated by these challenges, this study proposes a novel floating breakwater integrated with flexible wave-dissipating elements inspired by natural vegetation, as shown in Figure 2. The flexible component is designed to mimic the morphology and dynamic behavior of marine plants while maintaining controllable geometry and material properties for engineering applications. The integration of flexible wave-dissipating structures (FWDS) beneath a floating breakwater aims primarily to (i) reduce the wave energy transmitted under the floating body to the lee side and (ii) reduce the amplitude of radiation waves emitted by the floating body through increased rotational inertia and added mass. In practice, the FWDS thus act both as a local dissipater (attenuating energy that would otherwise pass beneath the pontoon) and as a dynamic modifier of the pontoon motion (reducing surge/pitch amplitudes and associated radiated energy). We emphasize that the present study focuses on two major controllable design parameters—elastic modulus and vertical extent—and provides initial guidance on their effective ranges. Considerations such as durability, biofouling and life-cycle cost remain important for practical deployment and will be addressed in subsequent studies.

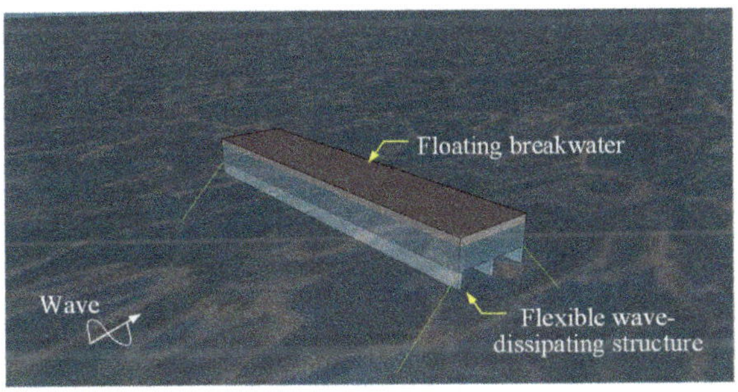

Figure 2. Conceptual illustration of a floating breakwater integrated with flexible wave-dissipating structures.

In addition to coastal protection, the proposed hybrid system may be employed to improve the operational stability of floating renewable energy platforms (e.g., FPV arrays and floating wind substructures), reduce mooring and structural loads on aquaculture installations, and serve as habitat-enhancing elements in integrated marine mitigation

https://doi.org/10.3390/jmse14010097

projects. Each application will require tailored design studies to account for platform geometry, payload, and operational constraints.

To capture the complex fluid–structure interactions involved, a two-dimensional numerical model is developed. This model is based on the Navier–Stokes (N-S) equations and incorporates multiple advanced techniques: the immersed boundary method (IBM) to handle solid boundaries, a finite element model (FEM) to simulate the deformation of the flexible structures, and the volume of fluid (VOF) method to track the free surface. Using this framework, the study systematically investigates the interaction mechanisms among the floating breakwater and the flexible wave-dissipating structures and water waves, evaluates their combined wave attenuation performance, and explores key design parameters for optimization.

Although prior studies have advanced understanding of flexible membranes, vegetation-inspired dissipators, and floating breakwaters, several recurring limitations remain. Many works emphasize material substitution or structural optimization without examining the coupled dynamics of flexible elements mounted beneath floating platforms. Validation efforts often stop at component-level benchmarks or linearized models rather than system-level experiments. Furthermore, a number of studies adopt two-dimensional or depth-integrated formulations that cannot capture three-dimensional processes such as spanwise vortex dynamics, lateral diffraction, or oblique-wave-induced yaw. The present study addresses a portion of these gaps by formulating a coupled immersed-boundary–finite-element–VOF framework and performing a focused parametric investigation of elastic modulus and vertical extent to reveal the dominant x–z plane interaction mechanisms. At the same time, it is important to state clearly that this work is confined to two-dimensional transverse-section simulations and that validation is performed at the component level (flexible-element FEM and IB–VOF coupling). Integrated system-scale experiments and three-dimensional simulations are required to fully assess practical applicability and are planned as follow-up work.

2. Numerical Model

The proposed system consists of two primary components: a floating breakwater and three flexible wave-dissipating structures. The floating breakwater is composed of a rectangular floating body moored by a tension-leg system, allowing it to move with three degrees of freedom (DOFs) under wave action: surge, heave, and pitch. Attached beneath the floating body are three identical flexible wave-dissipating structures, which are fixed to the lower surface of the breakwater. These flexible structures deform and bend under wave loading, thereby dissipating part of the incident wave energy and enhancing the overall wave attenuation performance of the coupled system.

Considering the geometric regularity of the configuration and the objective of developing a preliminary yet efficient numerical framework, the system is simplified to a two-dimensional (2D) model in the x–z plane. This simplification enables a detailed investigation of the coupled hydrodynamic behavior between the floating breakwater and the flexible wave-dissipating structures while significantly reducing computational cost and allowing refined local flow-field analysis.

2.1. Governing Equations and Wave Generation

The numerical simulations are conducted in a 2D numerical wave tank (NWT) based on an incompressible Navier–Stokes (N–S) solver (Figure 3). The flow is governed by the continuity and modified N–S equations as follows:

$$\frac{\partial u_i}{\partial x_i} = Q \tag{1}$$

https://doi.org/10.3390/jmse14010097

$$\frac{Du_i}{D_t} = -\frac{1}{\rho}\frac{\partial p}{\partial x_i} + 2v\frac{\partial D_{ij}}{\partial x_j} - \frac{2v}{3}\frac{\partial Q}{\partial x_i} - g_i - \gamma u_i \delta_{i2} + f_i \quad (2)$$

where x_i represents the orthogonal Cartesian coordinates (x, z), and $u_i = (u, w)$ are the corresponding velocity components, p represents the pressure, ν is the kinematic viscosity of water, γ is a wave-damping coefficient (non-zero only within sponge layers), δ_{i2} equals 0 and 1 in the x- and z-directions, respectively, D_{ij} is the velocity stress tensor, f_i is the body-force term (non-zero only near the immersed boundary), and Q is the wave-source term applied at $x = x_s$. The source is defined as $Q = q_{source}(z,t)/\Delta x_s$ where Δx_s is the grid spacing at the source, and q_{source} represents the flux density that gradually increases during the first three wave periods at the start of wave generation [25]. The source term is expressed as:

$$q_{source}(z,t) = \begin{cases} \left\{1 - \exp\left(-\frac{2t}{T_i}\right)\right\} 2U_0 \frac{\eta_0 + h}{\eta_s + h}; & \frac{t}{T_i} \leq 3 \\ 2U_0 \frac{\eta_0 + h}{\eta_s + h}; & \frac{t}{T_i} > 3 \end{cases} \quad (3)$$

where t is the simulation time, U_0 is the horizontal velocity amplitude, and η_s and η_0 are the free-surface displacements at the source and the theoretical wave elevation from third-order Stokes theory, respectively. To minimize wave reflection at the generation boundary, the depth-integrated flux q_{source} is modified by $(\eta_s + h)/(\eta_0 + h)$, following Ohyama and Nadaoka [26].

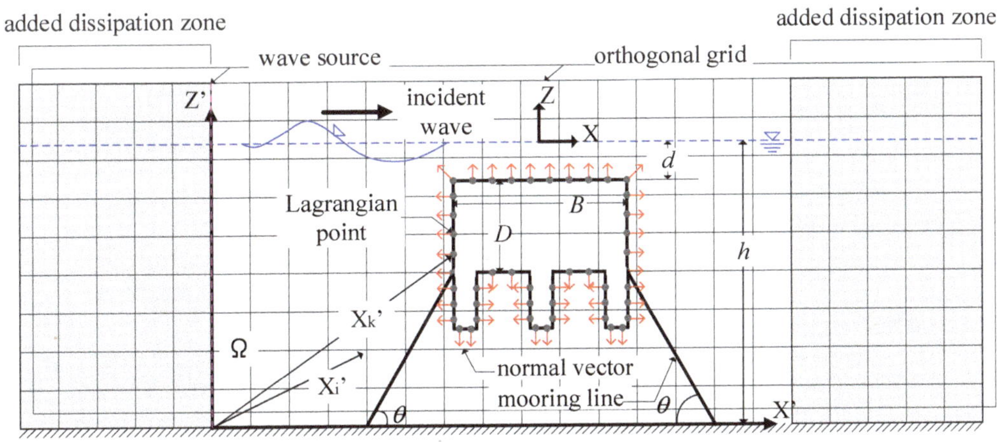

Figure 3. Computational domain.

The free surface is captured using the Volume of Fluid (VOF) method, governed by the advection equation of the VOF function F:

$$\frac{\partial F}{\partial t} + \frac{\partial (u_i F)}{\partial x} = FQ \quad (4)$$

where $F = 1$ and $F = 0$ represent the liquid and gas phases, respectively, and $0 < F < 1$ denotes the interface region.

The governing equations are solved numerically using the two-step projection method proposed by Chorin [27,28]. The computational domain is discretized on a rectangular staggered grid, where velocity components are defined at cell faces and pressure at cell centers. The velocity field is first advanced explicitly to an intermediate value, followed by solving the Poisson Pressure Equation (PPE) to obtain the pressure correction. Subsequently, the velocity is updated to the next time level. The Eulerian force term f_i acting near the immersed boundaries is evaluated using the Physical Virtual Model (PVM) developed

https://doi.org/10.3390/jmse14010097

by Lima E Silva et al. [29]. In the PVM framework, the total virtual force is explicitly decomposed into acceleration, inertial, viscous, and pressure components, ensuring stable and accurate representation of fluid–structure interactions. Further details of the numerical implementation can be found in refs. [30–33].

Although real floating breakwater systems are inherently three-dimensional, the present configuration—comprising a pontoon and vertically oriented flexible wave-dissipating elements—extends continuously in the transverse direction and is subjected to predominantly unidirectional incoming waves. Under this geometric arrangement, the dominant hydrodynamic responses (surge, heave, and pitch of the pontoon and the deformation dynamics of the flexible elements) occur primarily within the vertical x–z plane. For these reasons, a two-dimensional (2D) formulation is suitable for resolving the key fluid–structure interaction mechanisms, consistent with previous studies employing 2D models to capture highly resolved local flow features and vortex dynamics. However, this simplification inherently neglects spanwise (y) variations and other three-dimensional (3D) processes.

The present study focuses exclusively on normal-incidence waves and does not account for multi-directional seas or oblique incidence, which may induce yaw, asymmetric loading, and spanwise variability. Such 3D directional effects can significantly influence energy partitioning and flexible-structure deformation in realistic ocean conditions. Extending the present modelling framework to 3D simulations with directional wave spectra is therefore an essential next step.

While the 2D model effectively captures the principal x–z-plane physics for laterally extensive systems, several 3D mechanisms are not represented. These include edge-vortex formation at finite-width structures, spanwise flow redistribution that may alter vortex shedding and energy dissipation, lateral diffraction, and enhanced local dissipation due to 3D wave breaking. Accordingly, the quantitative results reported herein should be interpreted within the scope of the assumed 2D configuration. The qualitative mechanisms identified—deformation-induced dissipation and vortex-mediated energy loss—are expected to persist in three dimensions, but accurate engineering design for finite-width systems will require dedicated 3D CFD–FEM simulations and scale-model tests. These efforts are planned as part of our future research.

Moreover, this study focuses on regular (monochromatic) wave forcing to isolate and clarify the underlying fluid–structure interaction mechanisms and to facilitate robust validation. The response to irregular seas, wave groups and multi-directional spectra will be investigated in future work as part of a comprehensive parametric study.

2.2. Hydrodynamic and Motion Model of the Floating Breakwater System

The floating breakwater (FBW) is modeled as a rigid body supported by a tension-leg mooring system and coupled with three flexible wave-dissipating structures. The buoyancy forces acting on the floating body are significantly greater than its self-weight, ensuring that the mooring lines remain in tension throughout the simulation and no slack conditions occur. We assume that the tension-leg mooring lines remain taut throughout the simulations; this assumption is supported by the predominance of buoyancy over self-weight for the chosen floating body and by the intended mooring configuration. The present work therefore neglects slack-line behavior, dynamic tension variability and mooring failure modes. A detailed study of mooring dynamics, including line elasticity, pretension variation, and possible slackening or failure, will be performed in future work to assess their influence on system stability and performance.

The dynamic behavior of the submerged FBW under incident wave action is illustrated schematically in Figure 4. Wave-induced forces acting on the structure are decomposed

https://doi.org/10.3390/jmse14010097

into horizontal (H) and vertical (V) components applied to the surfaces denoted as H_1–H_4 and V_1–V_4, respectively. The total hydrodynamic forces and moments on the breakwater can be obtained by integrating the pressure distributions over these surfaces. Considering the rigid-body motion of the floating breakwater in the x–z plane, the equations of motion can be derived from Newton's second law for translation and rotation as follows:

$$\sum M_{c.g} = \sum M_{cg.x} + \sum M_{cg.z} + \sum M_{cg.T} - J_{FB}\alpha_{FB} \tag{5}$$

$$\sum F_X = m\alpha_x = H_1 + H_3 - H_2 - H_4 + 2(T_2\cos\theta_2 - T_1\cos\theta_1) \tag{6}$$

$$\sum F_Z = m\alpha_z = V_1 + V_3 - W - V_2 - V_4 - 2(T_2\sin\theta_2 + T_1\sin\theta_1) \tag{7}$$

where m is the mass of the floating body, α_x, and α_z are the translational accelerations in the horizontal and vertical directions, respectively, α_{FB} is the angular acceleration about the center of gravity (C.G.), J_{FB} is the mass moment of inertia of the floating body about its C.G., H_i and V_i denote the resultant horizontal and vertical wave forces on each surface, T_1 and T_2 are the tensile forces acting on the mooring lines, $\sum M_{cg.x}$, $\sum M_{cg.z}$, and $\sum M_{cg.T}$ are the moments about the C.G. generated by the horizontal forces, vertical forces, and mooring tensions, respectively.

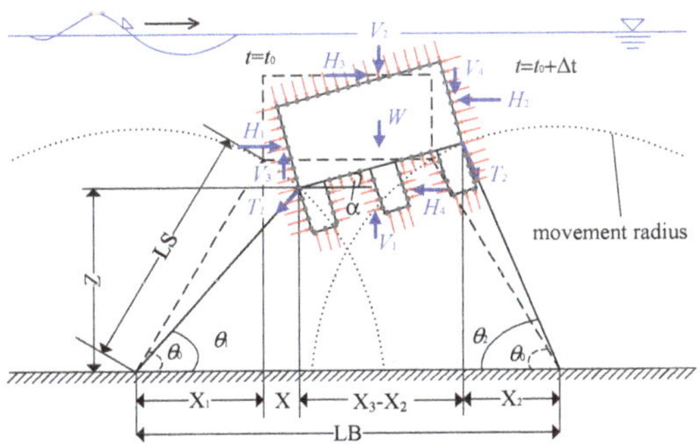

Figure 4. Dynamics of the floating breakwater integrated with flexible wave-dissipating structures.

These coupled equations govern the translational (surge and heave) and rotational (pitch) responses of the floating breakwater under wave excitation. The motion of the floating body in turn affects the surrounding flow field and the deformation of the flexible wave-dissipating structures, forming a two-way fluid–structure interaction (FSI) system.

To ensure numerical stability, the motion equations are solved using a time-marching approach synchronized with the Navier–Stokes solver. At each time step, hydrodynamic pressures on the floating body surfaces are integrated to yield the instantaneous wave forces and moments, which are then substituted into Equations (5)–(7) to update the body's position and orientation. The updated geometry is subsequently applied to the flow solver via the immersed boundary method (IBM), allowing dynamic coupling between the rigid floating body and the surrounding fluid field.

Potential operational issues such as debris accumulation, entanglement and progressive structural degradation or failure are not treated in the current simulations. These hazards will be examined in future experimental tests and long-term numerical studies that consider debris loading, impact loads and maintenance strategies.

2.3. Finite Element Modeling of the Flexible Wave-Dissipating Structures

The hybrid system considered in this study (see Figures 2–4) consists of two main components: (1) a floating breakwater (FBW) modeled as a rigid body, and (2) three flexible wave-dissipating structures (FWDS) attached to the underside of the FBW, as illustrated in

https://doi.org/10.3390/jmse14010097

Figure 5. Each flexible wave-dissipating structure is modeled as a thin rectangular plate with dimensions of 0.075 m × 0.02 m, uniformly distributed along the bottom surface of the breakwater. The deformation of these flexible structures under wave-induced hydrodynamic loads is simulated using the finite element method (FEM).

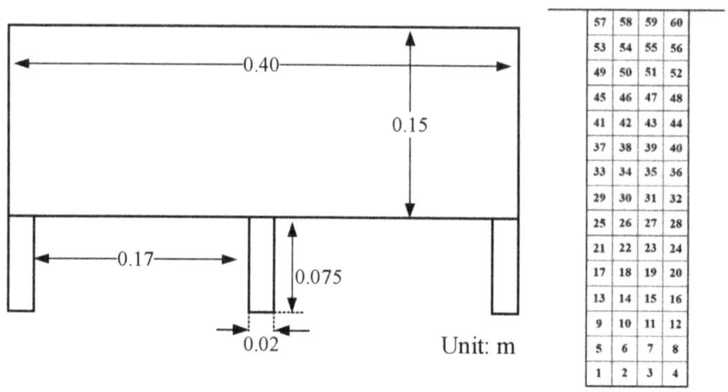

(a) (b)

Figure 5. (**a**) Two-dimensional schematic diagram of the floating breakwater coupled with flexible wave-dissipating structures, with main geometric dimensions indicated. (**b**) Finite element mesh discretization and node numbering of a single flexible wave-dissipating structure.

Each flexible plate is discretized into 60 four-node rectangular elements and 80 nodes in total. The nodes are numbered sequentially from left to right and bottom to top, ensuring consistent element connectivity. The four-node rectangular element is chosen because of its simplicity, numerical stability, and ability to accurately capture the bending and deformation behavior of thin flexible structures under moderate strains.

For each element, the displacement field is approximated by an interpolation (shape) function derived from the nodal coordinates. This function expresses the displacement components within an element in terms of the nodal displacements, allowing the strain–displacement relationships to be formulated directly. Based on these relationships, the element stiffness matrix $[K^e]$ is obtained through the standard strain energy formulation, while the mass matrix $[M^e]$ is derived from the kinetic energy expression. A lumped (concentrated) mass matrix is employed to reduce computational cost and improve numerical efficiency without compromising accuracy for the present problem.

The nodal displacement vector of an individual element can be written as:

$$\mathbf{q}_t^e(t) = \left[u_1(t)v_1(t)w_1(t)\ \ldots\ u_n(t)v_n(t)\ \ w_n(t)\right]^{\mathrm{T}} \tag{8}$$

where $u_i(t)$, $v_i(t)$, and $w_i(t)$ represent the displacement components in the local x-, y-, and z-directions at node i, respectively.

Additionally, applying the principle of virtual work to each finite element yields the following dynamic equilibrium equation:

$$[M^e]\ddot{\mathbf{q}}_t^{\,e} + [C^e]\dot{\mathbf{q}}_t^{\,e} + [K^e]\mathbf{q}_t^e = \mathbf{F}_t^e \tag{9}$$

where $[M^e]$, $[C^e]$, and $[K^e]$ are the mass, damping, and stiffness matrices of the element, respectively; $\mathbf{F}_t^e$ is the external load vector containing the hydrodynamic forces exerted by the surrounding fluid.

https://doi.org/10.3390/jmse14010097

The damping matrix $[C^e]$ can be determined using Rayleigh damping, expressed as a linear combination of the mass and stiffness matrices, i.e., $[C^e] = \alpha[M^e] + \beta[K^e]$, where α and β are mass- and stiffness-proportional damping coefficients, respectively.

The global stiffness, mass, and damping matrices of the flexible structure are assembled by summing the corresponding element matrices according to the global nodal numbering system:

$$\begin{aligned}[K] &= \sum_{e=1}^{n}[K^e] \\ [M] &= \sum_{e=1}^{n}[M^e] \\ [C] &= \sum_{e=1}^{n}[C^e]\end{aligned} \tag{10}$$

where n denotes the total number of elements.

After the application of boundary conditions (e.g., fixed connections between the upper edge of each flexible structure and the bottom of the floating breakwater), the global dynamic equation of motion for the entire flexible system can be expressed as:

$$[M]\ddot{\mathbf{q}}_t + [C]\dot{\mathbf{q}}_t + [K]\mathbf{q}_t = \mathbf{F}_t \tag{11}$$

Here, $\mathbf{q}_t$, $\dot{\mathbf{q}}_t$ and $\ddot{\mathbf{q}}_t$ denote the global displacement, velocity, and acceleration vectors of all nodes, respectively. The right-hand-side vector $\mathbf{F}_t$ represents the distributed hydrodynamic loading on the flexible structures, transferred from the flow field through the fluid–structure interaction module.

Once the global motion equation is established, the transient structural responses—nodal displacements, velocities, and accelerations—can be obtained through time-domain integration. Alternatively, for efficiency and modal insight, the system can be transformed into modal coordinates using the mode superposition method, which decouples the global system into a set of independent single-degree-of-freedom equations. The resulting modal responses provide the basis for analyzing the deformation characteristics and energy dissipation behavior of the flexible wave-dissipating structures under different hydrodynamic conditions.

2.4. Overall Numerical Procedure

The coupled solver integrates (i) an incompressible Navier–Stokes solver using the two-step projection method for velocity–pressure coupling, (ii) the Volume-of-Fluid (VOF) method for free-surface tracking, (iii) an immersed boundary method (IBM) to impose moving solid boundaries in an Eulerian fluid grid, and (iv) a finite-element structural solver to compute deformation of the flexible elements. The IBM provides two-way interaction by exchanging forces and kinematic constraints between the fluid and Lagrangian structural representation, while the FEM computes structural internal forces and modal responses for the FWDS.

The complete computational procedure of the present coupled model is summarized in Figure 6, and the main steps are described as follows:

1. Wave force computation:

   Calculate the external hydrodynamic forces acting on the breakwater surfaces (H_1–H_4 and V_1–V_4) by integrating the local pressure field around the floating body.

2. Floating body motion update:

   Determine the translational and rotational displacements of the floating breakwater, and subsequently update the positions of the corresponding Lagrangian points.

3. Flexible structure deformation:

https://doi.org/10.3390/jmse14010097

Compute the deformation of the three flexible wave-dissipating structures using the modal superposition method, and update their instantaneous configurations to reflect the new equilibrium positions.

4. Immersed boundary forcing:

Evaluate the forcing term f_i to impose the solid boundary condition on the fluid field, following the immersed boundary approach described in refs. [30,31].

5. Fluid field update:

Advance the fluid velocity and pressure fields using the two-step projection method, incorporating the computed virtual forces to ensure dynamic consistency between the fluid and structures.

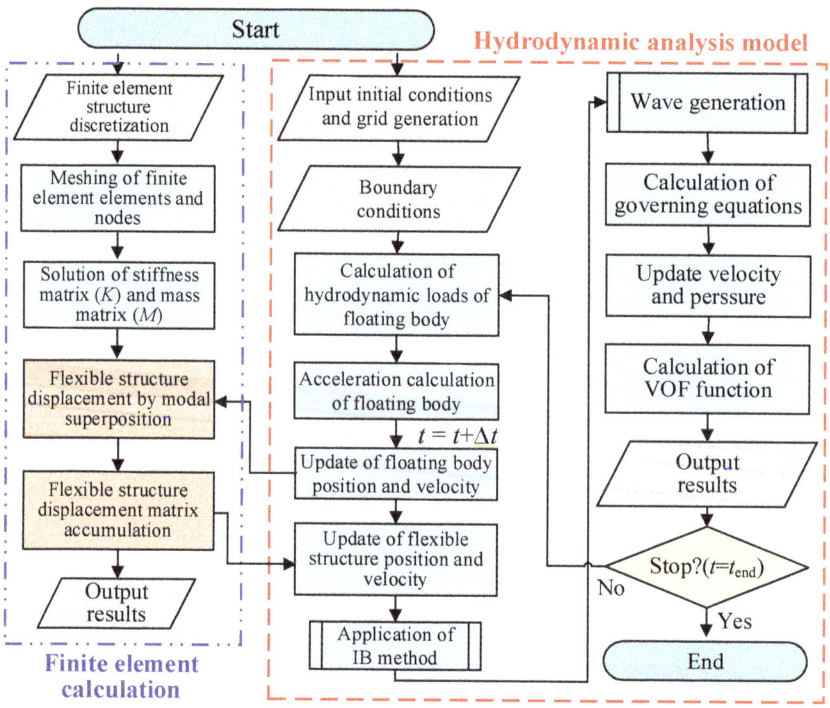

Figure 6. Flowchart of the numerical procedure for the coupled fluid–structure interaction model.

This iterative procedure continues at each time step until convergence is achieved for both the fluid and structural responses. The framework allows for strong two-way coupling between the flow solver, the floating body motion module, and the finite element model of the flexible structures, ensuring accurate representation of the nonlinear fluid–structure interaction dynamics. Relevant numerical and modeling studies on complex coupled flows and advanced numerical techniques are discussed to provide broader methodological context [34–36]. These methods illustrate the diversity of numerical approaches available for strongly coupled multi-physics problems.

In the present 2D simulations we intentionally used a high spatial resolution ($\Delta x = \Delta z = 0.01$ m in the model domain) so that the dominant vortex formation and shedding phenomena at the studied Reynolds numbers are directly resolved, consistent with the approach and validation discussed in refs. [30–33]. This direct-resolution strategy reduces dependence on turbulence closures for the current cases. However, this does not substitute for three-dimensional turbulence dynamics that arise at higher Reynolds numbers and in

https://doi.org/10.3390/jmse14010097

full-scale systems; therefore, for future 3D and prototype-scale studies, we plan to adopt appropriate turbulence models (LES or hybrid RANS–LES) and grid-adaptive strategies to ensure fidelity across scales.

3. Model Validation

The numerical model developed in this study builds upon the authors' previous work, which has already been validated for simulating the hydrodynamic interactions between rigid floating breakwaters and surface waves. Therefore, the present validation focuses exclusively on the model's capability to simulate the motion and deformation of flexible structures under both static and dynamic loading conditions. Two benchmark cases were conducted: (i) the deflection of a cantilever beam under concentrated loads, and (ii) the motion of a single-stem flexible vegetation model subjected to regular waves.

The validation work presented in Section 3 focuses on the fidelity of the key numerical components employed in this study. Specifically, the finite-element structural model has been verified through static deflection tests of a cantilever beam subjected to concentrated loads, while the IB–VOF fluid solver and the fluid–structure coupling strategy have been validated against experimental measurements of a single flexible stem oscillating under regular wave excitation. These component-level benchmarks confirm the accuracy of the numerical building blocks used in the coupled simulations.

However, it is acknowledged that the present study does not include a system-level experimental validation of the fully integrated floating breakwater equipped with multiple FWDS units. The combined hydrodynamic–structural behavior—such as integrated wave attenuation, global platform motions (surge, heave, and pitch), FWDS deformation, and vortex-shedding patterns—has therefore not yet been assessed through physical testing.

To address this limitation, a scaled laboratory campaign of the complete floating breakwater–FWDS assembly is planned for future work. The forthcoming experiments will quantify wave attenuation, structural responses, and flow features under controlled wave-tank conditions, providing system-level benchmarks for the present numerical framework. In parallel, extended numerical verifications involving multiple interacting flexible elements will be performed to further substantiate model robustness.

Overall, while the current validation confirms the reliability of the structural solver, the IB–VOF fluid solver, and their coupling for flexible-element FSI at the component scale, full experimental validation of the integrated system remains a key direction for future research. The planned laboratory and numerical efforts will serve to refine and strengthen the predictive capability of the model for practical floating breakwater applications.

3.1. Cantilever Beam Test

To verify the accuracy of the flexible structural model under static conditions, a cantilever beam subjected to a tip load was analyzed and compared with the theoretical solution for beam deflection. The theoretical expression for the deflection curve is given as:

$$y = -\frac{Px^2}{6EI}(3l - x) \tag{12}$$

where y is the vertical displacement, P is the concentrated load, E is the elastic modulus, I is the second moment of inertia, l is the beam length, and x is the distance from the fixed end.

In the simulation, the cantilever beam had a rectangular cross-section of 0.075 m $\times$ 0.02 m, with $E = 2.17 \times 10^6$ Pa, $\rho = 1.37 \times 10^3$ kg/m^3, Poisson's ratio $\nu = 0.394$, and $I = 3 \times 10^{-7}$ m^4. Concentrated loads of $F = 0.50$ N, 0.75 N, and 1.00 N were applied at the free end. The beam was discretized into 60 four-node rectangular finite elements and 80 nodes, ensuring adequate spatial resolution to capture deformation.

https://doi.org/10.3390/jmse14010097

Figure 7 presents the comparison between the simulated and theoretical deflection profiles for the three loading cases. The results show an excellent agreement between the two, with only negligible discrepancies. This demonstrates that the proposed finite element model accurately reproduces the static deformation behavior of flexible beams, providing a solid basis for subsequent dynamic analyses under wave-induced loading.

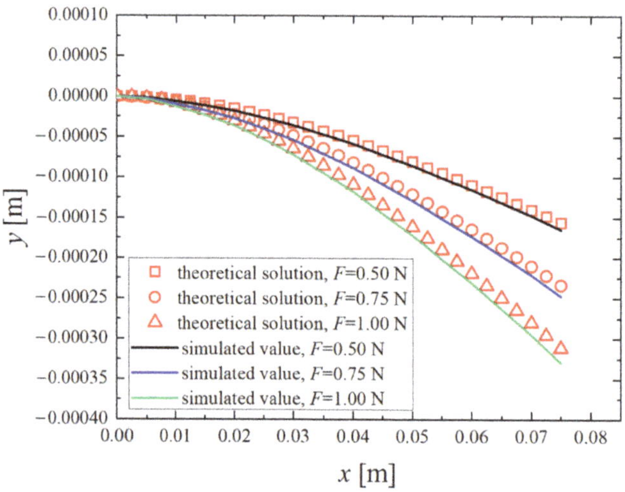

Figure 7. Comparison of simulated and theoretical deflections of a cantilever beam under concentrated loads of 0.50 N, 0.75 N, and 1.00 N.

3.2. Single-Stem Flexible Vegetation Test

A second validation case was conducted to assess the model's capability in simulating the dynamic response of flexible structures in oscillatory flow. The simulation setup follows the experimental conditions of ref. [37], where regular waves with a period $T = 4.0$ s, height $H = 0.2$ m, and water depth $h = 0.8$ m were generated. The computational domain was discretized using a uniform orthogonal grid with $\Delta x = \Delta z = 0.01$ m to ensure numerical stability and resolution.

The flexible vegetation model was positioned downstream of the wave source at $x = 1.20$ m. It was discretized into 60 finite elements and 80 Lagrangian nodes, enabling accurate tracking of deformation under wave forcing. The time step was set to $\Delta t = 0.005$ s, and the total simulation time was 20 s. The structural properties were consistent with those used in Section 3.1, with $E = 2.17 \times 10^6$ Pa, $\rho = 1.37 \times 10^3$ kg/m^3, and $\nu = 0.394$.

To ensure a realistic initialization, the flexible structure was first subjected to hydrostatic pressure in still water, allowing it to reach a quasi-equilibrium configuration before wave loading commenced. The computed extreme positions of the structure during one wave cycle were compared with the experimental data from refs. [37,38]. Figure 8 illustrates the comparison, showing excellent agreement between the simulated and measured tip displacements. These results demonstrate that the present model can accurately reproduce the motion characteristics of flexible vegetation under wave action. The successful validation under both static and dynamic conditions confirms the robustness and reliability of the coupled finite element–fluid interaction framework, providing a firm foundation for subsequent investigations of flexible wave-dissipating structures integrated with floating breakwaters.

https://doi.org/10.3390/jmse14010097

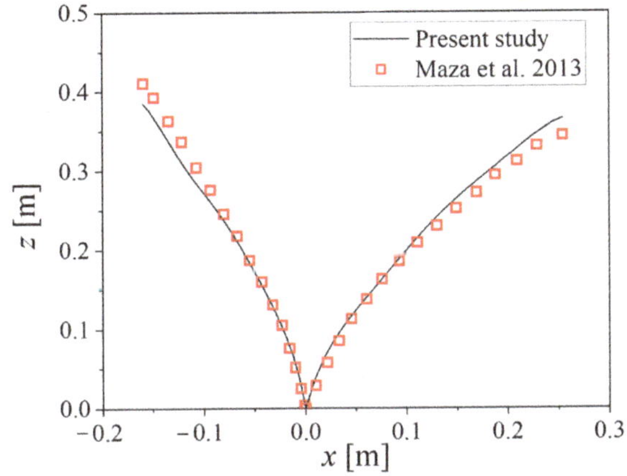

Figure 8. Comparison of the simulated and experimental extreme displacements [37] of a single flexible stem during one wave cycle.

4. Wave Attenuation Performance Evaluation Method

Based on the wave elevation data recorded by Gauge 1, Gauge 2, Gauge 3, and Gauge 5 (see Figure 9), the wave reflection and transmission characteristics of the floating breakwater coupled with flexible wave-dissipating structures were evaluated. The least-squares method [39] in combination with the Fast Fourier Transform (FFT) algorithm was employed to decompose the measured free surface elevation into incident and reflected components. The simulated wave elevations were utilized to calculate the wave reflection coefficient $K_r = H_r/H$, where H_r and H denote the reflected and incident wave heights, respectively, and the transmission coefficient $K_t = \sqrt{E_t/E_i}$ where E_t and E_i represent the transmitted and incident wave energies. The wave energy at each location was obtained by integrating the spectral energy over all frequency components, with the amplitude of each frequency extracted through FFT analysis.

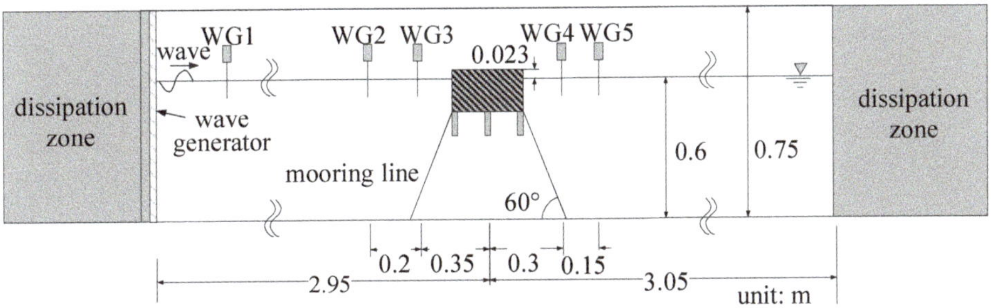

Figure 9. Schematic of the computational domain, including the wavemaker, floating breakwater, absorbing zones, and the positions of the five wave gauges. For clarity, the horizontal distances between the wavemaker and the floating breakwater and between the breakwater and the outlet boundary are not drawn to scale, as the figure is intended to highlight the placement of the breakwater and measurement points rather than depict exact geometric proportions.

The energy loss coefficient, K_{loss}, was defined to quantify the proportion of wave energy dissipated through viscous effects such as vortex formation, flow separation, and shear-induced damping near the edges of the floating and flexible components. According

https://doi.org/10.3390/jmse14010097

to the principle of energy conservation, the following relationship can be established among the dimensionless coefficients:

$$K_{\text{loss}} = 1 - K_t^2 - K_r^2 \tag{13}$$

In summary, the reflection coefficient (K_r), transmission coefficient (K_t), and energy loss coefficient (K_{loss}) together provide a comprehensive framework for assessing the wave attenuation performance of the coupled breakwater–flexible structure system. Specifically, K_r quantifies the portion of wave energy reflected by the structure, K_t represents the fraction of wave energy transmitted past it, and K_{loss} reflects the energy dissipated through viscous damping, vortex shedding, and internal deformation within the flexible components. By systematically analyzing these coefficients under various wave conditions and structural configurations, the underlying mechanisms of wave attenuation—reflection, transmission, and viscous dissipation—can be quantitatively distinguished, providing a solid foundation for subsequent optimization and hydrodynamic performance evaluation of the proposed system.

The decomposition $K_{loss} = 1 - K_t^2 - K_r^2$ provides a convenient energy balance and remains informative even under nonlinear conditions; however, this formulation implicitly lumps all non-recovered energy into K_{loss}, including viscous dissipation, turbulence, nonlinear wave–structure energy exchange, and numerical dissipation. When strong nonlinearities, breaking or high-amplitude vortex shedding occur, the estimate of spectral energies via FFT and the least-squares separation may introduce additional uncertainty. We therefore interpret K_{loss} as a bulk measure of irreversible energy loss rather than a strict separation of distinct physical mechanisms. Where possible, we corroborate energy-loss trends with flow-field diagnostics (vorticity, vortex shedding) to strengthen physical interpretation.

5. Results and Discussion

The present simulations represent a transverse section of a long (or very wide) floating platform and therefore neglect spanwise edge effects and lateral diffraction associated with finite-length bodies. The objective of the current work is to reveal the fundamental coupling mechanisms and dissipation pathways induced by flexible elements in the x–z plane; engineering applicability for finite-size systems will require subsequent three-dimensional studies and model tests, which are identified as immediate future research tasks.

A two-dimensional numerical wave tank was established to investigate the hydrodynamic performance of the floating breakwater system. The rectangular computational domain measured 6.00 m × 0.75 m, with 0.80 m-long sponge layers implemented at both ends to effectively minimize wave reflections and boundary interference. The present domain (6.00 m × 0.75 m) with 0.80 m-long sponge layers at each end follows the configuration used in prior validation studies that demonstrated effective reflection damping for the considered wave conditions (see refs. [30–33]). The wave generation source was positioned at $x = 0$ m, while the center of the floating breakwater system was located at $x = 2.95$ m, $z = 0.548$ m. The numerical simulations were carried out with a time step of $\Delta t = 0.005$ s and a total simulation time of 20 s.

To capture the wave evolution throughout the computational domain, five wave gauges were evenly distributed along the tank to record the free surface elevation in real time. The detailed configuration of the computational domain, including the locations of the wave gauges, the wave maker, the floating breakwater, and the absorbing zones, is shown in Figure 9. For clarity, some structural dimensions in the figure are not drawn to scale in order to highlight key regions of interest. The selected computational setup and parameters ensure that the numerical results accurately represent the hydrodynamic behavior of the floating breakwater system under wave excitation.

https://doi.org/10.3390/jmse14010097

The floating breakwater (FBW) was modeled as a rigid body with a planform size of 0.40 m × 0.15 m. The long and short sides of the breakwater were discretized by 80 and 30 Lagrangian nodes, respectively. These nodes were coupled to the flow field via the immersed boundary method (IBM), enabling accurate capture of the breakwater's translational and rotational motions under wave action.

The flexible wave-dissipating structures (FWDS) were attached to the bottom of the floating breakwater. Each flexible structure had a height of 0.075 m and a width of 0.02 m, and all three were evenly spaced along the underside of the breakwater. In the flow field, the flexible structures were represented as solid boundaries using Lagrangian points, consistent with the numerical treatment adopted for the breakwater (see Figures 3–5). This configuration allows for the accurate coupling between fluid motion and structural deformation, ensuring that the dynamic response and energy dissipation behavior are faithfully reproduced.

The tension-leg mooring system was modeled with mooring lines anchored to the seabed at an inclination angle of 60°, ensuring that the lines remained taut throughout the wave–structure interaction process. This configuration provided sufficient restoring stiffness to constrain the motion of the floating breakwater, preventing excessive drift or instability.

In all simulations, the still-water depth was set to h = 0.60 m and the wave steepness was fixed at H/L = 0.03. Regular (monochromatic) waves with model-scale periods T = 1.0–2.0 s were used. Under the geometric scale adopted in this study (1:25), these model periods correspond approximately to prototype periods of 5.0–10.0 s, which span a wide range of common nearshore sea states. The chosen steepness represents appreciable nonlinearity while remaining below typical breaking thresholds for the water depths considered; therefore the selected conditions are appropriate for probing the device response in energetic but non-breaking seas and for isolating the principal fluid–structure interaction mechanisms.

We recognize, however, that real ocean conditions are often irregular and may include directional spreading, steep breaking waves, wave groups, and wave–current interactions that can substantially affect performance. The present regular-wave parametric study therefore provides controlled, mechanistic insight rather than comprehensive performance limits. Extension to a broader parameter space—including irregular (spectral) seas, higher steepness and breaking conditions, focused wave groups and wave–current coupling—is planned as part of our future work to assess device behavior under realistic and extreme sea states and to support eventual three-dimensional and experimental validation.

All simulations assume constant fluid properties representative of seawater (1025 kg m^{-3}). The effects of variations in density and viscosity (e.g., freshwater vs. seawater, temperature dependence) are expected to be secondary for the present geometric scales but will be quantified in future sensitivity studies.

To evaluate the influence of the flexible component's material properties, a series of cases with varying elastic modulus (E) values were conducted, while other parameters—namely the density (ρ = 1.37 × 10^3 kg/m^3) and Poisson's ratio (ν = 0.394)—were kept constant. This approach isolates the effect of material stiffness on the overall hydrodynamic performance of the coupled floating breakwater–flexible structure system, providing valuable insights into the energy dissipation mechanisms and motion responses induced by wave loading.

The rectangular computational domain is discretized with a uniform orthogonal grid with $\Delta x = \Delta z$ = 0.01 m. Including the sponge layers and all computational cells, the total mesh size in the present 2D runs is approximately 51,000 cells. With a time step Δt = 0.005 s and a total simulated physical time of 20 s (4000 time steps), a single simulation required about 6.5 h of wall-clock time on a workstation with an Intel Core i9-9980HK (8 physical

https://doi.org/10.3390/jmse14010097

cores). These timings reflect the current serial/modestly parallel implementation; for future 3D and large-scale studies, we will employ domain decomposition and MPI/OpenMP parallelization, adaptive mesh refinement in regions of interest (near FWDS tips), and modal reduction techniques for the structural solver to reduce computational cost.

5.1. Wave-Structure Interactions

This section analyzes the interaction process between waves and two types of structures: a conventional floating pontoon breakwater and a composite floating breakwater equipped with Flexible Wave-Dissipating Structures (FWDS). The focus is placed on evaluating the influence of the flexible components on the overall hydrodynamic performance of the floating system.

Figure 10 presents the time series of free surface elevation recorded by wave gauge WG5, located 0.25 m behind the floating breakwater, for both cases with and without the FWDS. As shown, the free surface elevation is significantly reduced when the FWDS are attached. At WG5, the average wave height behind the structure without FWDS is 0.0437 m, while with FWDS it decreases to 0.0269 m. Compared with the incident wave height, the wave attenuation ratio increases from 38.10% (without FWDS) to 48.30% (with FWDS), representing an improvement of 10.20%. This observation indicates that the FWDS effectively modulate the wave energy distribution and propagation path through complex wave–structure interactions, thereby enhancing the overall wave-dissipating performance of the floating breakwater.

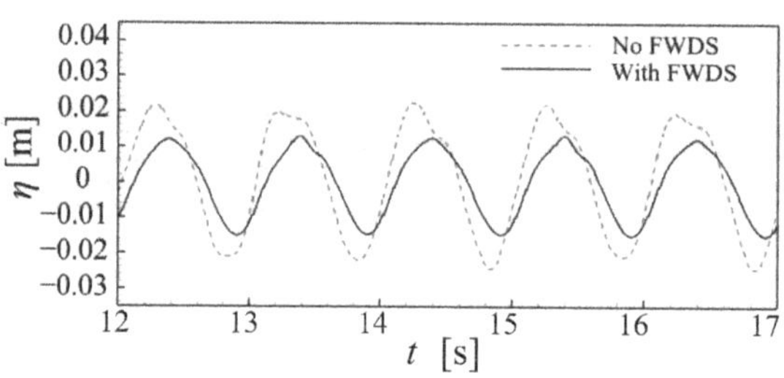

Figure 10. Comparison of free-surface elevation at the leeward side (WG5) for cases with and without Flexible Wave-Dissipating Structures (FWDS). (h = 0.6 m, T = 1.0 s, H = 0.046 m, H/L = 0.03).

By adding submerged flexible elements beneath the floating breakwater, the system's effective added mass and rotational inertia increase. This additional inertia reduces the amplitude of surge and pitch motions for a given incident wave, thereby decreasing the amplitude of radiation waves generated by the floating body. In this sense, FWDS contribute to the dynamic stability of the FB and reduce motion-induced transmission of wave energy.

Figures 11 and 12 further illustrate the spatial distributions of the velocity field and vorticity field, respectively, under regular wave conditions (h = 0.6 m, T = 1.0 s, H = 0.046 m, H/L = 0.03). The velocity field represents the instantaneous flow motion around the structure, described by the velocity components in the x and z directions (units: m/s). The vorticity, ζ, which quantifies local flow rotation, is defined as follows:

$$\zeta = \frac{\partial w}{\partial x} - \frac{\partial u}{\partial z} \approx \frac{\Delta w}{\Delta x} - \frac{\Delta u}{\Delta z} \tag{14}$$

where u and w are the velocity components in the x- and z-directions, respectively, and ζ has the dimension of s^{-1}. In the vorticity contour plots, black arrows denote the instantaneous velocity vectors, while colors represent the vorticity magnitude and sign—positive (red) indicating counterclockwise rotation and negative (blue) indicating clockwise rotation.

https://doi.org/10.3390/jmse14010097

As shown in Figure 11, the interaction between the rectangular floating breakwater and incident waves exhibits strong periodicity. The generation, evolution, and shedding of vortices around the breakwater have been extensively discussed in previous studies (see ref. [31]). Comparatively, Figure 12 reveals a more complex flow pattern in the case with FWDS. When waves propagate from the left, at t = 8.1 s, the floating system remains nearly at equilibrium, with the pontoon level and the flexible structures vertically downward. As the waves advance (Figure 12b), the structure begins to move rightward under wave excitation, forming positive vorticity zones on both the front and rear surfaces due to wave reflection. Meanwhile, negative vortices emerge near the tips of the flexible structures as a result of strong local disturbances.

As the wave crest passes (Figure 12c,d), the pontoon tilts and partial overtopping occurs, but with a significantly smaller volume of overtopped water than that observed in the rigid case (Figure 11c,d). The motion amplitude of the structure reaches its maximum, accompanied by vortex shedding beneath the FWDS tips, which leads to energy dissipation through vortex evolution and detachment. When the structure returns to its equilibrium position (Figure 12f), the vorticity near the leeward side intensifies and detaches from the surface (Figure 12g). The structure then rotates slightly clockwise, producing strong wave reflection on the front face and generating a negative vorticity region behind it. The high negative vorticity zones around the flexible tips transform into positive vorticity regions, consistent with the dynamic deformation of the FWDS. At t = 8.7–8.8 s, the local flow velocity near the flexible components increases, forming a more complex localized flow pattern. When the wave crest passes (Figure 12i,j), the leeward flow field exhibits backflow regions and strong vortex diffusion and shedding, indicating wave energy dissipation as the vorticity propagates outward and decays in the far field.

The flexibility and deformability of the FWDS enable them to adapt to incident wave motion, effectively mitigating wave impact forces and spreading wave energy over a wider area. The velocity field distribution in Figures 11 and 12 confirms that the presence of FWDS reduces peak flow velocities and smooths their spatial distribution. This demonstrates that the dynamic motion of the flexible structures disperses concentrated wave energy, alleviating local flow acceleration and suppressing high-energy zones. In contrast, the rigid pontoon (without FWDS) produces concentrated velocity regions (Figure 11f,g), implying limited energy dissipation and localized hydrodynamic impacts.

Similarly, significant differences are observed in the vorticity field distributions. Without FWDS, strong vortices form near the upper right corner of the structure, caused by viscous flow separation at sharp edges, consistent with findings in previous literature. These intense vortices concentrate wave energy locally and enhance nonlinear flow phenomena such as wave breaking and vortex shedding. When FWDS are included, the vorticity distribution becomes more uniform and spatially extended. Notably, high-vorticity regions shift away from the structural corners toward the sides, reflecting the fluid–structure coupling effect of the flexible components. Through deformation, the FWDS alleviate local velocity gradients, reduce shear effects, and promote a smoother vorticity distribution. This uniformization reduces local hydrodynamic loads, enhances wave energy dissipation, and improves the overall wave-attenuation efficiency of the composite floating breakwater.

The cyclic deformation of FWDS produces alternating flow separation and tip vortices. These vortices detach and evolve into turbulent structures that act to dissipate coherent wave energy via viscous dissipation and turbulent mixing. As a result, vortex generation and shedding are major contributors to K_{loss}, and the spatial distribution and strength of vorticity are directly correlated with the observed increase in energy dissipation for more flexible or taller FWDS.

https://doi.org/10.3390/jmse14010097

It should be noted that the present study intentionally focuses on the essential hydrodynamic interaction between the breakwater and the flexible wave-dissipating structures within a two-dimensional numerical framework. Additional physical processes and extended structural or environmental factors, though important, fall beyond the current scope and will be explored in future extensions of this work.

Further quantitative evaluation is performed through the calculation of reflection (K_r), transmission (K_t), and energy dissipation coefficients (K_{loss}) for cases with and without FWDS, as shown in Figure 13. Six regular wave conditions are tested with periods ranging from T = 1.0 s to 2.0 s at an interval of 0.2 s, while water depth and wave steepness remain constant. The reflection coefficient K_r shows little dependence on wave period, remaining between 0.2 and 0.3, but increases on average by 51.91% due to the presence of FWDS—particularly under long-period waves. Conversely, the transmission coefficient K_t increases with wave period in both configurations. For shorter waves, limited propagation and weaker excitation lead to low transmission. As the period increases, longer waves induce stronger excitation and larger structural motions, and near the system's natural frequency, resonance amplifies oscillations, reducing the shielding effect and increasing transmission. Consequently, energy dissipation efficiency decreases with longer wave periods. When T increases from 1.0 s to 1.6 s, the system exhibits the highest dissipation efficiency, as the incident wave energy is concentrated near the free surface and more effectively attenuated by the structure. For longer waves, energy spreads vertically, increasing transmission and causing nonlinear variations in K_t and K_{loss}. This indicates that the wave-attenuation performance of the FWDS is strongly nonlinear with respect to wave period, showing high sensitivity within specific ranges.

5.2. Influence of Elastic Modulus on the Hydrodynamic Performance of the Composite System

The material properties of the Flexible Wave-Dissipating Structures (FWDS) are expected to have a substantial impact on their overall wave attenuation performance. In this section, the influence of the elastic modulus (E) of the FWDS is investigated in detail. The elastic modulus, representing the stiffness of the material, serves as a key parameter controlling the deformation response of the flexible elements under wave loading. Based on the range reported by Sun et al. [19], six elastic modulus values are considered, from 1.085×10^5 Pa to 2.17×10^{10} Pa, the latter approximating a rigid material. Other material parameters, including density ($\rho = 1.37 \times 10^3$ kg/m^3) and Poisson's ratio (ν = 0.394), are kept constant throughout the analysis.

A higher elastic modulus corresponds to a stiffer material that undergoes limited deformation, whereas a lower modulus provides greater flexibility, allowing the structure to deform more readily under hydrodynamic excitation. By systematically varying E, the dynamic characteristics of the FWDS can be tuned to optimize wave energy dissipation and control wave propagation.

https://doi.org/10.3390/jmse14010097

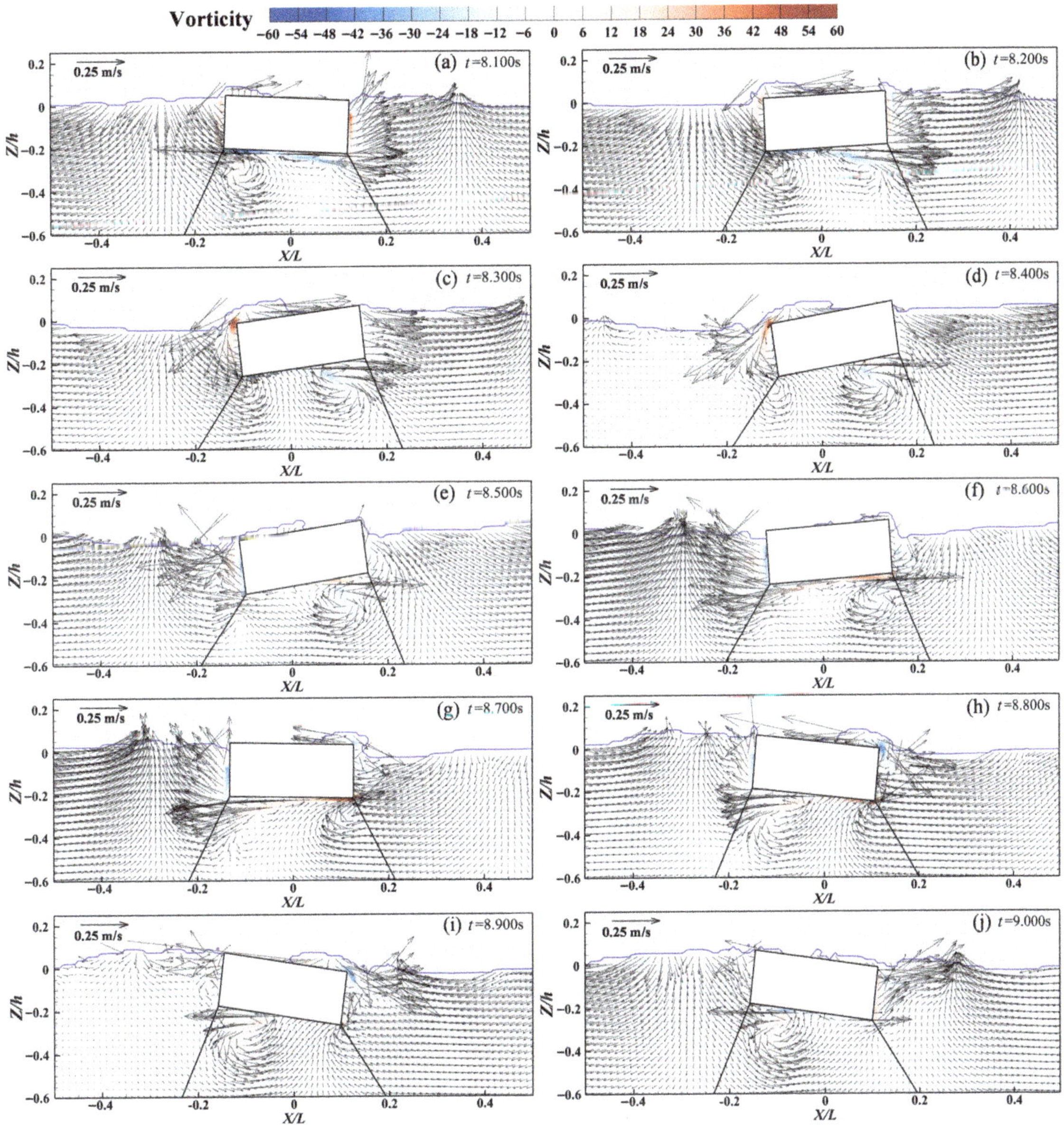

Figure 11. Instantaneous velocity field and vorticity distribution of the floating breakwater without FWDS under regular waves. The colour bar represents the vorticity field ζ in units of s^{-1}, where positive values indicate counter-clockwise rotation and negative values indicate clockwise rotation. The arrows denote the instantaneous water-particle velocity vectors, with arrow length scaled according to the reference magnitude of 0.25 m/s, as indicated in the legend. The free surface profile is also shown for reference. (h = 0.6 m, T = 1.0 s, H = 0.046 m, H/L = 0.03).

https://doi.org/10.3390/jmse14010097

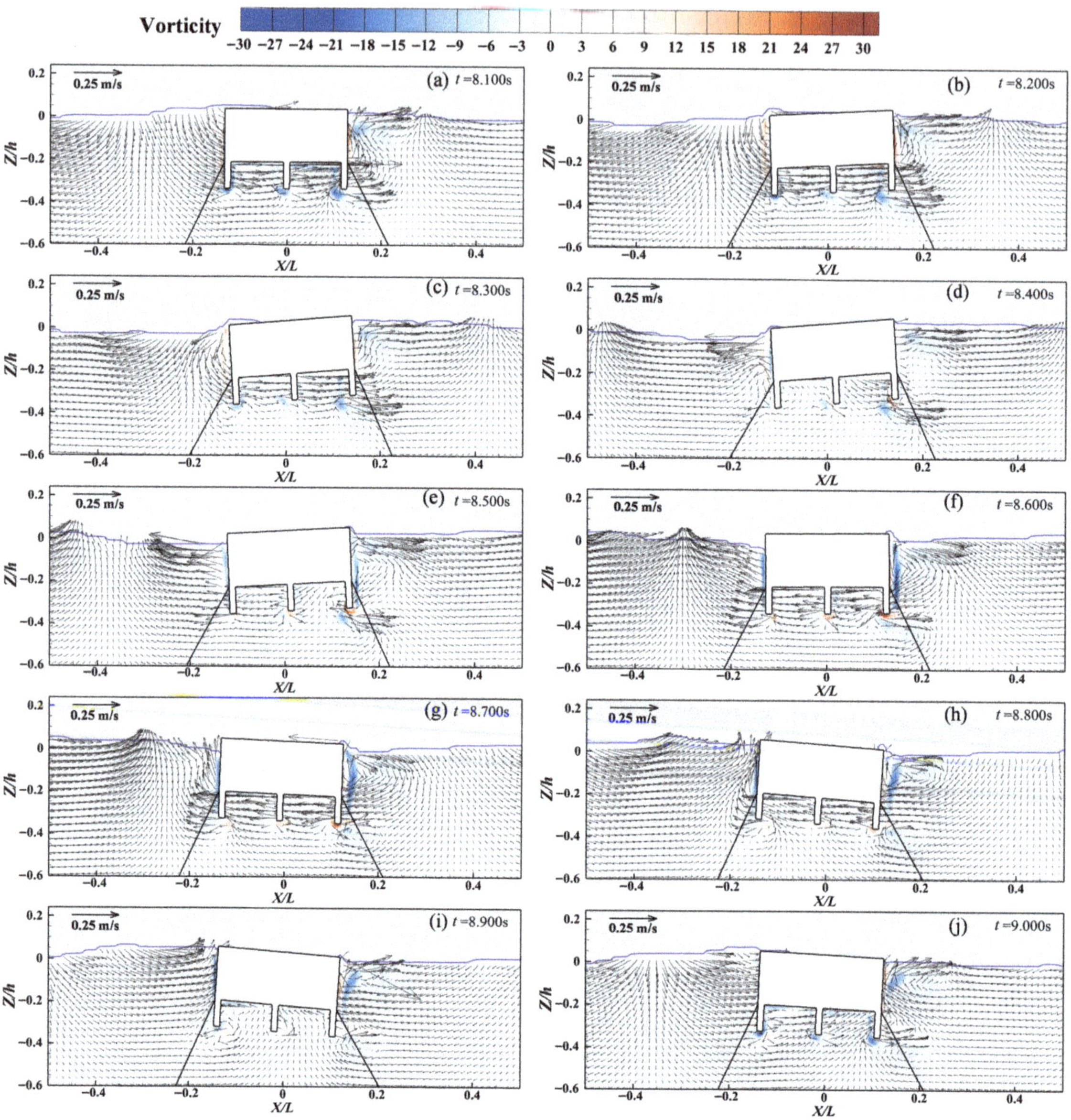

Figure 12. Instantaneous velocity field and vorticity distribution of the floating breakwater with FWDS under regular waves. The colour bar represents the vorticity field ζ in units of s^{-1}, where positive values indicate counter-clockwise rotation and negative values indicate clockwise rotation. The arrows denote the instantaneous water-particle velocity vectors, with arrow length scaled according to the reference magnitude of 0.25 m/s, as indicated in the legend. The free surface profile is also shown for reference. (h = 0.6 m, T = 1.0 s, H = 0.046 m, H/L = 0.03).

https://doi.org/10.3390/jmse14010097

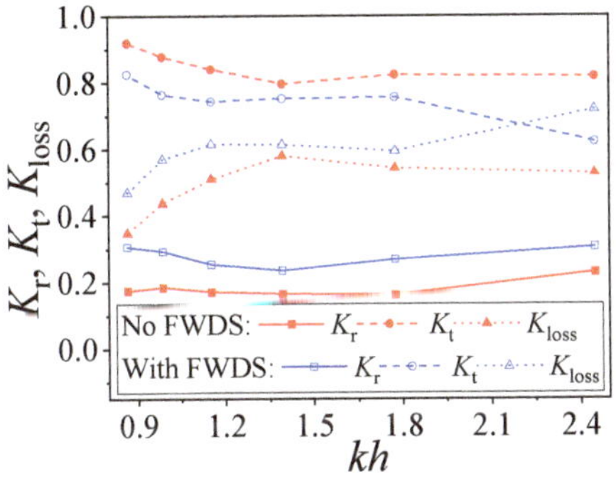

Figure 13. Reflection, transmission, and energy dissipation coefficients (K_r, K_t, K_{loss}) of the floating breakwater with and without FWDS under different wave periods.

Figure 14 presents the time histories of the free surface elevation measured at WG5 (located on the leeward side of the structure) for different elastic modulus values. As shown, the overall wave elevation patterns remain similar for most cases, except when the elastic modulus is at its minimum value of 1.085 × 10^5 Pa. In this highly flexible condition, a secondary wave crest emerges following the primary peak, indicating a stronger nonlinear response. This phenomenon suggests that, for low elastic modulus values, the flexible structures exhibit pronounced dynamic deformations and potentially induce vortex-induced vibrations (VIV), leading to complex wave–structure interaction behaviors. Conversely, when the elastic modulus is larger, the deformation of the FWDS is restricted, enhancing the system's ability to reflect and radiate incident wave energy due to increased structural rigidity.

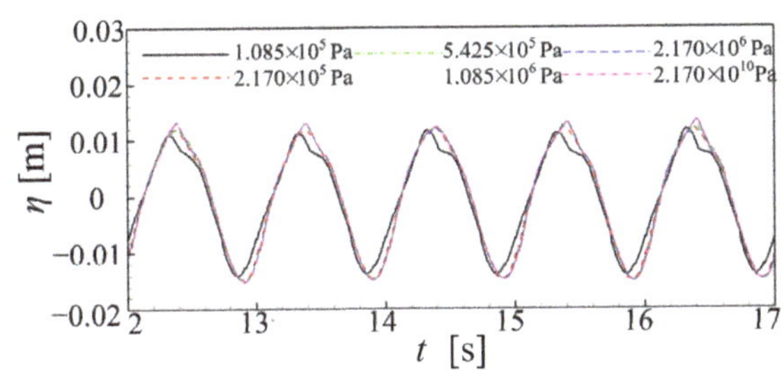

Figure 14. Time histories of free surface elevation at WG5 for different elastic modulus values of the Flexible Wave-Dissipating Structures (FWDS). (h = 0.6 m, T = 1.0 s, H = 0.046 m, H/L = 0.03).

To further elucidate the hydrodynamic mechanisms associated with the material stiffness, Figures 15 and 16 show the velocity and vorticity field distributions within one wave period for two representative cases: $E = 2.17 \times 10^{10}$ Pa (rigid-like) and $E = 1.085 \times 10^5$ Pa (highly flexible), respectively.

Despite similarities in the overall flow evolution patterns, the two cases reveal distinct flow characteristics and energy dissipation mechanisms. For the rigid case (Figure 15), limited deformation leads to stronger wave reflection and localized flow disturbances. The surrounding flow field exhibits higher velocity gradients and more pronounced nonlinear behaviors, particularly near the structural edges. The strong reflection and radiation of wave energy generate compact vortex structures along the structure's front and rear faces, where the formation and detachment of vortices significantly affect local flow velocity and energy dissipation patterns.

https://doi.org/10.3390/jmse14010097

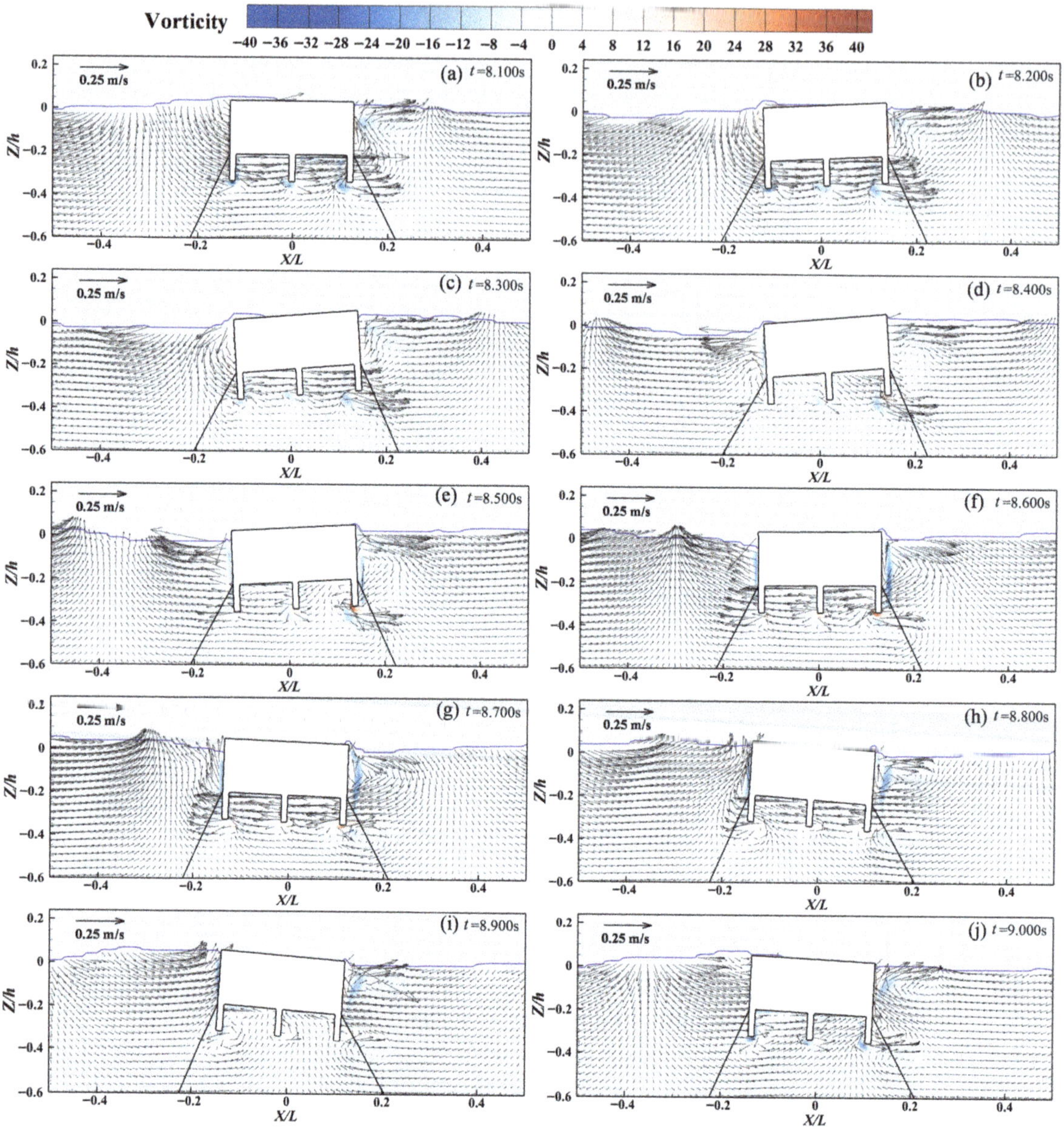

Figure 15. Instantaneous velocity and vorticity fields around the composite floating breakwater with rigid FWDS ($E = 2.17 \times 10^{10}$ Pa). The colour bar represents the vorticity field ζ in units of s^{-1}, where positive values indicate counter-clockwise rotation and negative values indicate clockwise rotation. The arrows denote the instantaneous water-particle velocity vectors, with arrow length scaled according to the reference magnitude of 0.25 m/s, as indicated in the legend. The free surface profile is also shown for reference. (h = 0.6 m, T = 1.0 s, H = 0.046 m, H/L = 0.03).

https://doi.org/10.3390/jmse14010097

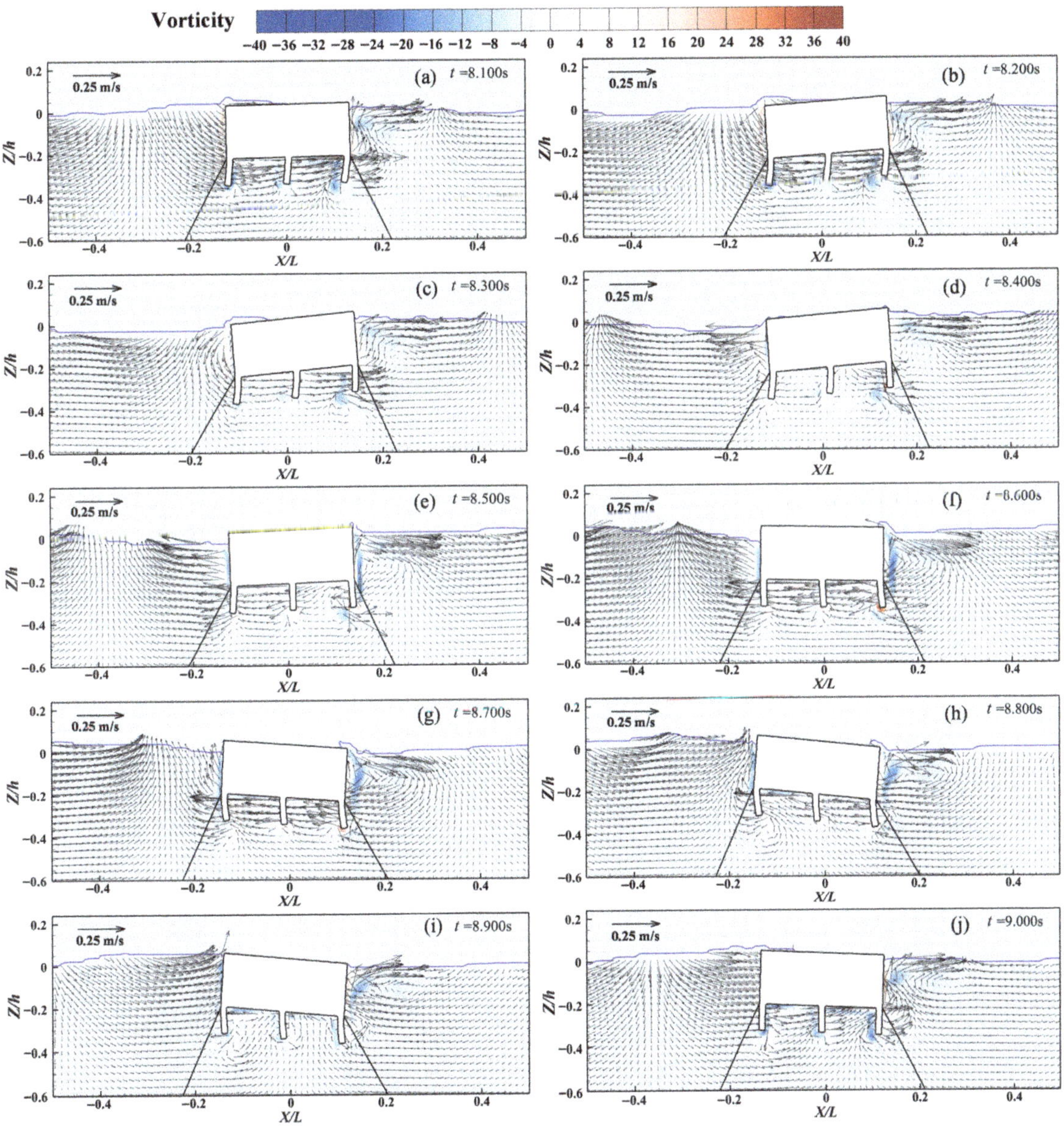

Figure 16. Instantaneous velocity and vorticity fields around the composite floating breakwater with highly flexible FWDS (E = 1.085 × 10^5 Pa). The colour bar represents the vorticity field ζ in units of s^{-1}, where positive values indicate counter-clockwise rotation and negative values indicate clockwise rotation. The arrows denote the instantaneous water-particle velocity vectors, with arrow length scaled according to the reference magnitude of 0.25 m/s, as indicated in the legend. The free surface profile is also shown for reference. (h = 0.6 m, T = 1.0 s, H = 0.046 m, H/L = 0.03).

In contrast, for the flexible case (Figure 16), the FWDS experience substantial cyclic deformation under wave excitation, resulting in large-scale oscillations and tip motions. At specific moments, such as t = 8.3 s (Figure 16c) and t = 8.7 s (Figure 16g), the flexible components undergo considerable bending and twisting, producing intense vorticity concentrations near the tips. These high-vorticity regions extend along the surface of the FWDS and evolve dynamically as the structures oscillate. The periodic motion of the flexible tips induces alternating flow shear and vortex shedding, promoting a more distributed energy dissipation process. Compared with the rigid case, the vorticity field for the flexible configuration is broader and more diffused (Figure 16h), indicating that energy dissipation occurs over a wider spatial range rather than being confined near the structural boundaries.

Although the flexible structures exhibit a weaker direct reflection capacity compared to the rigid configuration, their ability to absorb and redistribute wave energy through large-scale deformation leads to a more uniform and stable wave attenuation effect.

Overall, the contrasting flow patterns between the rigid and flexible FWDS highlight two distinct energy dissipation mechanisms. Rigid FWDS: Dominated by strong wave reflection and localized vortex generation near the structure, leading to sharp but confined energy losses. Flexible FWDS: Characterized by distributed deformation, wide-ranging vortex evolution, and smoother energy dispersion, resulting in more stable and sustained wave attenuation. The flow field and vorticity evolution patterns directly determine the hydrodynamic performance and wave-dissipating capability of each configuration under different wave conditions.

Figure 17 quantifies the influence of the elastic modulus on three key dimensionless coefficients: the reflection coefficient (K_r), transmission coefficient (K_t), and energy dissipation coefficient (K_{loss}). These coefficients are calculated using data from wave gauges WG2, WG3, and WG5, representing reflected, incident, and transmitted waves, respectively. Six elastic modulus values are examined: 1.085×10^5 Pa, 2.17×10^5 Pa, 5.425×10^5 Pa, 1.085×10^6 Pa, 2.17×10^6 Pa, and 2.17×10^{10} Pa. The investigated elastic-modulus range (1.085×10^5–2.17×10^{10} Pa) was chosen to span orders of magnitude from highly compliant materials (soft polymers, elastomers) to near-rigid structural materials (metals, rigid composites), thereby revealing the full spectrum of deformation-mediated hydrodynamic responses. Representative examples include soft elastomers and flexible polymers in the lower range ($\approx 10^5$–10^7 Pa), fiber-reinforced composites and stiff polymers in the mid-range ($\approx 10^7$–10^9 Pa), and structural metals/composites that approximate the high end ($>10^9$ Pa). We emphasize that elastic modulus is only one selection criterion; practical material choice must balance stiffness with durability, corrosion and fouling resistance, manufacturability and cost. The present parametric results supply guidance on the stiffness–performance relationship, which can then be combined with engineering constraints to select candidate materials.

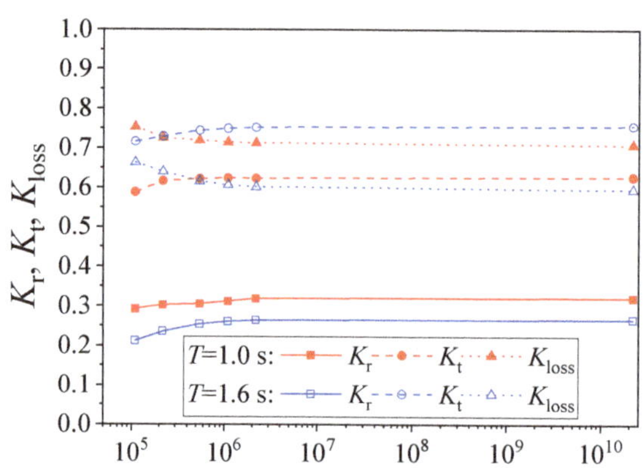

Figure 17. Variation of reflection (K_r), transmission (K_t), and energy dissipation (K_{loss}) coefficients with different elastic modulus values of the FWDS.

https://doi.org/10.3390/jmse14010097

As shown, both K_r and K_t increase with the elastic modulus, especially when E rises from 1.085×10^5 Pa to 5.425×10^5 Pa. In this range, the flexible structures are highly sensitive to wave-induced deformation, resulting in pronounced variations in wave reflection and transmission. Beyond this range, the growth rates of K_r and K_t level off, suggesting that the effect of increasing rigidity becomes saturated. When the structure becomes sufficiently stiff, further increases in E contribute little to additional reflection, as deformation-induced energy absorption diminishes.

This observation has important design implications: for practical engineering applications, an excessively high elastic modulus not only reduces the adaptive capacity of the FWDS but also increases material and construction costs without significant improvement in wave reflection efficiency. Similarly, the transmission coefficient K_t shows an asymptotic behavior, indicating that once the structure achieves a certain stiffness, additional rigidity does not significantly enhance wave transmission, and the transmitted energy remains limited.

The energy dissipation coefficient (K_{loss}) exhibits the opposite trend, decreasing rapidly with increasing E and then stabilizing. At lower elastic modulus values, the greater deformability of the FWDS allows for enhanced energy absorption and viscous dissipation through complex flow interactions. As the structure becomes more rigid, deformation diminishes, and energy dissipation stabilizes at a lower level.

Integrating the results of free surface deformation, structural response, and energy coefficients, it can be concluded that the elastic modulus exerts a significant influence on wave attenuation performance. Higher stiffness enhances reflection and transmission but reduces energy dissipation, while lower stiffness promotes broader energy dispersion and adaptive damping of incident waves. Although excessively flexible structures may induce local instability or reduced reflection, they exhibit superior adaptability to varying sea states and complex wave environments, providing potential long-term advantages in practical applications.

It is also worth noting that the FWDS in this study possess a finite thickness of 0.02 m, and the lowest tested modulus (1.085×10^5 Pa) still represents a material with moderate stiffness. If either the thickness or the modulus were further reduced, the interaction with the local flow could weaken sharply, resulting in decreased energy dissipation. Therefore, although K_{loss} increases with decreasing E within the tested range, this does not imply that the smallest possible elastic modulus always yields the best performance. Further parametric studies and high-resolution fluid–structure interaction analyses will be conducted in future work to clarify the broader effects of extreme material flexibility on wave–structure interactions. The above trends can be used as objective functions in an optimization framework (e.g., maximize K_{loss} while constraining K_r and structural stress). Future work will formulate such design optimizations to produce actionable design maps for practitioners.

The observed trend—decreasing K_r and K_t and increasing K_{loss} with decreasing elastic modulus—can be explained by the deformation-mediated redistribution of wave energy. Softer materials permit larger cyclic bending and tip motions, which (i) produce time-varying body shapes that absorb incident energy through internal strain and (ii) generate more extended vortex shedding and shear layers that convert coherent wave energy into turbulent and viscous dissipation. By contrast, stiffer FWDS deform little, acting more like rigid obstacles that reflect incident energy and produce localized, strong vortices but less distributed dissipation. Therefore, elastic modulus controls the relative importance of reflection versus distributed viscous dissipation via its effect on deformation amplitude and vortex dynamics.

https://doi.org/10.3390/jmse14010097

5.3. Influence of the Height of the Flexible Wave-Dissipating Structures

In the previous section, the height of each flexible wave-dissipating structure was fixed at 0.075 m. To further investigate the influence of structural geometry on hydrodynamic performance, this section examines the case where the height of the flexible structures is increased to 0.15 m, while all other simulation conditions remain the same. The objective is to clarify the effect of structural height on wave attenuation efficiency and to provide theoretical insights for the optimal design of flexible wave-dissipating systems.

Figure 18 illustrates the evolution of the velocity and vorticity fields around the coupled system when the flexible structures have a height of 0.15 m. It can be observed that the taller structures exert a stronger influence on the surrounding flow field. The enhanced wave–structure interaction induces pronounced velocity gradients near the edges and side surfaces of the flexible elements. As the incoming flow encounters the taller structures, local flow separation occurs, forming high-intensity vortices, particularly around the corners and tips (see Figure 18b,f). These vortices not only modify the local hydrodynamics but also enhance the dissipation of wave energy by promoting turbulent mixing and viscous losses.

Moreover, the cyclic excitation from incident waves intensifies the unsteady flow patterns around the structure, producing multiple small-scale vortices at the upper tips of the flexible elements (see Figure 18g,h). The formation and evolution of these vortices indicate a more complex energy conversion process, in which part of the wave energy is transformed into rotational kinetic energy of the fluid. This dynamic process enhances the overall energy dissipation capability of the system and contributes to the reduction of transmitted wave amplitude.

Figure 19 compares the reflection coefficient (K_r), transmission coefficient (K_t), and energy dissipation coefficient (K_{loss}) for structures with different heights. Overall, as the flexible structure height increases, K_r rises, K_t decreases, and K_{loss} increases. This trend suggests that taller flexible structures are more effective in reflecting and dissipating incoming wave energy. The increased vertical extent allows the structures to intercept a larger portion of the incident wave front, thereby reducing the transmitted wave energy and enhancing overall wave attenuation performance.

However, the improvement in wave dissipation efficiency becomes less significant under long-period wave conditions. This implies that when the wavelength is much greater than the structure height, the interaction between the wave and the flexible element becomes less pronounced, limiting the incremental benefit of increasing structural height. Therefore, while taller flexible structures can enhance energy dissipation under shorter wave conditions, their efficiency gain diminishes in longer waves, where the hydrodynamic loading is more evenly distributed along the structure. The simulations indicate a diminishing marginal benefit of increased FWDS height for longer period waves: in the present parameter set, this occurs for $T \geq 1.6$ s. The physical explanation is that for long wavelengths (relative to structure height), the incident energy is less concentrated near the free surface and therefore the additional vertical extent yields proportionally smaller interaction and dissipation. Increasing height further would shift this threshold toward longer periods, but at the expense of structural cost and potential operational constraints.

Increasing the vertical extent of FWDS strengthens interaction with incident waves by intercepting a larger portion of the wave front and by promoting stronger flow separation and tip-vortex formation. Consequently, reflection K_r and dissipation K_{loss} increase while transmission K_t decreases. The effect is most pronounced for shorter and intermediate periods (in our scaled tests for $T \leq 1.6$ s), whereas for longer waves the incremental benefit of height diminishes because the incoming energy is distributed over greater depths and the relative blockage effect is reduced.

https://doi.org/10.3390/jmse14010097

In summary, increasing the height of the flexible wave-dissipating structures can strengthen their interaction with waves and improve the overall dissipation of wave energy, especially for intermediate and short waves. Nevertheless, the optimal design should balance structural dimensions with target wave conditions to achieve effective energy attenuation without excessive material use or hydrodynamic resistance. Although increasing structural stiffness enhances reflection, the extended structural height compensates by amplifying viscous and vortical dissipation, leading to a net increase in energy loss.

The present parametric study focused on regular waves; the effectiveness and potential vulnerability of flexible elements under random seas, energetic wave groups and storm conditions will be the subject of ongoing research. We will investigate transient focusing events, large-amplitude forcing, and associated fatigue and stability issues in our subsequent three-dimensional and experimental campaigns.

5.4. Design Implications and Practical Guidance

To enhance the practical relevance of the parametric findings and address the engineering implications highlighted by the reviewers, Section 5.4 summarizes design-oriented insights derived from the numerical analysis. These guidelines provide qualitative direction for early-stage engineering design and identify the key trade-offs among material stiffness, geometry, and wave dissipation performance.

The parametric investigations conducted in this study reveal several trends useful for the preliminary design of flexible wave-dissipating structures. Although the findings are based on the present two-dimensional numerical framework, they provide qualitative guidance for material selection and geometric configuration in early-stage engineering design.

1. Applications prioritizing wave-energy dissipation.

For configurations where reducing the transmitted wave energy is the primary objective (e.g., protection of downstream offshore installations), flexible materials with a moderate-to-low elastic modulus—within the tested range—are preferable. Such materials facilitate larger structural deformation and promote the formation of vortices around the flexible elements, enhancing distributed energy dissipation. A sufficient vertical extent (height) also supports the development of strong vortex shedding and improves the overall dissipation efficiency.

2. Applications where increased reflection is acceptable or desirable.

In scenarios where upstream wave reflection is desired (e.g., shoreline protection), stiffer materials and larger frontal areas are more effective. These configurations increase the reflection coefficient K_r but tend to reduce the dissipation coefficient K_{loss}, indicating a shift from dissipative to reflective performance.

3. Avoiding excessively stiff materials.

Very high stiffness does not proportionally improve wave reflection and may lead to diminishing returns while increasing structural costs. Overly stiff configurations also reduce deformation-induced dissipation, limiting their hydrodynamic adaptability.

4. Considering long-term durability and cost trade-offs.

Beyond hydrodynamic performance, practical material selection should account for durability, fatigue resistance, and maintenance cost. Composite materials or coated polymers may offer a balanced combination of moderate stiffness, deformation capacity, and long-term durability.

https://doi.org/10.3390/jmse14010097

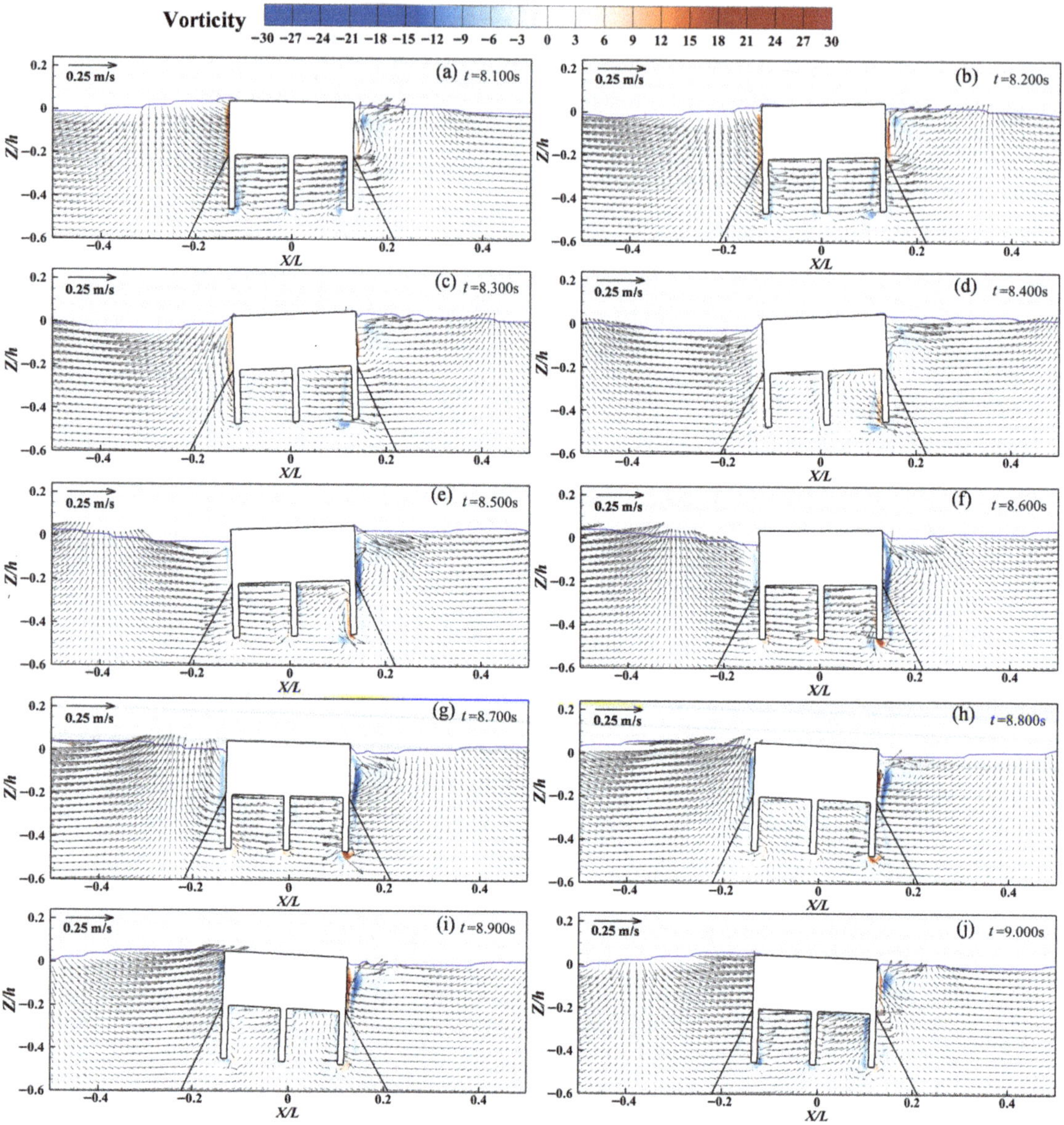

Figure 18. Instantaneous velocity and vorticity fields around the coupled floating breakwater system with flexible wave-dissipating structures of height 0.15 m at different time instants. The colour bar represents the vorticity field ζ in units of s^{-1}, where positive values indicate counter-clockwise rotation and negative values indicate clockwise rotation. The arrows denote the instantaneous water-particle velocity vectors, with arrow length scaled according to the reference magnitude of 0.25 m/s, as indicated in the legend. The free surface profile is also shown for reference. (h = 0.6 m, T = 1.0 s, H = 0.046 m, H/L = 0.03).

https://doi.org/10.3390/jmse14010097

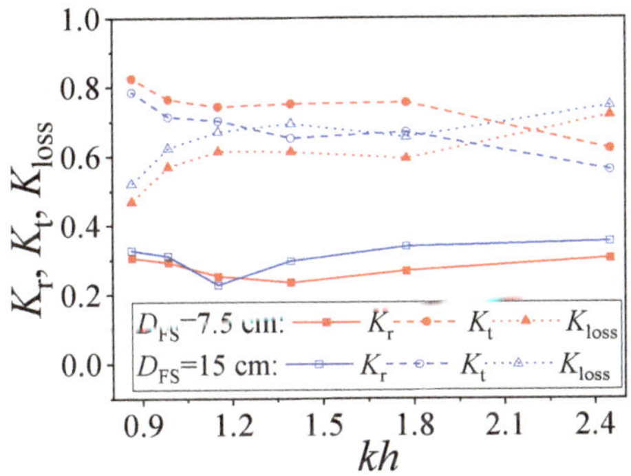

Figure 19. Variation of the reflection coefficient (K_r), transmission coefficient (K_t), and energy dissipation coefficient (K_{loss}) with different heights of the flexible wave-dissipating structures.

These qualitative guidelines are intended to support preliminary decision-making. In future work, we will develop a formal multi-objective optimization framework to generate quantitative design charts that simultaneously consider K_{loss}, K_r, structural stress, material fatigue, and cost.

6. Conclusions

In this study, a two-dimensional numerical model was developed to investigate the hydrodynamic performance of a floating breakwater coupled with flexible wave-dissipating structures (FWDS). The model combines the immersed boundary method with the finite element formulation for flexible structural motion, enabling accurate simulation of the fluid–structure interaction between incident waves, the floating body, and the flexible elements. The numerical scheme was first validated through benchmark tests on cantilever beam deflection under concentrated loads and flexible vegetation motion under wave action, both showing excellent agreement with theoretical and experimental results, thereby confirming the reliability of the proposed approach. The model was then applied to evaluate the effects of structural flexibility and geometric parameters on wave attenuation performance. The key findings can be summarized as follows:

The elastic modulus of the flexible wave-dissipating structures significantly influences the overall hydrodynamic behavior of the coupled system. With decreasing elastic modulus, the FWDS exhibit greater deformation and energy absorption capacity, leading to enhanced wave attenuation. In contrast, stiffer structures generate stronger wave reflection but reduced energy dissipation. The results indicate that an optimal elastic modulus exists that balances reflection, transmission, and dissipation for efficient wave energy reduction.

Analysis of the instantaneous velocity and vorticity fields revealed distinct flow mechanisms associated with structural flexibility. Stiffer configurations produce localized high-vorticity zones and pronounced reflected waves, while softer structures induce larger-scale flow disturbances and distributed energy dissipation through cyclic deformation and vortex shedding near their tips.

Increasing the height of the flexible elements improves wave reflection and reduces wave transmission, particularly for shorter wave periods. However, under long-period wave conditions, the improvement becomes less significant, suggesting that the influence of height is constrained by the wavelength-to-structure-height ratio. Taller flexible structures nevertheless promote stronger flow separation and vortex formation, enhancing localized energy dissipation.

Overall, the proposed coupled model effectively captures the nonlinear hydrodynamic interactions between flexible and rigid components in a floating breakwater system. The

https://doi.org/10.3390/jmse14010097

results demonstrate that both the elastic modulus and structural height of the flexible wave-dissipating elements play crucial roles in governing wave attenuation performance. These findings provide valuable insights for the optimal design of hybrid floating breakwaters, enabling adaptive control of wave reflection and transmission through structural flexibility. Future research will further explore three-dimensional effects, irregular wave conditions, and the long-term fatigue behavior of flexible components to support their practical application in coastal and offshore engineering.

To address the inherent limitations of the present two-dimensional and component-validated study, future work will extend the numerical framework to fully three-dimensional simulations to resolve spanwise flow, lateral diffraction, and oblique-wave incidence. In parallel, scaled laboratory experiments of the integrated floating-breakwater–FWDS system will be carried out to measure wave attenuation, platform motions, FWDS deformations, and vortex-shedding characteristics under controlled conditions. These combined efforts will provide system-level validation and support the development of practical design guidelines for real-world deployment.

Building upon this, the numerical model will be expanded to include three-dimensional directional wave spectra to quantify yaw motions, spanwise variability, and multi-directional wave loading. Such analyses are essential for assessing the robustness of design recommendations under realistic ocean conditions where waves approach from varying directions.

Moreover, the present study considered only regular waves within a moderate steepness range. Future simulations will incorporate irregular seas, broadband spectra, storm-like wave groups, and extreme events to evaluate device performance under realistic and energetic ocean states. These extensions will allow assessment of reliability, survivability, and hydrodynamic efficiency across a wider range of operational and extreme scenarios.

Beyond hydrodynamic performance, environmental interactions will also be examined. Planned numerical–experimental campaigns in collaboration with coastal geomorphologists and marine ecologists will investigate sediment transport, scour potential, and biological colonization around the integrated system, providing a holistic understanding of ecological impacts and eco-engineering opportunities.

The current configuration, consisting of three identical FWDS units arranged with uniform spacing, was selected to isolate baseline hydrodynamic mechanisms. Future parametric studies will systematically explore the influence of FWDS number, spacing (lateral and longitudinal), orientation (inclined or staggered arrangements), and permeable/porous membrane-type designs on wave reflection, transmission, and dissipation. These investigations aim to generate comprehensive engineering guidance for practical array layouts.

The present simulations assume idealized material properties and do not account for long-term degradation mechanisms such as fatigue, biofouling, UV exposure, or chemical aging. These processes can modify mass, stiffness, and surface roughness, thereby altering hydrodynamic response and dissipation characteristics. Future research will include controlled aging experiments, biofouling tests, and coupled structural-degradation models to evaluate durability, maintenance intervals, and lifecycle performance.

Finally, the model did not include breaking waves, storm-scale forcing, or large variations in water depth. These conditions will be addressed in forthcoming three-dimensional simulations and physical experiments to evaluate extreme loads, survival behavior, and performance sensitivity to depth variations. Collectively, the planned extensions will enable a comprehensive assessment of the system's behavior under realistic and extreme sea states, supporting safe and efficient design for practical deployment.

https://doi.org/10.3390/jmse14010097

Author Contributions: Conceptualization, X.J. and W.P.; methodology, X.J., S.G. and X.Y.; validation, K.W. and S.F.; writing—original draft preparation, X.Y.; writing—review and editing, S.G., K.W., S.F. and W.P. All authors have read and agreed to the published version of the manuscript.

Funding: The research was funded by the National Key R&D Program of China (Grant No. 2024YFB4207000). The authors acknowledge and express their gratitude to the reviewers of this paper for their valuable and constructive comments that helped to improve the quality of the manuscript.

Data Availability Statement: The data presented in this study are available on request from the corresponding author.

Conflicts of Interest: Authors Xianlin Jia, Kangjie Wang, and Sai Fu are employed by the company Power China Hua Dong Engineering Co., Ltd. The remaining authors declare that the research was conducted in the absence of any commercial or financial relationships that could be construed as a potential conflict of interest.

References

1. Lindeque, P.K.; Cole, M.; Coppock, R.L.; Lewis, C.N.; Miller, R.Z.; Watts, A.J.; Wilson-McNeal, A.; Wright, S.L.; Galloway, T.S. Are we underestimating microplastic abundance in the marine environment? A comparison of microplastic capture with nets of different mesh-size. *Environ. Pollut.* **2020**, *265*, 114721. [CrossRef]
2. Zhao, C.; Tang, J.; Shen, Y.; Wang, Y. Study on wave attenuation in following and opposing currents due to rigid vegetation. *Ocean Eng.* **2021**, *236*, 109574. [CrossRef]
3. Knutson, P.L.; Brochu, R.A.; Seelig, W.N.; Inskeep, M. Wave damping in Spartina alterniflora marshes. *Wetlands* **1982**, *2*, 87–104. [CrossRef]
4. Jadhav, R.S.; Chen, Q. Probability distribution of wave heights attenuated by salt marsh vegetation during tropical cyclone. *Coast. Eng.* **2013**, *82*, 47–55. [CrossRef]
5. Jadhav, R.; Chen, Q. Field investigation of wave dissipation over salt marsh vegetation during tropical cyclone. *Coast. Eng. Proc.* **2012**, *1*, 41–51. [CrossRef]
6. Jadhav, R.S.; Chen, Q.; Smith, J.M. Spectral distribution of wave energy dissipation by salt marsh vegetation. *Coast. Eng.* **2013**, *77*, 99–107. [CrossRef]
7. Dalrymple, R.A.; Kirby, J.T.; Hwang, P.A. Wave diffraction due to areas of energy dissipation. *J. Waterw. Port Coast. Ocean Eng.* **1984**, *110*, 67–79. [CrossRef]
8. Mendez, F.J.; Losada, I.J. An empirical model to estimate the propagation of random breaking and nonbreaking waves over vegetation fields. *Coast. Eng.* **2004**, *51*, 103–118. [CrossRef]
9. Hu, Z.; Suzuki, T.; Zitman, T.; Uittewaal, W.; Stive, M. Laboratory study on wave dissipation by vegetation in combined current–wave flow. *Coast. Eng.* **2014**, *88*, 131–142. [CrossRef]
10. Hu, Z.; Lian, S.; Zitman, T.; Wang, H.; He, Z.; Wei, H.; Ren, L.; Uijttewaal, W.; Suzuki, T. Wave breaking induced by opposing currents in submerged vegetation canopies. *Water Resour. Res.* **2022**, *58*, e2021WR031121. [CrossRef]
11. Magdalena, I.; Andadari, G.R.; Reeve, D.E. An integrated study of wave attenuation by vegetation. *Wave Motion* **2022**, *110*, 102878. [CrossRef]
12. Liu, H.; Fang, H.; Lin, P. A theoretical model for wave attenuation by vegetation considering current effects. *Coast. Eng.* **2024**, *190*, 104508. [CrossRef]
13. Stamos, D.G.; Hajj, M.R.; Telionis, D.P. Performance of hemi-cylindrical and rectangular submerged breakwaters. *Ocean Eng.* **2003**, *30*, 813–828. [CrossRef]
14. Diamantoulaki, I.; Angelides, D.C.; Manolis, G.D. Performance of pile-restrained flexible floating breakwaters. *Appl. Ocean Res.* **2008**, *30*, 243–255. [CrossRef]
15. Koley, S.; Sahoo, T. Scattering of oblique waves by permeable vertical flexible membrane wave barriers. *Appl. Ocean Res.* **2017**, *62*, 156–168. [CrossRef]
16. Guo, Y.C.; Mohapatra, S.C.; Soares, C.G. Experimental study on the performance of an array of vertical flexible porous membrane type breakwater under regular waves. *Ocean Eng.* **2022**, *264*, 112328. [CrossRef]
17. Guo, Y.C.; Mohapatra, S.C.; Soares, C.G. An experimental study on the performance of combined membrane breakwaters against incident waves. *Ocean Eng.* **2023**, *280*, 114793. [CrossRef]
18. Sun, W.; Nakamura, T.; Cho, Y.; Mizutani, N. Numerical Investigation of Solitary Wave Attenuation by a Vertical Plate-Type Flexible Breakwater Constructed Using Hyperelastic Neo-Hookean Material. *J. Mar. Sci. Eng.* **2024**, *12*, 1004. [CrossRef]

https://doi.org/10.3390/jmse14010097

19. Sun, B.; Li, C.; Yang, S.; Zhang, H.; Song, Z. Experimental and numerical study on the wave attenuation performance and dynamic response of kelp-box type floating breakwater. *Ocean Eng.* **2022**, *263*, 112374. [CrossRef]
20. Wu, J.-P.; Li, C.-R.; Qi, J.-J.; Xu, X.-Y.; Li, J.-N.; Zhang, X.-X.; Qin, Q.; Zhou, Z.-R. Experimental study on the hydrodynamic performance of a side-by-side box-membrane breakwater. *Ocean Eng.* **2024**, *302*, 117658. [CrossRef]
21. Meng, Y.; Li, X.; He, M.; Sui, Y.; Zhang, X. Study on wave dissipation performance of ecological breakwater with seagrasses planted on submerged horizontal plate. *Ocean Eng.* **2025**, *43*, 189–199. [CrossRef]
22. Zhao, X.; Zhou, Y.; Zong, Y.; Yang, Z.; Luo, M. A CIP-based numerical simulation of wave interaction with a fluid-filled membrane submerged breakwater. *Ocean Eng.* **2022**, *260*, 111819. [CrossRef]
23. Li, A.J.; Liu, Y.; Li, H.J.; Fang, H. Analysis of water wave interaction with a flexible submerged perforated semi-circular breakwater. *Eur. J. Mech. B-Fluids* **2020**, *79*, 345–356. [CrossRef]
24. Li, A.J.; Liu, Y.; Li, H.J.; Fang, H. Analysis of water wave interaction with a submerged fluid-filled semi-circular membrane breakwater. *Ocean Eng.* **2020**, *197*, 106901. [CrossRef]
25. Brorsen, M.; Larsen, J. Source generation of nonlinear gravity waves with the boundary integral equation method. *Coast. Eng.* **1987**, *11*, 93–113. [CrossRef]
26. Ohyama, T.; Nadaoka, K. Development of a numerical wave tank for analysis of nonlinear and irregular wave field. *Fluid Dyn. Res.* **1991**, *8*, 231–251. [CrossRef]
27. Chorin, A.J. On the convergence of discrete approximations to the Navier-Stokes equations. *Math. Comput.* **1969**, *23*, 341–353. [CrossRef]
28. Chorin, A.J. Numerical solution of the Navier-Stokes equations. *Math. Comput.* **1968**, *22*, 745–762. [CrossRef]
29. Lima, E.; Silva, A.L.F.; Silveira-Neto, A.; Damasceno, J.J.R. Numerical simulation of two-dimensional flows over a circular cylinder using the immersed boundary method. *J. Comput. Phys.* **2003**, *189*, 351–370. [CrossRef]
30. Lee, K.H.; Mizutani, N. A numerical wave tank using direct-forcing immersed boundary method and its application to wave force on a horizontal cylinder. *Coast. Eng. J.* **2009**, *51*, 27–48. [CrossRef]
31. Peng, W.; Lee, K.-H.; Shin, S.-H.; Mizutani, N. Numerical simulation of interactions between water waves and inclined-moored submerged floating breakwaters. *Coast. Eng.* **2013**, *82*, 76–87. [CrossRef]
32. Peng, W.; Huang, X.; Fan, Y.; Zhang, J.; Ren, X. Numerical analysis and performance optimization of a submerged wave energy converting device based on the floating breakwater. *J. Renew. Sustain. Energy* **2017**, *9*, 044503. [CrossRef]
33. Peng, W.; Fan, Y.; Zhang, J. Numerical study on optimization of the float in a shoreline wave energy converter. *J. Coast. Res.* **2018**, *85*, 1296–1300. [CrossRef]
34. Soomro, F.A.; Alamir, M.A.; El-Sapa, S.; Haq, R.U.; Soomro, M.A. Artificial neural network modeling of MHD slip-flow over a permeable stretching surface. *Arch. Appl. Mech.* **2022**, *92*, 2179–2189. [CrossRef]
35. Raizah, Z.; El-Sapa, S.; Aly, A.M. ISPH simulations of thermosolutal convection in an annulus amongst an inner prismatic shape and outer cavity including three circular cylinders. *Case Stud. Therm. Eng.* **2022**, *30*, 101736. [CrossRef]
36. Aly, A.M.; El-Sapa, S. Double rotations of cylinders on thermosolutal convection of a wavy porous medium inside a cavity mobilized by a nanofluid and impacted by a magnetic field. *Int. J. Numer. Methods Heat Fluid Flow* **2022**, *32*, 2383–2405. [CrossRef]
37. Maza, M.; Lara, J.L.; Losada, I.J. A coupled model of submerged vegetation under oscillatory flow using Navier–Stokes equations. *Coast. Eng.* **2013**, *80*, 16–34. [CrossRef]
38. Chen, H.; Zou, Q.-P. Eulerian-Lagrangian flow-vegetation interaction model using immersed boundary method and OpenFOAM. *Adv. Water Res.* **2019**, *126*, 176–192. [CrossRef]
39. Mansard, E.P.D.; Funke, E.R. The measurement of incident and reflected spectra using a least squares method. In Proceedings of the 17th International Conference on Coastal Engineering, Sydney, Australia, 23–28 March 1980; pp. 154–172.

Disclaimer/Publisher's Note: The statements, opinions and data contained in all publications are solely those of the individual author(s) and contributor(s) and not of MDPI and/or the editor(s). MDPI and/or the editor(s) disclaim responsibility for any injury to people or property resulting from any ideas, methods, instructions or products referred to in the content.

https://doi.org/10.3390/jmse14010097

Journal of
Marine Science and Engineering

MDPI

Article

Hydrodynamic Performance and Mooring Safety Assessment of an Offshore Floating Movable Fish Cage

Sung-Jae Kim [1], Seong-Jae Jeong [1,*] and Sung-Ju Park [2]

1 Fisheries Engineering Division, National Institute of Fisheries Science, Busan 46083, Republic of Korea; sjkim8@korea.kr
2 School of Electrical and Control Engineering, Tongmyong University, Busan 48520, Republic of Korea; parksj0314@tu.ac.kr
* Correspondence: denkmal@korea.kr

Abstract: This study evaluates the hydrodynamic performance of a movable fish cage equipped with a spread mooring system in offshore condition. It investigates the global behavior and safety of a mooring system under environmental influences such as waves, currents, and biofouling. A numerical model was developed using the Cummins equation and a lumped-mass line model to capture the coupling effects between the floating structure and mooring lines. The steel frame was modeled using Morison members, whereas fishing nets were represented by a screen model incorporating drag forces. Parametric studies were performed to assess the effects of varying mooring line lengths, current speeds, and biofouling on cage behavior. Evidently, heavier chains reduced excursions but increased tension, whereas high current speeds increased the line tension (owing to increased drift) and mooring line stiffness by up to 66%. Biofouling increased the maximum excursion by 6% and line tension by up to 17%. Safety evaluations based on the American Bureau of Shipping rules examined intact and damaged conditions, comparing estimated line tensions with allowable values. The findings confirm that the mooring system ensures reliable station-keeping performance even under challenging conditions, validating its suitability for offshore deployment and ensuring the safety and stability of floating fish cage systems.

Keywords: fish cage; spread mooring; hydrodynamics; mooring tension; extreme environment; current

1. Introduction

Fish cages are aquaculture structures designed to create controlled environments for cultivating various fish species in open water. Fish cage farming has attracted increasing attention owing to depleting fishery resources and the decline in labor-intensive fishing practices. Among the various types of fish cages, floating fish cages offer advantages, such as seamless access for feeding, monitoring, and maintenance, as well as improved water circulation. Moreover, it promotes a healthier aquatic environment for fish. However, offshore fish cage farming has limitations. These facilities are generally open systems, rendering them vulnerable to environmental loads, including typhoons, algal blooms, and high water temperatures. Exposure to such natural hazards increases the risk of damage to the structures and loss of stock, necessitating robust design and effective mooring systems to ensure stability and safety under extreme environmental conditions [1].

The number of offshore fish cage aquaculture facilities has significantly increased. However, a growing trend exists toward the use of semi-submersible structures instead of traditional collar-type facilities, owing to their enhanced safety and stability. Such structures must adhere to marine design standards, particularly those established by Det Norske Veritas (DNV) or the American Bureau of Shipping (ABS) [2]. Most floating structures, including fish farms and oil and gas platforms, are analyzed solving potential theory-based 3D radiation/diffraction problems to assess hydrodynamic performance while incorporating the drag and lift effects of nets [3–16]. Finite element analysis (FEA) is suitable for evaluating mooring line behavior and performing floater–net–mooring system coupled analyses [4,17]. A common model for offshore applications is the semi-submersible offshore fishing farm (SOFF) model. Yu et al. [8] and Miao et al. [9] conducted experimental and numerical studies to analyze the hydrodynamic performance of a SOFF under regular waves, with and without uniform currents. They employed the potential theory and a simple spring-based mooring system for simplicity. Ma et al. [10] and Yue et al. [11] applied vessel-shaped floating aquaculture platforms with single-point mooring. Moreover, they modeled the system using the potential theory and the lumped-mass line model to investigate the global behavior and coupling effects under regular waves. Pang et al. [13] validated the dynamics of a vessel-shaped structure with Y-shaped moorings. Chen et al. [15] examined the hydrodynamic performance of a spread-moored semi-submersible structure certified by the China Classification Society. Zhang et al. [16] proposed a new aquaculture platform combining an upper rigid structure with lower collar-type nets, assessing the hydrodynamic impact of the netting system under irregular waves. Mohapatra and Soares [18,19] adopted a semi-analytic solution, based on the potential and linear wave theories, to evaluate the hydro-elastic behavior of an array of vertical cylindrical flexible fish cages under regular waves and ocean currents. Therefore, most studies have primarily focused on evaluating the hydrodynamics of floating aquaculture structures, with limited attention given to assessing mooring line behavior.

Considering the effects of biofouling on fish cages and the safety of mooring systems under a one-line failure scenario is critical in the design of floating fish cages. Previous studies on the biofouling effects on nets have primarily focused on evaluating drag forces, mass variation, and blockage effects [20–22], without addressing the hydrodynamic performance of fish cages. Additionally, studies on damaged mooring conditions have been solely focused on the assessment of the safety of mooring systems for arrays of collar-type fish cages [23–25], and other floating structures have rarely been explored. To the best of the authors' knowledge, comprehensive studies evaluating the performance of offshore floating fish cages—encompassing the global performance of floating aquaculture structures, the safety of mooring systems in accordance with ship registry guidelines, and the impact of biofouling on nets—remain scarce.

This study evaluated the hydrodynamic performance of a floating movable fish cage (X-AQUDS), assessing the safety of its mooring system based on the ABS rules [26] and guidelines [27]. A movable fish cage is appealing for rearing and transporting farmed species, facilitating the cultivation of high-value and diverse species, including migratory fish. Its mobility also helps avoid natural disasters like typhoons and red tides. In 2014, the X-AQUDS was designed by the National Institute of Fisheries Science (NIFS), as shown in Figure 1, in accordance with the IMO regulations [28,29]. These include typhoons, algal blooms, and elevated sea temperatures. Kim and Jeong [30] conducted resistance tests on a fish cage in a NIFS towing tank to optimize the shield design for the netting system. In 2014, a prototype model was constructed. From 2015 to 2017, a real sea test was performed in the coastal area of Tongyeong, Republic of Korea, achieving a fish survival rate exceeding 95% during transport tests to evade extreme environmental loads. This study explores the potential applications of this cage in offshore installations. A dynamic analysis model incorporating the potential theory and the lumped-mass line model was established to evaluate the global behavior of the fish cage and its interactions with the mooring lines. The effects of the netting system were modeled by the screen model approach of Kristensen and Faltinsen's screen-type approach [31]. The study optimized the spread mooring system, analyzed the impact of current speed and netting (including biofouling effects), and assessed the mooring system's safety, following the ABS standards and guidelines, based on this numerical model [26,27].

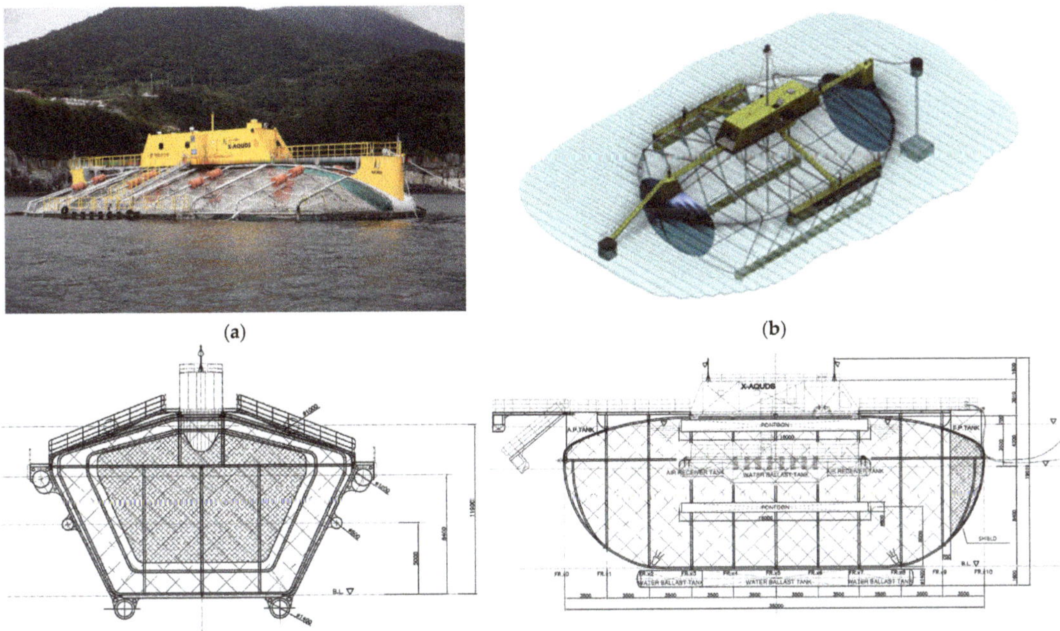

Figure 1. Floating movable fish cage (X-AQUDS; (**a**) real sea test; (**b**) 3D model; (**c**) GA in front view; and (**d**) GA in side view).

This study provides valuable guidelines for optimizing mooring systems, advancing sustainable offshore aquaculture infrastructure, and ensuring system stability in challenging environments. The remainder of this study is organized as follows: Section 2 presents the mathematical formulation. Section 3 presents the numerical model and environmental conditions. Section 4 discusses the results. Section 5 concludes the study.

2. Mathematical Formulation

A floating moored fish cage comprises pontoons with steel frames, netting, and mooring lines, necessitating careful consideration of the coupling effects among these components. Although steel frames and mooring systems have slender body, pontoons are relatively blunt and generate radiated and diffracted waves. The dynamics of the fish cage can be expressed by Equation (1), based on the Cummins equation [32]. This study evaluates the global six degree-of-freedom (DOF) motions of a floating fish cage and the safety of its mooring system, with the assumption that the pontoons and steel frames function as a single rigid body without local deformation.

$$(\mathbf{M}+\mathbf{M}_{a,\infty})\ddot{\boldsymbol{\xi}}(t)+\mathbf{K}\boldsymbol{\xi}(t)=\mathbf{F_w}(t)+\mathbf{F_C}(t)+\mathbf{F_{Drag}}(t)+\mathbf{F_M}(t) \quad (1)$$

where $\mathbf{M}$, $\mathbf{M}_{a,\infty}$, and $\mathbf{K}$ represent the mass matrix (including pontoons and steel frames), added mass matrix at infinite frequency, and restoring coefficient matrix, respectively. These matrices have 6×6 dimensions. The $\mathbf{F_w}$, $\mathbf{F_C}$, $\mathbf{F_{Drag}}$, and $\mathbf{F_M}$ correspond to the wave excitation load, radiation damping force, drag force, and connection force of the mooring system, respectively. The position vector $\boldsymbol{\xi}$ combines three translational motion components (surge, sway, and heave) and three rotational motion components (roll, pitch, and yaw). The overdot notation indicates time derivatives; for instance, $\dot{\boldsymbol{\xi}}$ and $\ddot{\boldsymbol{\xi}}$ are the velocity and acceleration vectors, respectively, of the six DOF motions. The restoring coefficients are calculated based on the hydrostatic pressure and submerged volume of the fish cage.

Radiation and diffraction problems must be solved to evaluate the added masses and radiation damping coefficients. The radiation problem involves assessing the effects of radiated waves generated by an oscillating body in calm water. However, the diffraction problem examines the wave excitation loads on a fixed body subjected to incident waves. The radiation damping force vector can be expressed by Equation (2), using a memory function derived from frequency-dependent radiation damping coefficients (**C**) through the Fourier cosine transform expressed in Equation (3). Wave excitation loads from the diffraction problem are calculated using a 3D hydrodynamic analysis program based on potential theory and the boundary element method, as shown in Equation (4).

$$\mathbf{F_C}(t) = \int_0^t \mathbf{R}(\tau - t)\dot{\boldsymbol{\xi}}(\tau)d\tau \quad (2)$$

$$\mathbf{R}(t) = -\frac{2}{\pi}\int_0^{\infty} \mathbf{C}(\omega)\cos(\omega t)d\omega \quad (3)$$

$$\mathbf{F_w}(t) = \mathrm{Re}\left[\sum_{j=1}^{Nw} A_j \mathbf{L}(\omega_j)e^{-i\omega_j t}\right] \quad (4)$$

where **R** and **L** represent the memory function, also known as the impulse response function, and the linear transfer function for the wave excitation loads, respectively. The variable N_w denotes the number of wave-frequency components used to generate irregular waves. However, A_j and ω_j represent the j-th wave amplitude and its corresponding wave frequency, respectively. The wave amplitude at each wave frequency can be determined based on the wave spectrum and statistical characteristics of the design waves.

The drag force terms in the Morison equation were used to express the drag effects induced by the pontoons, steel frames, and netting. These drag forces can be expressed as shown in Equation (5), which evaluates the forces based on the drag coefficient ($\mathbf{C_D}$) and the projected area ($\mathbf{A_P}$).

$$\mathbf{F_{Drag}} = \frac{1}{2}\rho \mathbf{C_D}\mathbf{A_P}\left(\mathbf{u} - \dot{\boldsymbol{\xi}}\right)\left|\mathbf{u} - \dot{\boldsymbol{\xi}}\right| \quad (5)$$

where ρ and **u** denote fluid density and velocity, respectively. Since this study focuses on the net effect on the global performance of the fish cage and the mooring system safety, the net effect is applied by adopting a drag screen model. The drag coefficient is determined based on the solidity index (S_n), which is the ratio of the effective area (A_{pn}) occupied by the net's threads and knots to the total area (A_t) covered by the net, as shown in Equation (6), as well as on the Reynolds number (R_e).

$$S_n = \frac{A_{pn}}{A_t} \quad (6)$$

The connection force vector can be evaluated iteratively by considering the interaction between the rigid body and the mooring lines. This study adopted two distinct approaches to estimate the connection forces: theoretical and numerical.

The connection forces in the theoretical approach are estimated using the catenary equation, which iteratively calculates the force based on simple line properties, such as weight, buoyancy, and axial stiffness, at each time step. Equations (7)–(9) outline the catenary equation for a freely hanging line in a vertical plane, extending between its bottom (x_b, z_b) and top ends (x_t, z_t).

$$t_t = t_b + \frac{wl_u}{T_h} \quad (7)$$

$$x_t = x_b + l_u\left\{\frac{1}{t_t - t_b}\ln\left(\frac{t_t + \sqrt{1+t_t^2}}{t_b + \sqrt{1+t_b^2}}\right) + \frac{T_h}{EA}\right\} \quad (8)$$

$$z_t = z_b + l_u \left\{ \frac{\sqrt{1+t_t^2} - \sqrt{1+t_b^2}}{t_t - t_b} + \frac{T_h}{EA} \frac{t_b + t_t}{2} \right\} \quad (9)$$

where t_b and t_t are the angles of the line tangents at the bottom and top ends. w, l_u, T_h, and EA denote the net weight of the line per unit length, unstretched length of the suspended line segment, horizontal component of tension, and axial stiffness of the line, respectively. A theoretical approach was used to identify the optimal configuration for the mooring system, minimizing the computation time. However, it does not account for environmental factors, including wave- and current-induced loading acting on the lines or bottom friction. Once the optimal condition was determined, a numerical approach was applied to analyze the detailed dynamics of the mooring system. In this model, the line is divided into discrete elements, with each element represented by two masses located at its ends and connected by massless springs and dampers, as illustrated in Figure 2. The hydrodynamic load acting on the lines from these factors can be calculated using Equation (10) because the dynamics of the mooring lines are influenced by waves and currents.

$$\mathbf{F_{hy}} = \mathbf{C_M}\Delta\mathbf{a_f} - \mathbf{C_a}\Delta\mathbf{a_r} + \frac{1}{2}\rho D l \mathbf{C_D}\mathbf{v}|\mathbf{v}| \quad (10)$$

where $\mathbf{C_M}$, $\mathbf{C_a}$, Δ, D, and l represent the inertia coefficient matrix, added mass coefficient matrix, mass, diameter, and sectional length of each line element, respectively. $\mathbf{a_f}$ and $\mathbf{a_r}$ denote the fluid acceleration and the relative acceleration between the fluid and line elements.

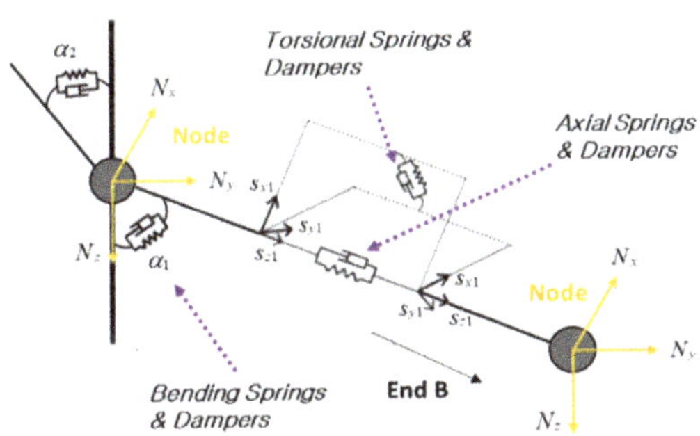

Figure 2. Lumped-mass line model [33].

Finally, the dynamics of a floating moored fish cage can be estimated by fully considering the coupling effects between the rigid body and the mooring system using an implicit calculation method to achieve accurate simulation results.

3. Numerical Model and Environmental Condition

The floating movable fish cage (X-AQUDS) comprised six pontoons and numerous steel frames to maintain cage space. The spaces between the frames were covered with nets to ensure the safety of the aquaculture fish, as shown in Figure 1. Six horizontal cylindrical pontoons were employed to provide buoyancy to the cages. Three pontoons existed on each side of the x-axis. Table 1 outlines the fish cage's specifications, and the mass and restoring characteristics of the cage are presented in Table 2.

Table 1. Dimensions of the X-AQUDS.

Item	Unit	Dimension
Length (L)	m	35.0
Breadth (B)	m	23.0
Depth (D)	m	14.0
Draft (T)	m	8.5
Volume (V)	m^3	5500
Center of gravity above Base (KG)	m	5.2
Pontoon (Material: SS510)		
Upper pontoon diameter (D_{up})	m	1.2
Upper pontoon length (L_{up})	m	16.0
Middle pontoon diameter (D_{md})	m	0.8
Middle pontoon length (L_{md})	m	16.0
Lower pontoon diameter (D_{lw})	m	1.4
Lower pontoon length (L_{lw})	m	23
Steel frame (Material: STPG)		
Outer diameter	mm	216.3
Wall thickness	mm	8.2

Table 2. Mass and restoring characteristics of the X-AQUDS.

Item	Unit	Dimension
Total mass (M)	ton	114.03
Roll radius of gyration at CG (R_{xx})	m	4.92
Pitch radius of gyration at CG (R_{yy})	m	5.00
Yaw radius of gyration at CG (R_{zz})	m	5.68
Heave restoring coefficient (K_{33})	kN/m	364.1
Roll restoring coefficient (K_{44})	kN m/rad	4199.7
Pitch restoring coefficient (K_{55})	kN m/rad	40,529.2

Pontoons have relatively blunt bodies; hence, the 3D radiation/diffraction problem was solved in the frequency domain by the commercial software, ANSYS AQWA 18.2. Figure 3 shows the mesh system for the six pontoons comprising 14,992 mesh elements. The torsional effects of the fish cage were neglected, and the entire structure was treated as a single rigid body. The mesh system was validated by comparing the root mean square error (RMSE) of the added mass and radiation damping coefficients for the 14,992 nodes (Table 3). The results showed a significant decrease in the RMSE with increasing node numbers. With an RMSE below 1% for all coefficients, the current mesh system was considered suitable for the present study. Figure 4 shows the frequency-dependent added masses, moments of inertia, and radiation damping coefficients at a water depth of 50 m. The added masses for the sway and heave motions were more significant than those for the surge motion, owing to the symmetrical installation of pontoons along the x-axis. The moments of inertia for the roll motion were superior to those for the pitch and yaw motions. These tendencies were observable for the radiation damping coefficients. This configuration enables seamless cage movement to a sheltered location during typhoons.

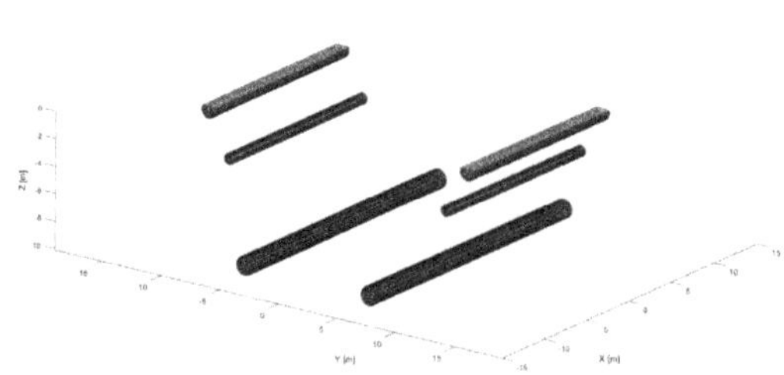

Figure 3. Numerical meshes for pontoons.

Table 3. RMSEs under different node number conditions (relative to 14,992 nodes).

# of Nodes	RMSE (Heave–Heave Added Mass)	RMSE (Pitch–Pitch Added Moment of Inertia)	RMSE (Heave–Heave Radiation Damping)	RMSE (Pitch–Pitch Radiation Damping)
5184	12.08	595.7	10.95	172.4
7800	7.44	413.7	6.59	90.5
12,664	0.46	72.0	1.02	12.2

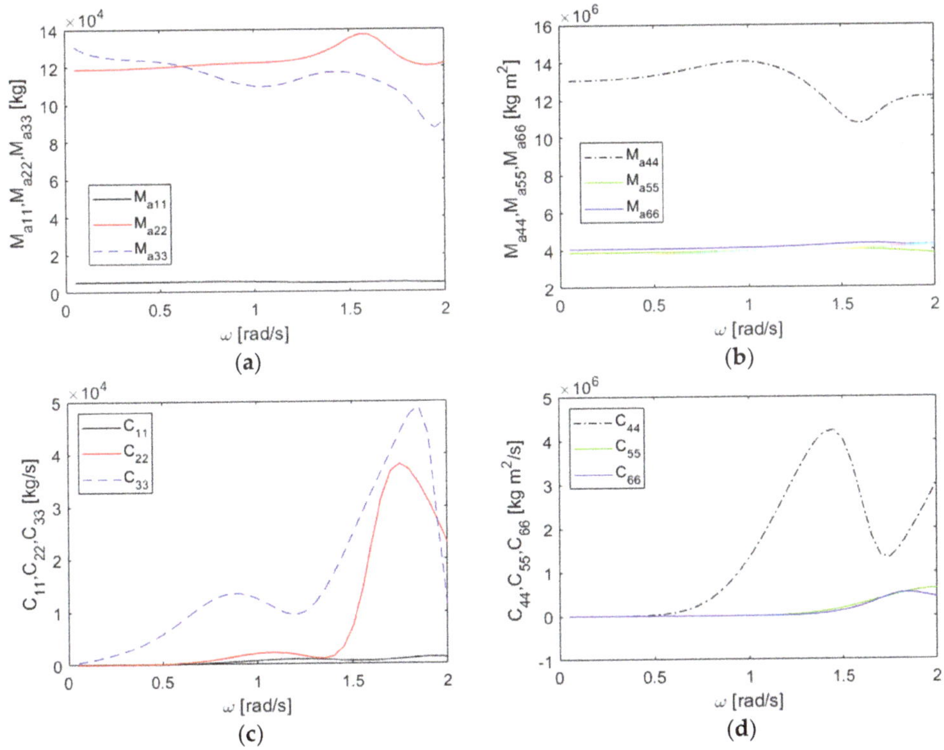

Figure 4. Added masses and moments of inertia (**a**,**b**) and radiation damping coefficients (**c**,**d**).

Figure 5 illustrates the numerical model of the moored fish cage (X-AQUDS). Figure 6 shows the layout of the spread mooring system. A numerical model was constructed by adding steel frames to six pontoons, totaling 94 steel frames, each with a diameter of 216.3 mm. We modeled pontoons and steel frames as rigid bodies, focusing solely on the inertia and drag effects. Fish cages were anchored to four groups of mooring lines, with each group comprising two lines. Each group was positioned at a 55° angle relative to the x-axis, with a 10° separation between the lines within each group (Figure 6). The mooring lines had a horizontal radius of 290 m from the origin. In this study, we examined four lengths of these lines, ranging from 285 m to 300 m at 5 m intervals. The mooring system was composed of studless chains, as illustrated in Table 4 [34]. The added mooring lines mass and drag coefficients were 1.0 and 1.2, respectively. The drag coefficient of the steel frame of the fish cage was 1.2.

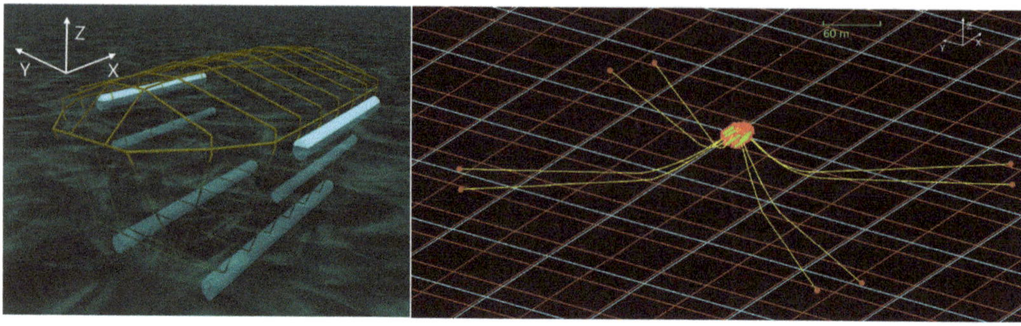

Figure 5. Numerical model for the fish cage with a spread mooring system.

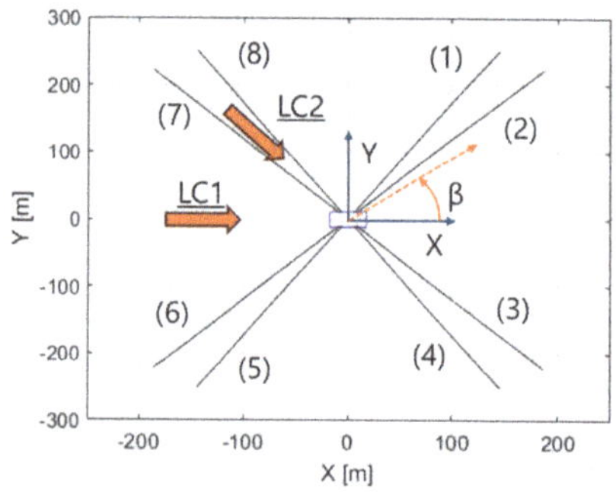

Figure 6. Layout of the spread mooring system for the fish cage (The numbers in the bracket indicate the mooring line number).

Table 4. Specification of the mooring chain.

Chain Diameter [mm]	Wet Weight [N/m]	Axial Stiffness [MN]	Minimum Breaking Load (MBL) [kN]
32	192	92.16	895
36	243	116.64	1125

The drag coefficient for the netting was determined based on the solidity ratio (S_n) and Reynolds number (*Re*). This study employed the empirical formula using S_n and *Re* proposed by Ninu et al. [35]. They conducted experiments to determine the drag coefficients of typical nylon fishing nets with and without biofouling and provided the empirical formula as Equation (11).

$$C_D = a_n R_e + b_n \tag{11}$$

where a_n and b_n are constants obtained by the experiment. Table 5 illustrates the solidity ratios, the constants, and the drag coefficients for different net conditions. This study used nylon nets with a similar solidity ratio as in the work of Ninu et al. [35]. Hence, the drag coefficient values can be directly applied to estimate the effects of nets and biofouling. A drag screen model was employed, and the instantaneous drag coefficient of the nets was evaluated by considering the Reynolds number and wetted surface area of each part at each timestep.

Table 5. Solidity ratio and drag coefficients with and without biofouling.

Condition	Solidity Ratio	a_n	b_n	Drag Coefficient (Re = 2000)	Drag Coefficient (Re = 4000)
No biofouling	0.22	-1.12×10^{-5}	0.20	0.1776	0.1552
With biofouling	0.31	-1.75×10^{-5}	0.27	0.2350	0.2000

The environmental design conditions for this study were selected based on the study by Hong et al. [36]. They built a numerical estimation model using SWAN based on the design environmental conditions, providing offshore design wave conditions for the Korean Peninsula. In accordance with the ABS guidelines [27], a 50-year return period of wave and current conditions can be used as design environmental condition for mobile mooring systems. Table 6 illustrates the design of the environmental conditions. The JONSWAP wave spectrum was adopted to generate irregular waves, with the enhancement factor calculated in accordance with the DNV GL rules [37]. Figure 7 displays the time series and power spectral densities of wave elevation and the vertical current profile. The calculated wave statistics showed a significant wave height of 6.5498 m and a peak wave period of 10.76 s. The reconstructed wave spectrum aligned closely with the theoretical solution, differing by less than 3%. A vertical current profile was calculated using the seventh power law with no directional variation along the z-axis. The wave and current directions were aligned to represent the worst-case scenario for a floating structure. Two environmental conditions were applied: a maximum excursion condition (LC1) and a maximum tension condition (LC2). The maximum excursion condition occurs when incoming waves and coplanar currents impact the front of the structure. However, the maximum tension condition occurs when environmental loads are applied to the center of the mooring line group, as shown in Figure 6.

Table 6. Environmental condition (50-year return period).

Significant Wave Height [m]	Peak Period [s]	Enhancement Factor	Surface Current Speed [m/s]
6.55	11.19	2.06	0.71

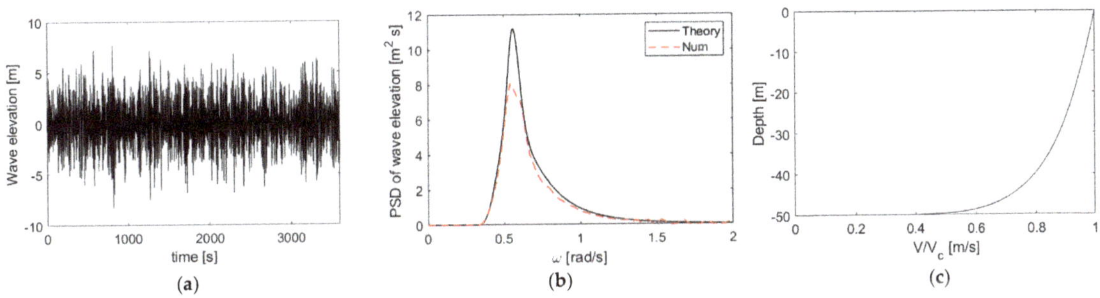

Figure 7. Time histories of the designed waves (**a**), their wave spectrum (**b**), and the vertical current profile (**c**).

4. Results

This study aimed to evaluate the dynamics of a floating moveable fish cage with a spread mooring system situated far from the seashore, assessing the safety of the mooring system according to the ABS rules [18,19].

4.1. Motion Response Amplitude Operators of the Freely Floating Fish Cage

The motion response amplitude operators (RAOs) of the freely floating fish cages were evaluated based on the numerical model (Section 2). Three different calculation scenarios were considered to assess the effects of the steel frames and netting system: (1) pontoons only, (2) pontoons with steel frames, and (3) pontoons with steel frames and nets. However, the local deformation in the numerical model was not considered. The analysis considered a freely floating body with incoming waves at an incident angle of 0°. Thus, we compared the displacements of the heave and pitch motions. While Figure 8 shows the time series of these motion displacements, Figure 9 shows the motion RAOs. When analyzing the time series, the drag effects from the steel frames and nets yielded smaller motion displacements and phase angle differences. We compared the motion RAOs from the frequency and time domain analyses to validate the numerical model, as depicted in Figure 9. The frequency domain results were obtained by solving the 3D radiation/diffraction problem. These results are in good agreement. Excessive motion reduction due to the drag effects from additional structures (steel frames and nets) is observed near the natural frequency. This drag damping typically reduces motion at the natural frequency, and the results align with these trends. The damped natural frequencies for the heave and pitch motions were 1.05 and 0.7 rad/s, respectively.

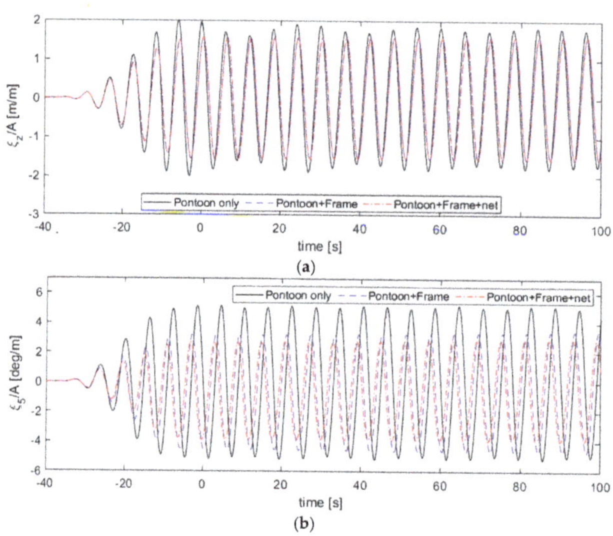

Figure 8. Time histories of heave (**a**) and pitch (**b**) motion displacements of a freely floating fish cage (β = 0° and ω = 1.05 rad/s).

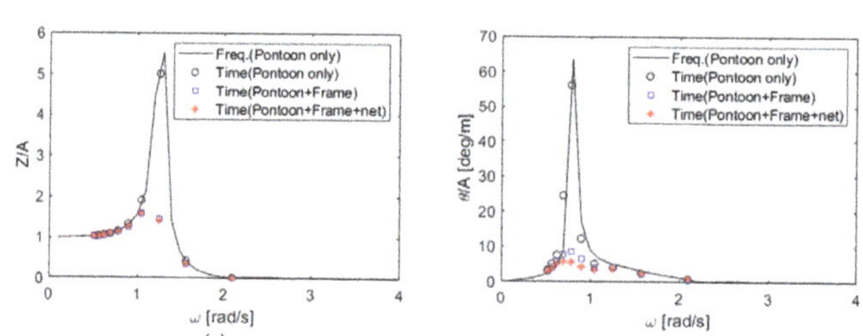

Figure 9. Heave (**a**) and pitch (**b**) motion RAOs without mooring systems (β = 0°).

4.2. Dynamics of a Moored Fish Cage

For station-keeping, a spread mooring system was applied to the cage by employing eight mooring lines. Two wave and current heading angles were applied to assess the maximum excursion and mooring line tension. First, we considered the maximum excursion condition, whereby the incident angle was 0°. Moreover, the motion displacement and mooring line tensions were observable. All the simulations involving irregular waves were conducted for 1 h with a time step of 0.025 s. For a one-hour simulation, the theoretical approach to calculating connection forces requires a computational cost of 6 h. Table 6 illustrates the environmental conditions. Two mooring line chains were used with a mooring line length of 290 m (Table 2). Figure 10 illustrates the three degree-of-freedom motion (surge, heave, and pitch) displacements. When the heave and pitch motions were compared, no significant differences were observed depending on the chain type. The maximum excursion was more noticeable with the lighter chain than with the heavier chain (owing to its greater stiffness). The low-frequency surge motion was more evident with the lighter chain than with the heavier chain, as observed in the power spectral densities. Consequently, the maximum excursion was significantly affected by the chain weight, whereas the other motion modes were less affected. Additionally, an analysis of the motion response spectrum indicated that the peak frequency of the primary responses aligned with the wave peak frequency because the incident waves had a longer peak period than the natural period of the cage.

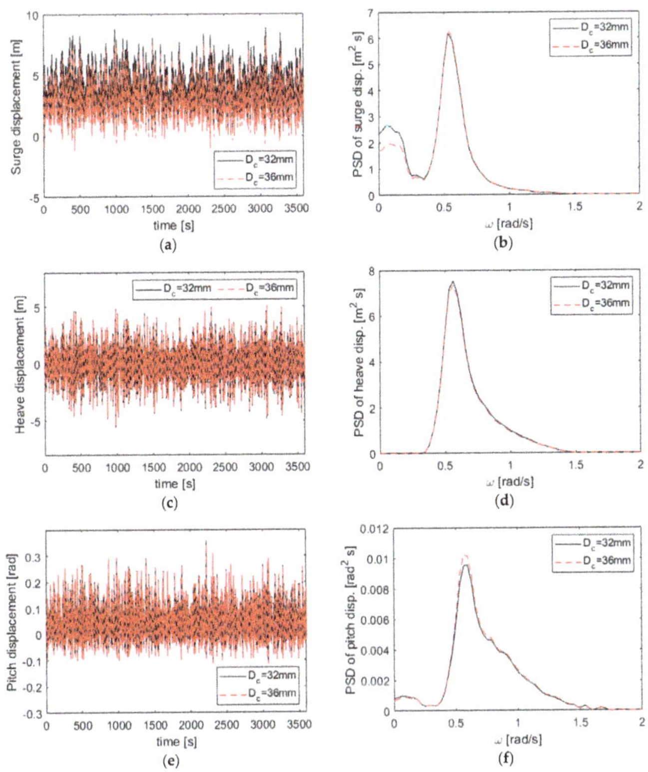

Figure 10. Time histories (**a**,**c**,**e**) and power spectral densities (**b**,**d**,**f**) of motion responses of the moored fish cage under the maximum excursion conditions (β = 0°; (**a**,**b**) surge: (**c**,**d**) heave; and (**e**,**f**) pitch).

The mooring line tensions were compared under two different chain conditions, as shown in Figure 11. A heavier chain yielded a more significant static tension across all the mooring lines. Owing to the drifting of the moored fish cage in the positive x-direction under the influence of surface currents, mooring lines #7 and #8 experienced higher tension than mooring lines #1 and #2, with line #8 exhibiting the maximum tension. No significant differences were observable in the motion responses at the wave peak frequency depending on the chain type. The heavier chain demonstrated higher line tension owing to its greater stiffness.

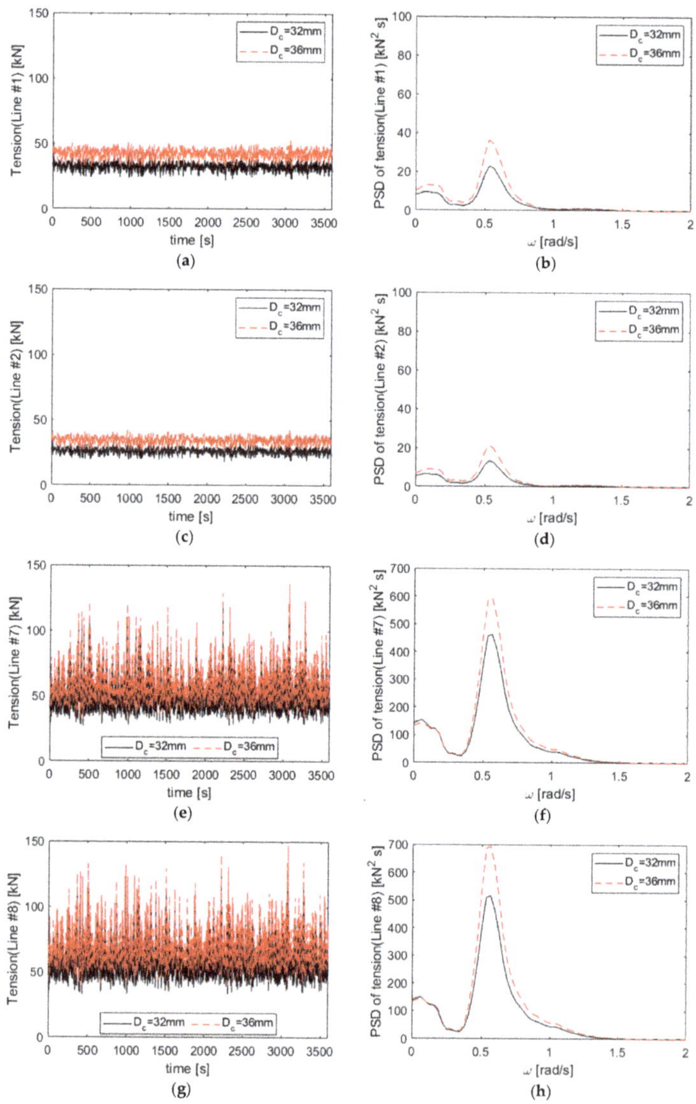

Figure 11. Time histories (**a**,**c**,**e**,**g**) and power spectral densities (**b**,**d**,**f**,**h**) of mooring line tensions of the moored fish cage under the maximum excursion conditions ($\beta = 0°$; (**a**,**b**) line #1; (**c**,**d**) line #2; (**e**,**f**) line #7; and (**g**,**h**) line #8).

Subsequently, the maximum tension condition was applied to assess the safety of the mooring system. Compared with the maximum excursion condition, where only three DOF motions were considered, this scenario required the consideration of six DOF motions because environmental loading was applied along the mooring lines. Figure 12 illustrates the trajectory of the fish cage. Table 7 illustrates the statistical values of the global motion of the cage, including the mean and standard deviation of the six DOF motions. The excursion range was broader when using the lighter chain owing to its weaker static tension than that of the heavier chain, as shown in Figure 10. Evaluation of the hydrodynamic characteristics revealed that the mean surge and sway motions were attributable to the chain diameter. However, the standard deviations of the surge, sway, and heave motions were nearly identical despite the chain diameter. The differences in the rotational motion displacements were also insignificant. Figure 13 shows the time histories and power spectral densities of the line tensions for the two mooring lines that experienced the highest tension. The maximum tension was observed in line #8 and was significantly higher than that in the maximum excursion scenario. The case with a chain diameter of 32 mm exhibited a more significant drift, yielding a higher line tension level. The spectral analysis of line tension indicated that its spectral tendencies aligned closely with those of the motion responses. The wave-frequency motion responses predominantly affected line tension, with a minor contribution from the low-frequency motion responses.

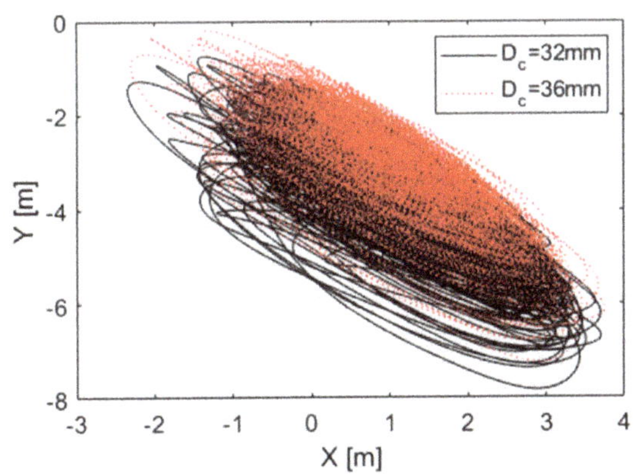

Figure 12. Trajectory of the fish cage in the global coordinate system under the maximum tension conditions.

Table 7. Hydrodynamic characteristics of the cage under the maximum tension conditions.

Item	Chain Diameter = 32 mm		Chain Diameter = 36 mm	
	Mean	STD	Mean	STD
Surge	1.095	0.923	0.980	0.922
Sway	−3.537	1.091	−2.954	1.061
Heave	−0.045	1.349	−0.041	1.332
Roll	0.015	0.063	0.015	0.062
Pitch	0.016	0.035	0.016	0.036
Yaw	−0.004	0.025	−0.003	0.024

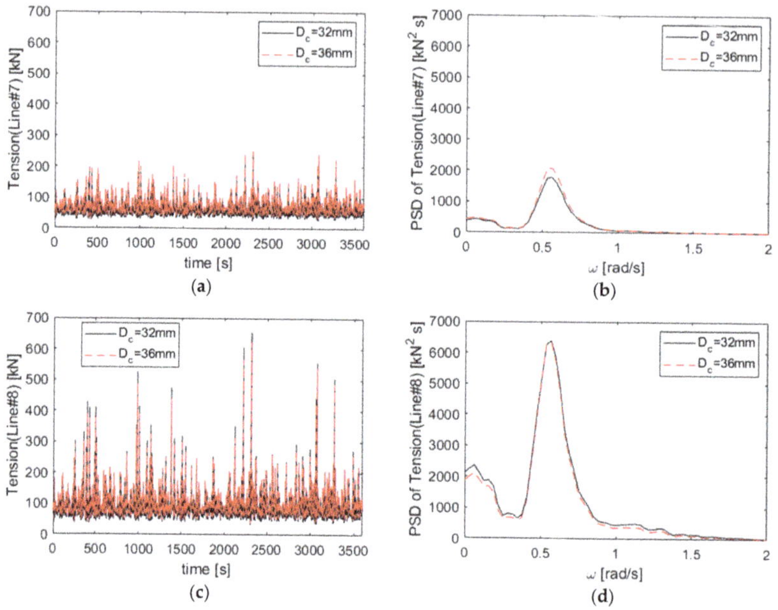

Figure 13. Time histories (**a**,**c**) and power spectral densities (**b**,**d**) of mooring line tensions of the moored fish cage under the maximum tension conditions ($\beta = -55°$; (**a**,**b**) line #7; and (**c**,**d**) line #8).

4.3. Effect of Mooring Line Length

We conducted a parametric study to examine the effect of mooring line length. The weight, pretension, and fairlead angle of the mooring lines were controlled by adjusting the line length while keeping the anchor and fairlead points fixed. Therefore, four different line lengths ranging from 285 to 300 m were applied. We used a catenary equation-based theoretical approach to minimize computational cost. Table 8 illustrates the hydrodynamic characteristics of the fish cage under the maximum excursion conditions. Table 9 illustrates the statistical values of the line tensions under the maximum tension conditions. Longer mooring lines yielded a superior maximum excursion and a more significant heave motion owing to the reduced pretension in the mooring system. Conversely, the pitch motion decreased as the mooring line length increased because a stiffer mooring line induced superior rotational motion. Regarding the longest mooring line, the maximum excursion was approximately half the water depth, suitable for designing the mooring system of the cage.

Table 8. Hydrodynamic characteristics of the fish cage with various mooring line length conditions under the maximum excursion conditions.

Chain Diameter [mm]	Line Length [m]	Max. Excursion [m]	STD of Heave [m]	STD of Pitch [deg]
32	285	4.340	1.467	4.279
	290	8.863	1.534	3.733
	295	15.192	1.554	3.623
	300	22.305	1.563	3.633
36	285	3.835	1.429	4.122
	290	8.030	1.513	3.819
	295	13.754	1.539	3.639
	300	20.606	1.551	3.638

Table 9. Statistical characteristics of mooring line tensions for the fish cage with various mooring line length conditions under the maximum tension conditions.

Chain Diameter [mm]	Length [m]	Max. Excursion [m]	Line Tension (#7) [kN] Static	Max	STD	Line Tension (#8) [kN] Static	Max	STD
32	285	3.651	87.17	349.60	37.07	128.75	835.03	94.90
	290	8.371	87.17	244.27	22.83	128.75	654.70	46.95
	295	13.432	87.17	188.26	17.87	128.75	550.12	39.15
	300	18.755	87.17	165.37	15.71	128.75	542.36	39.04
36	285	3.317	107.89	357.49	39.39	158.30	908.96	95.36
	290	7.834	107.89	250.95	24.33	158.30	623.83	45.18
	295	12.685	107.89	200.18	18.15	158.30	484.93	35.10
	300	17.925	107.89	173.03	15.87	158.30	451.37	34.18

The safety of the mooring system was evaluated by comparing the maximum measured tension with the maximum allowable tension, calculated by dividing the minimum breaking load by the factor of safety (FoS). For intact mobile and permanent mooring systems under extreme design conditions, the FoS was 1.67 in accordance with the ABS rules. For a damaged floating structure, such as a one-line broken structure under similar environmental conditions, the FoS was 1.25. The maximum allowable tensions for the two chain diameters under intact conditions were 535.9 and 673.6 kN, respectively. A mooring system with heavier chain lines was suitable for this fish cage because the maximum tension of all mooring line lengths with lighter chains surpassed the allowable tension owing to the lower pretension in the lines. Moreover, longer mooring lines yielded a lower maximum tension and reduced tension variability related to fatigue damage. Therefore, longer mooring lines enhanced the safety of fish cage systems. Moreover, a mooring line length of 290 m with a chain diameter of 36 mm was selected owing to its insignificant excursion and the validated safety of the mooring lines at a low cost for the mooring system.

4.4. Effect of Surface Current Speed

The optimal mooring conditions selected in the previous section were applied for further analysis. This study conducted a finite element analysis employing the lumped-mass line model approach to analyze the line dynamics and coupling effect between a fish cage and mooring lines. The theoretical approach and the FE-based numderical approach required 6 and 12 h of computational times, respectively. The maximum tension condition (LC2) requires approximately 24 h of computational time owing to its dependency on environmental conditions. Figure 14 shows the motion responses at different surface current speeds under the maximum excursion conditions. The surge motion responses were significantly affected by the current speed. The fish cage experienced greater drift, stretching the mooring lines under high current speeds. Moreover, the stiffness of the mooring system increased, reducing the motion responses around the wave peak frequency. However, the heave and pitch motions were not significantly affected by changes in the current speed. Figure 15 illustrates the mooring line tension at various surface current speeds. Although the smaller wave-frequency surge motion responses were observable at higher current speeds, the mooring line tension was higher. This was attributable to the stiffened mooring lines inducing significant tension at the wave peak frequency despite the insignificant corresponding motion responses. Consequently, higher current speeds negatively affected the safety of the mooring system by increasing the mooring line tensions.

The maximum tension condition was used as an input to evaluate the hydrodynamic performance of the fish cage and the safety of the mooring system. Figure 16 illustrates the trajectory of the cage under various current speed conditions. With an incident wave and current angle of −55°, the cage moved southeast. As the current speed increases, the mean position moves farther from the origin, as shown in Figure 14a. Table 10 illustrates the maximum excursion and statistical values of the line tensions. As the current speed increased, the maximum excursion and statistical values of the line tensions increased.

At a current speed of 1.055 m/s, the tension in line #8 surpassed the permissible limit of 673.6 kN, which is determined by dividing the minimum breaking load by the safety factor. This result suggests that the structural stability of the movable fish cage is compromised at speeds above 1 m/s, which must be maintained to ensure integrity. Since maximum tension and its variability are closely linked to mooring lines' structural failure and fatigue damage, the current speed must be carefully accounted for in the design of mooring systems.

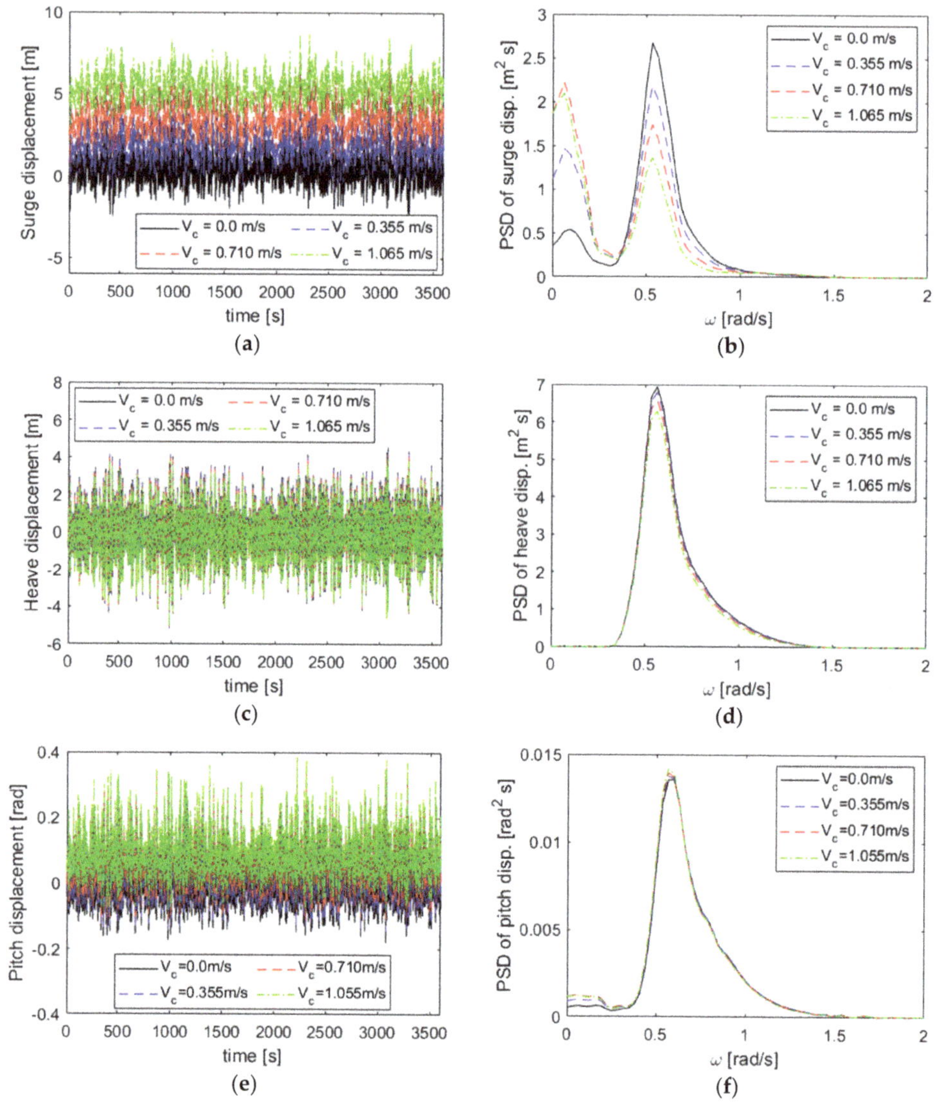

Figure 14. Time histories (**a**,**c**,**e**) and power spectral densities (**b**,**d**,**f**) of motion displacements of the moored fish cage with various surface current speed conditions under the maximum excursion conditions ($\beta = 0°$; (**a**,**b**) surge; (**c**,**d**) heave; and (**e**,**f**) pitch).

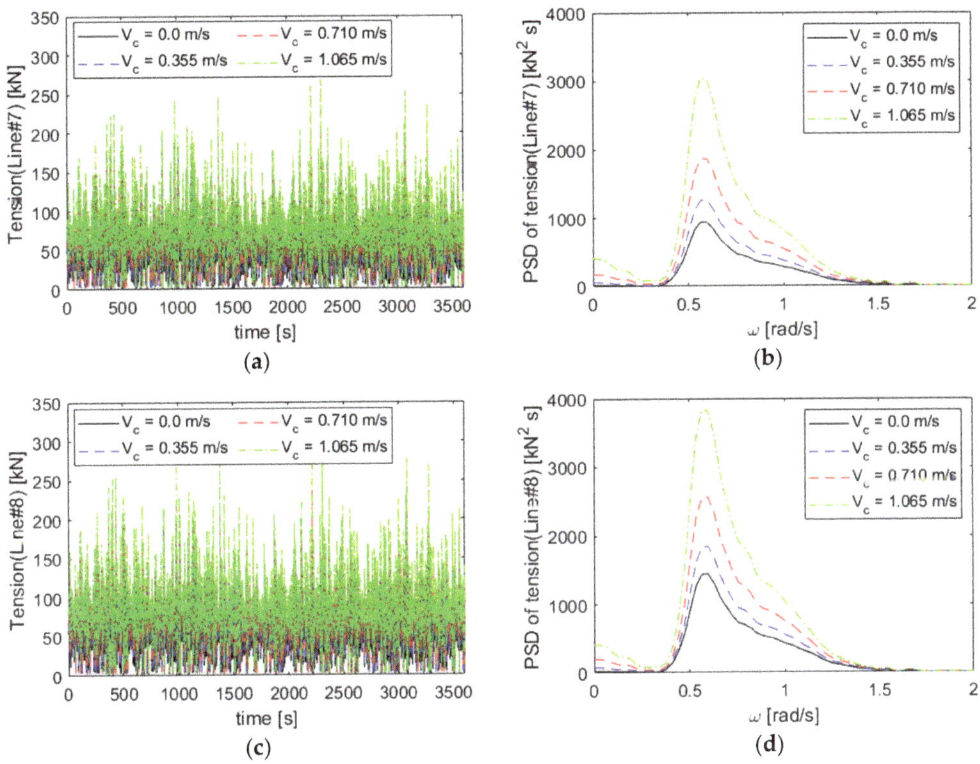

Figure 15. Time histories (**a**,**c**) and power spectral densities (**b**,**d**) of mooring line tensions of the moored fish cage with various surface current speed conditions under the maximum excursion conditions ($\beta = 0°$; (**a**,**b**) line #7; and (**c**,**d**) line #8).

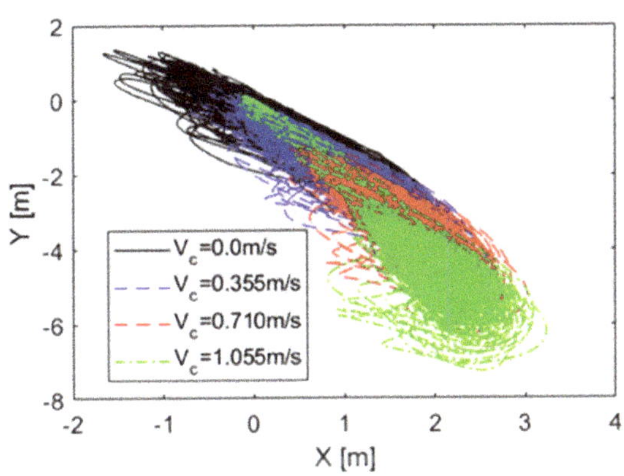

Figure 16. Trajectory of the fish cage with various surface current speed conditions under the maximum tension conditions ($\beta = -55°$).

Table 10. Statistical characteristics of mooring line tensions for the moored fish cage with various surface current speed conditions under the maximum tension conditions.

Current Speed [m/s]	Max. Excursion [m]	Line Tension (#7) [kN]			Line Tension (#8) [kN]		
		Static	Max	STD	Static	Max	STD
0.0	4.275	42.01	205.08	19.97	57.32	237.20	28.84
0.355	5.547	42.02	210.08	25.14	74.37	331.76	39.64
0.71	6.720	42.03	284.18	33.44	104.34	483.68	57.91
1.065	7.693	42.03	374.84	46.30	150.06	699.74	86.38

4.5. Effect of the Netting System

Biofouling is a critical factor when considering fish cages because it alters material properties, such as mass and drag effects. For simplicity, the mass variation was not considered in this study. However, the drag coefficient variation was considered using the experimentally obtained drag coefficients. Table 5 illustrates the solidity ratios and drag coefficients. First, the dynamic responses were compared regarding the net conditions, as shown in Figure 17. The heave and pitch motion responses were minimally affected; however, the surge motion yielded significant changes. A substantially increased low-frequency surge motion was observable owing to the increased drag force. However, the wave-frequency surge motion was slightly reduced with the application of the netting system. Under biofouling, the maximum excursion increased by approximately 6%.

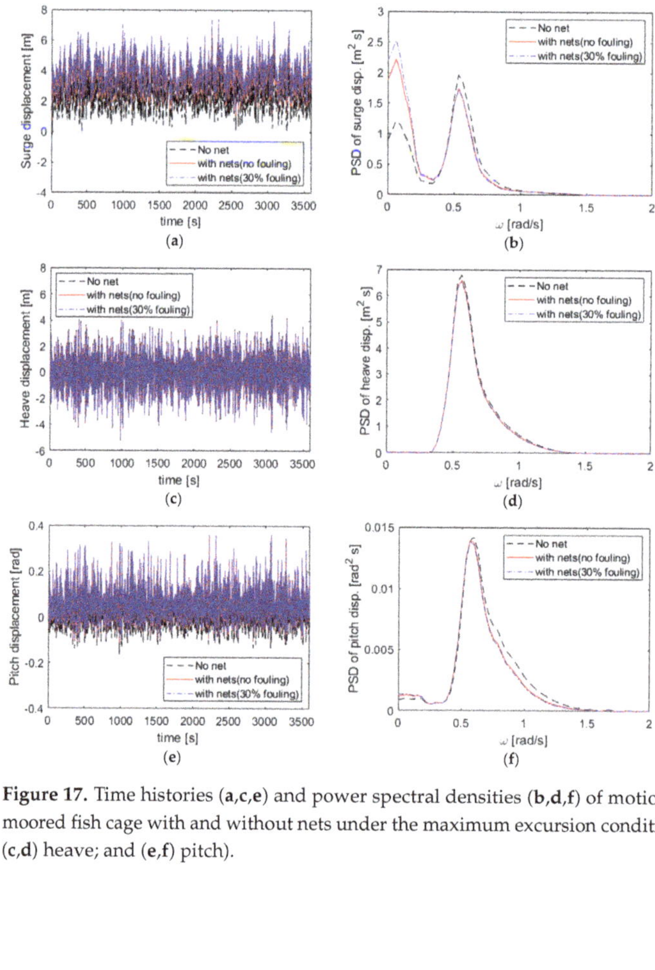

Figure 17. Time histories (**a**,**c**,**e**) and power spectral densities (**b**,**d**,**f**) of motion displacements of the moored fish cage with and without nets under the maximum excursion conditions (β = 0°; (**a**,**b**) surge; (**c**,**d**) heave; and (**e**,**f**) pitch).

The maximum tension condition ($\beta = -55°$) was applied to evaluate the safety of the mooring system under the influence of biofouling. Figure 18 shows the trajectories of the fish cages. As the drag coefficient increased owing to biofouling, the maximum excursion also increased, as shown in Figure 17. When examining the tensions of lines #7 and #8 (Figure 19), line #8 experienced significant tension. Increased line tension was observable at the wave peak frequency with a higher drag effect. Compared with Figure 17, the low-frequency surge motion is a crucial parameter related to the line tension. Although the maximum tension with biofouling is 17% higher than that without biofouling, it remained below the maximum allowable tension. Therefore, this moored cage system is considered well designed from a safety perspective despite having 30% biofouling.

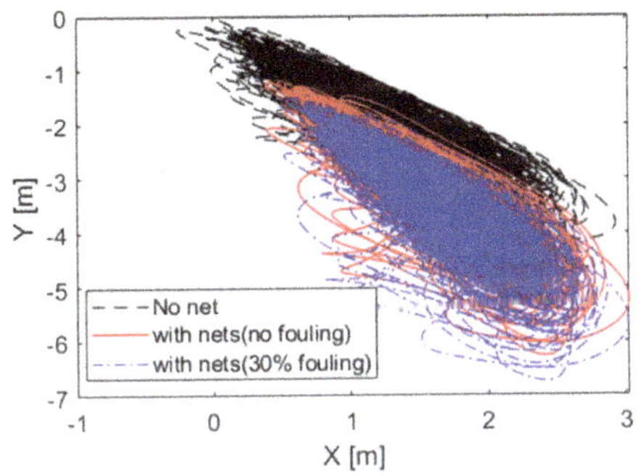

Figure 18. Trajectory of the fish cage with and without nets under the maximum tension conditions ($\beta = -55°$).

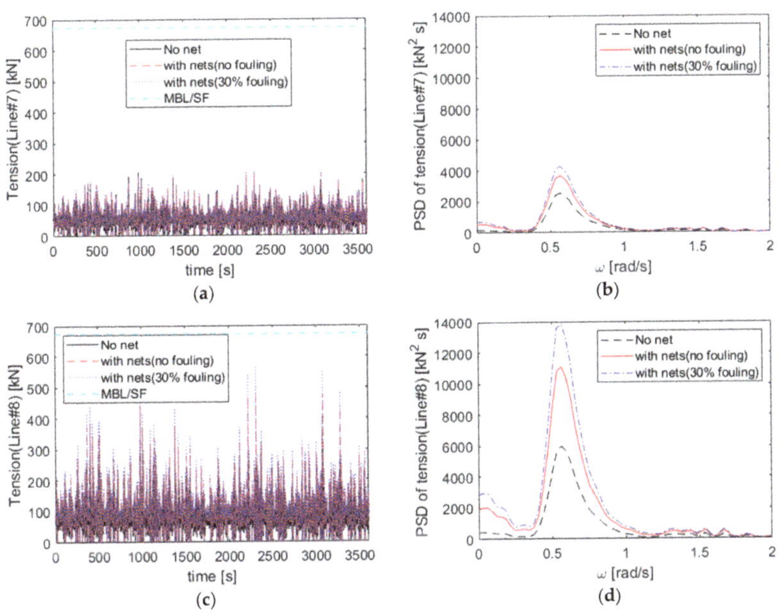

Figure 19. Time histories (**a**,**c**) and power spectral densities (**b**,**d**) of mooring line tensions of the moored fish cage with and without nets under the maximum tension conditions ($\beta = -55°$; (**a**,**b**) line #7; and (**c**,**d**) line #8).

4.6. Mooring Line Failure Condition

Finally, two worst-case scenarios were examined to assess the safety of the mooring system when one mooring line was broken. The first scenario (Damaged Case 1) involved the absence of mooring line #7. The second scenario (Damaged Case 2) involved the absence of mooring line #8. This analysis applied a single environmental load condition (maximum tension condition) to test the safety of the mooring system. Figure 20 illustrates the trajectory of the fish cage in all three cases (one intact and two damaged). The cage experienced a more significant drift when the line was broken. Owing to the high tension on line #8, the fish cage drifted considerably southeast when the line broke. When only line #7 was broken, line #8 reduced the drift to the west because its laid angle was deeper (60°) than that of line #7 (50°). Figure 21 shows the line tensions of mooring lines #7 and #8 under the two damage scenarios. The tension was significantly higher than in the intact condition. For the damaged cases, a safety factor of 1.25 was applied, resulting in a maximum allowable tension of 900 kN. Because the maximum tension for the damage scenarios did not exceed the allowable tension, the mooring system was considered suitable for the moveable fish cage at the selected offshore site.

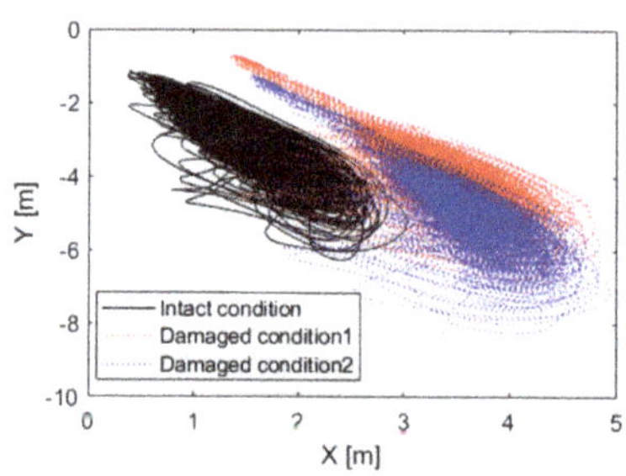

Figure 20. Trajectory of the fish cage under one intact case and two different damaged cases ($\beta = -55°$).

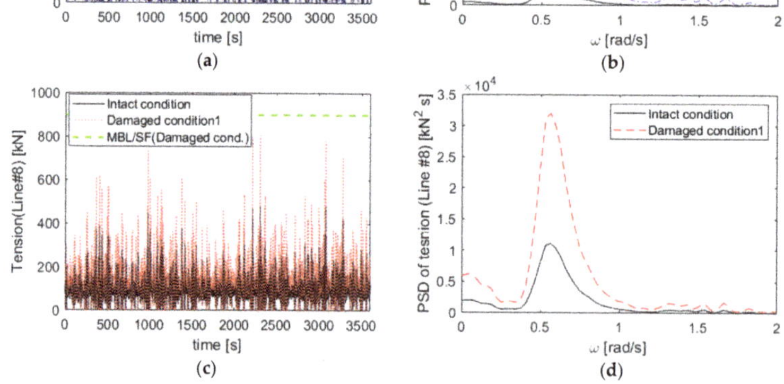

Figure 21. Time histories (**a**,**c**) and power spectral densities (**b**,**d**) of mooring line tensions of the moored fish cage under two different damaged conditions ($\beta = -55°$; (**a**,**b**) line #7 and (**c**,**d**) line #8).

5. Conclusions

This study evaluated the hydrodynamic performance and safety of a floating movable fish cage with a spread mooring system under offshore conditions. A numerical model was developed using potential theory with radiation/diffraction problems, the Cummins equation to represent the dynamics of the floating structure, and a lumped-mass line model to capture the detailed behavior and tension of the mooring lines. The steel frames were modeled using Morison members. However, fishing nets were incorporated using a screen model to account for the drag forces, which depend on the solidity ratio and Reynolds number. The mooring system comprises eight lines with four station-keeping groups. The numerical model was examined under environmental conditions with a 50-year return period to evaluate the maximum excursion and line tension of the mooring system. A parametric study was performed regarding mooring line length, current speed, and fishing nets with and without biofouling. The study findings are summarized as follows:

- A parametric study showed that longer mooring lines increased excursions but reduced the pitch motion and tension variability. Heavier chains (36 mm diameter) decreased excursions by up to 33%, improving station-keeping, but increased the maximum line tension by 17% compared with lighter chains. A 290 m line with a 36 mm chain diameter optimized excursions and mooring system safety.
- Higher current speeds significantly affected the surge motion, causing greater drift and stretching mooring lines, while minimally impacting the heave and pitch motions. Increased current speeds stiffened the mooring system, reducing motion responses near the wave peak frequency but increasing mooring line tension. At a 1.055 m/s current speed, mooring line tension exceeded safe limits, emphasizing the importance of considering current speed in mooring system design for stability and structural safety.
- Biofouling substantially altered the fish cage hydrodynamics by changing drag properties, leading to an increased low-frequency surge motion and a 6% rise in the maximum excursion, with minimal effects on the heave and pitch motions. Under the maximum tension conditions, biofouling induced the increased line tension at the wave peak frequency by 17%; however, the line tension remained within allowable limits.
- Two worst-case scenarios involving individual mooring line failures were investigated to evaluate system safety and resilience. Both scenarios led to increased drift and higher line tension levels compared to the intact condition. Nevertheless, the mooring system kept the line tension within the allowable safety threshold.

The fish cage mooring system installed offshore met the safety criteria established by the ABS rules in intact and damaged conditions. This comprehensive analysis confirms the structural and hydrodynamic integrity of the fish cage and its mooring system. However, further consideration of the characteristics of cultured species in fish cages, surrounding marine facilities, and seabed topography is crucial.

Author Contributions: Conceptualization: S.-J.K. and S.-J.J.; methodology, S.-J.K.; software, S.-J.P.; validation, S.-J.K. and S.-J.P.; formal analysis, S.-J.K.; investigation, S.-J.K. and S.-J.J.; resources, S.-J.J.; data curation, S.-J.J. and S.-J.P.; writing—original draft preparation, S.-J.K.; writing—review and editing, S.-J.J. and S.-J.P.; visualization, S.-J.J. and S.-J.P.; supervision, S.-J.K.; project administration, S.-J.J.; funding acquisition, S.-J.K. and S.-J.J. All authors have read and agreed to the published version of the manuscript.

Funding: This study was supported by a grant from the National Institute of Fisheries Science, Ministry of Oceans and Fisheries, Republic of Korea (grant number: R2024034).

Institutional Review Board Statement: Not applicable.

Informed Consent Statement: Not applicable.

Data Availability Statement: The original contributions presented in this study are included in the article. Further inquiries can be directed to the corresponding author.

Conflicts of Interest: The authors declare no conflicts of interest.

References

1. Chu, Y.I.; Wang, C.M.; Park, J.C.; Lader, P.F. Review of cage and containment tank designs for offshore fish farming. *Aquaculture* **2020**, *519*, 734928. [CrossRef]
2. Chu, Y.-I.; Wang, C.-M.; Zhang, H.; Abdussamie, N.; Karampour, H.; Jeng, D.-S.; Baumeister, J.; Aland, P.A. Offshore fish farms: A review of standards and guidelines for design and analysis. *J. Mar. Sci. Eng.* **2023**, *11*, 762. [CrossRef]
3. Kim, M.H.; Koo, B.; Mercier, R.M.; Ward, E.G. Vessel/mooring/riser coupled dynamic analysis of a turret-moored FPSO compared with OTRC experiment. *Ocean Eng.* **2005**, *32*, 1780–1802. [CrossRef]
4. Cifuentes, C.; Kim, M.H. Hydrodynamic response of a cage system under waves and currents using a Morison-force model. *Ocean Eng.* **2017**, *141*, 283–294. [CrossRef]
5. Jin, C.; Kang, H.; Kim, M.H.; Bakti, F.P. Performance evaluation of surface riding wave energy converter with linear electric generator. *Ocean Eng.* **2020**, *218*, 108141. [CrossRef]
6. Kim, S.-J.; Koo, W.; Kim, M.H. The effects of geometrical buoy shape with nonlinear Froude-Krylov force on a heaving buoy point absorber. *Int. J. Nav. Archit. Ocean Eng.* **2021**, *13*, 86–101. [CrossRef]
7. Kim, S.-J.; Koo, W. Numerical study on a multibuoy-type wave energy converter with hydraulic PTO system under real sea conditions. *IEEE J. Ocean. Eng.* **2021**, *46*, 573–582. [CrossRef]
8. Yu, S.; Qin, H.; Li, P.; Xu, Z. Nonlinear vertical accelerations and mooring loads of a semi-submersible offshore fish farm under extreme conditions. *Aquac. Eng.* **2021**, *95*, 102193. [CrossRef]
9. Miao, Y.-J.; Ding, J.; Tian, C.; Chen, X.-J.; Fan, Y.-L. Experimental and numerical study of a semi-submersible offshore fish farm under waves. *Ocean Eng.* **2021**, *225*, 108794. [CrossRef]
10. Ma, C.; Zhao, Y.-P.; Bi, C.-W. Numerical study on hydrodynamic responses of a single-point moored vessel-shaped floating aquaculture platform in waves. *Aquac. Eng.* **2022**, *96*, 102216. [CrossRef]
11. Yue, W.; Wang, W.; Sheng, S.; Ye, Y.; Hong, T. Analysis of the wave load and dynamic response of a new semi-submersible wave-energy-powered aquaculture platform. *Ocean Eng.* **2022**, *248*, 110346. [CrossRef]
12. Ding, W.-W.; Jiang, J.-Q.; Yue, W.-Z.; Li, Y.-Z.; Wang, W.-S.; Sheng, S.-W.; Chen, M. Numerical study on hydrodynamic performance of a new semi-submersible aquaculture platform. *Appl. Sci.* **2023**, *13*, 12652. [CrossRef]
13. Pang, G.; Zhang, S.; Liu, H.; Zhu, S.; Yuan, T.; Li, G.; Han, X.; Huang, X. Hydrodynamic response analysis for a new semi-submersible vessel-shaped fish farm platform based on numerical simulation. *Front. Mar. Sci.* **2023**, *10*, 1135757. [CrossRef]
14. Kim, S.-J.; Jin, C.; Kim, M. Time-dependent responses and mooring tensions of a moored floating structure in tsunami waves. *Mar. Struct.* **2024**, *93*, 103538. [CrossRef]
15. Chen, M.; Huang, L.; Cheng, H.; Wu, Q. Numerical investigation on mooring line configurations of a semisubmersible fish farm for global performance. *J. Mar. Sci. Eng.* **2024**, *12*, 880. [CrossRef]
16. Zhang, L.; Zhen, X.; Duan, Q.; Huang, Y.; Chen, C.; Li, Y. Hydrodynamic characteristics analysis and mooring system optimization of an innovative deep-sea aquaculture platform. *J. Mar. Sci. Eng.* **2024**, *12*, 972. [CrossRef]
17. Lee, C.-W.; Kim, Y.-B.; Lee, G.-H.; Choe, M.-Y.; Lee, M.-K.; Koo, K.-Y. Dynamic simulation of a fish cage system subjected to currents and waves. *Ocean Eng.* **2008**, *35*, 1521–1532. [CrossRef]
18. Mohapatra, S.C.; Guedes Soares, C. A semi-analytical model of an array of moored floating flexible offshore net cages under current loads. *Ocean Eng.* **2024**, *291*, 116309. [CrossRef]
19. Mohapatra, S.C.; Guedes Soares, C. A review of the hydroelastic theoretical models of floating porous nets and floaters for offshore aquaculture. *J. Mar. Sci. Eng.* **2024**, *12*, 1699. [CrossRef]
20. Reyes, R.; del Norte-Campos, A.; Añasco, N.C.; Santander-de Leon, S.M.S. Biofouling development in marine fish farm influenced by net colour, immersion period and environmental conditions. *Aquac. Res.* **2020**, *51*, 3129–3138. [CrossRef]
21. Cornejo, P.; Guerrero, N.M.; Montes, R.M.; Quiñones, R.A.; Sepúlveda, H.H. Hydrodynamic effect of biofouling in fish cage aquaculture netting. *Aquaculture* **2020**, *526*, 735367. [CrossRef]
22. Chen, Q.-P.; Bi, C.-W.; Zhang, Z.-X.; Zhao, Y.-P. Hydrodynamic effect of different biofouling types on aquaculture netting. *Ocean Eng.* **2023**, *279*, 114430. [CrossRef]
23. Tang, H.-J.; Yeh, P.-H.; Huang, C.-C.; Yang, R.-Y. Numerical study of the mooring system failure of aquaculture net cages under irregular waves and current. *Ocean Eng.* **2020**, *216*, 108110. [CrossRef]
24. Cheng, H.; Li, L.; Ong, M.C.; Aarsæther, K.G.; Sim, J. Effects of mooring line breakage on dynamic responses of grid moored fish farms under pure current conditions. *Ocean Eng.* **2021**, *237*, 109638. [CrossRef]
25. Hou, H.-M.; Liu, Y.; Dong, G.-H.; Xu, T.-J. Reliability assessment of mooring system for fish cage considering one damaged mooring line. *Ocean Eng.* **2022**, *257*, 111626. [CrossRef]
26. ABS. *Rules for Building and Classing Floating Production Installations*; American Bureau of Shipping: Houston, TX, USA, 2014.
27. ABS. *Guide for Building and Classing Aquaculture Installations*; American Bureau of Shipping: Spring, TX, USA, 2020.
28. IMO. *Standards for Ship Maneuverability*; MSC 76/23/Add. 1; International Maritime Organization: London, UK, 2002.
29. IMO. *Explanatory Notes to the Standards for Ship Maneuverability, MSC/Circ.1053*; International Maritime Organization: London, UK, 2002.

30. Kim, H.-J.; Jeong, S.-J. Resistance and stability evaluation of mobile fish-cage. *J. Korean Soc. Fish. Ocean Technol.* **2016**, *52*, 79–87. [CrossRef]
31. Kristiansen, T.; Faltinsen, O.M. Modeling of current loads on aquaculture net cages. *J. Fluids Struct* **2012**, *34*, 218–235. [CrossRef]
32. Cummins, W. The impulse response function and ship motion. *Schiffstechnik* **1962**, *47*, 101–109.
33. Chung, W.C.; Jin, C.; Kim, M.; Kim, S. Parametric study on mooring system design of submerged floating tunnel under extreme wave and seismic excitation. *J. Mar. Sci. Eng.* **2023**, *11*, 2357. [CrossRef]
34. Depalo, F.; Wang, S.; Xu, S.; Guedes Soares, C. Design and analysis of a mooring system for a wave energy converter. *J. Mar. Sci. Eng.* **2021**, *9*, 782. [CrossRef]
35. Niño, Y.; Vidal, K.; Tamburrino, A.; Zamorano, L.; Beltrán, J.F.; Estay, G.; Muñoz, A. Normal and tangential drag forces of nylon nets, clean and with fouling, in fish farming: An experimental study. *Water* **2020**, *12*, 2238. [CrossRef]
36. Hong, S.N.; Yun, U.H.; Kim, K.W.; Choi, C.Y.; Hwang, B.G.; Lee, S.C. *Development of Eco-Friendly Brass Fishing Nets and Cage Aquaculture Facilities*; Ministry of Oceans and Fisheries: Sejong-si, Republic of Korea, 2020.
37. DNVGL. *Environmental Conditions and Environmental Loads*; Det Norske Veritas: Oslo, Norway, 2000.

Disclaimer/Publisher's Note: The statements, opinions and data contained in all publications are solely those of the individual author(s) and contributor(s) and not of MDPI and/or the editor(s). MDPI and/or the editor(s) disclaim responsibility for any injury to people or property resulting from any ideas, methods, instructions or products referred to in the content.

MDPI AG
Grosspeteranlage 5
4052 Basel
Switzerland
Tel.: +41 61 683 77 34

Journal of Marine Science and Engineering Editorial Office
E-mail: jmse@mdpi.com
www.mdpi.com/journal/jmse

Disclaimer/Publisher's Note: The title and front matter of this reprint are at the discretion of the Guest Editors. The publisher is not responsible for their content or any associated concerns. The statements, opinions and data contained in all individual articles are solely those of the individual Editors and contributors and not of MDPI. MDPI disclaims responsibility for any injury to people or property resulting from any ideas, methods, instructions or products referred to in the content.

www.ingramcontent.com/pod-product-compliance
Lightning Source LLC
Chambersburg PA
CBHW040833250926
51850CB00007BA/1303
* 9 7 8 3 7 2 5 8 7 7 1 5 7 *